The Gender Vendors

The Gender Vendors

Sex and Lies from Abraham to Freud

A. L. Jones

LEXINGTON BOOKS
Lanham • Boulder • New York • London

Published by Lexington Books
An imprint of The Rowman & Littlefield Publishing Group, Inc.
4501 Forbes Boulevard, Suite 200, Lanham, Maryland 20706
www.rowman.com

16 Carlisle Street, London W1D 3BT, United Kingdom

Selections from *The Sacrifice of Isaac: The Aqedah (Genesis 22) and Its Interpretations*, edited by Ed Noort and Eibert Tigchelaar. Reproduced by permission of Koninklijke BRILL NV.

Selections from "My Daughter, Malala," by Ziauddin Yousafzai, March 2014. Reproduced by permission of TED.com.

British Library Cataloguing in Publication Information Available

Library of Congress Cataloging-in-Publication Data

Jones, A .L., 1949-
The gender vendors : sex and lies from Abraham to Freud / A. L. Jones.
pages cm.
Includes bibliographical references and index.
ISBN 978-0-7391-9096-8 (cloth) -- ISBN 978-0-7391-9097-5 (electronic)
1. Sex role—Philosophy. 2. Biology—Social aspects—Philosophy. 3. Nature and nurture—Philosophy. 4. Feminist theory. I. Title.
HQ1075.J66 2014
305.4201--dc23

2014020792

∞ ™ The paper used in this publication meets the minimum requirements of American National Standard for Information Sciences Permanence of Paper for Printed Library Materials, ANSI/NISO Z39.48-1992.

Printed in the United States of America

For Suzanne

Contents

Acknowledgments

This book would not have come into being without the generosity and kindness of my family, friends, and colleagues who engendered and/or nurtured its writing and offered invaluable comment and criticism. My heartfelt thanks go to Suzanne Holly Jones, Kim Kingston, Liz MacFarlane, Nick Chare, Maree Pardy, Fran Martin, Nigel Abbott, Carol Middleton, Carol Goudie, Mitchell Kelly, Dennis Mazalin, David Green, Wasana Weeraratne Green, Mal Sweetnam, and Helen Glezer. My deep gratitude goes to an initially anonymous peer reviewer, who waived her anonymity to invite me to respond to her critical review and discuss aspects of the work in more depth; the book has benefited enormously as a result of Carol Delaney's rigorous criticism and generous assistance. I am particularly grateful to my editors, Claire Duffy, Amy King, and Kelly Blackburn, for their unswerving encouragement, goodwill, and dedication to the work. It goes without saying that the book's shortcomings are entirely my responsibility. My deepest appreciation, as always, goes to my daughter and son.

Introduction: Sex and Lies, Seed and Soil

'The mother of what is called her child is not the parent,' says Aeschylus, 'but the nurse of the newly-sown embryo. The one who mounts is the parent.'[1] Amid countless differences, what do the following have in common? Genesis, Hesiod, Aeschylus, Aristotle, chastity codes, honor crimes, the Catholic Church, the early modern witch hunts, female genital mutilation, sexual violence, the medicalization and criminalization of gender nonconformity, and campaigns against women's suffrage. Neither the question nor the answer is new. Cumulative losses of historical insight since the 1970s, however, have arguably left twenty-first-century feminist theory ill-equipped to challenge neoliberal ideology. To my question about commonalities, the obvious answer is institutionalized sexism, but what do we know of its origins? Too often, the default response is 'It's just human nature.' This book's project is to deconstruct institutionalized sexism in order to contribute to its dismantling.

Among numerous tropes about male superiority, putatively the oldest and most potent is almost invisible in its apparent naturalness. My specific objective is to denaturalize this trope, to 'estrange' this proto-theory of the gendered body and procreation, namely, 'the seed and the soil,' first critically discussed by cultural anthropologist Carol Delaney in 1977.[2]

As in Aeschylus, seed-and-soil is a figurative expression for a proto-theory of male monogenesis, namely, that the father is the sole generating parent; the mother is not an equal partner in procreation but rather acts as nurturing medium, like soil, for the man's seed-child. In other words, men give life; women merely give birth.[3]

Why does an enquiry into seed-and-soil matter for feminism? For one thing, the metaphor highlights several characteristics of gender stratification itself, including its durability, unconsciousness, and toxic power relations in the guise of innocuousness. And, as Delaney shows, the trope is older than Western thought itself and inseparable from its ancient cosmologies and creation myths. Seed-and-soil is embedded so apparently naturally as to have become invisible.[4] The metaphor permeates the ancient Hebrew, Greco-Roman, Christian, and Islamic histories that constitute the foundational narratives of Western culture;[5] hence, its story matters to everyone who cares about gender equality and why it is taking so long to achieve. By coming to grips with the historical processes and events that led to the construction of gender stratification, we are better equipped to ask: What will it take to dismantle institutionalized sexism in its myriad forms? Though the question appears simplistic, I see it as neither simple nor even necessarily solvable. My aim is to identify some promising directions for feminist politics, in particular, for moving beyond the dualisms that plague it: culture/biology, essentialism/anti-essentialism, difference-based versus equality-based feminism, and the notion of woman-as-agent versus that of woman-as-victim. On the last binarism, ever since neoliberalism's self-reported global triumph,[6] the victors have been telling us that feminism, too, has won. To recalcitrant second-wave feminists, they say, 'You are tilting at windmills. Don't you realize your battle is over and your goals have been met? Why not just get on with enjoying the fruits of your labor?' The myth has proven seductive as well as generating much warranted criticism.[7] Suffice to say here is that, even if feminism looked no further than the interests of White, middle-class, heterosexual women, current statistics show that its goals are far from being met.[8]

As to my specific project, I build on Delaney's ethnographic study of seed-and-soil by examining some watersheds in the history of the trope, including case studies from the Hebrew Scriptures, ancient Greece, early Christianity, the medieval Church, the early modern witch hunts, and the campaigns of the nineteenth and twentieth centuries against women's suffrage. To clarify my stance on the contested issue of historical continuities, I do not claim that seed-and-soil constitutes a continuous narrative throughout history, but I have chosen to maintain the term throughout the book in order to draw attention to how entrenched interests often serve up old standards (e.g., the 'innate inferiority' of Woman) under new names (e.g., Woman as Eve, Woman as witch, or Woman as criminal deviant) to meet new political challenges (e.g., population decline). It goes without saying that, first, such historical continuities are chameleons on the shifting political landscape; second, their counter-narratives are of equal importance as objects of study; and third, the correlations outlined in this book are for the purpose of furthering scholarly enquiry. I make no claims of universal application and even less of causation. As my aim is to examine seed-and-soil at particular periods across millennia, of necessity I omit much of the detail, oversimplify the narratives, and leave stretches of history unsurveyed.

Psychology being my field, while my project is interdisciplinary its focus is predominantly psychological. I approach the task of estranging this largely unconscious metaphor with the deconstructive and dialectical tools of contemporary depth psychology, particularly Luce Irigaray's deconstructive, psychoanalytic approach to the unconscious dynamics of texts.[9] Immediately, the question arises as to how I ought to draw the line on transgressing the boundaries of other disciplines. I tread lightly in the less familiar territory of philosophy, history, sociology, and anthropology and when it comes to highly specialized debates I acknowledge these but refrain from commenting further.

This chapter introduces some abiding themes germane to the deconstruction of seed-and-soil. These include the historical conditions that produced it; its embeddedness in ancient Western cosmologies; its place in Aristotelian biology; and its link to ideologies of conquest. The chapter includes a section on why history matters to feminism and the tensions thus produced. Another section introduces some key ideas in the psychology of gender stratification and constructions of Woman. As in contemporary sociology, I use the term 'gender stratification' to mean the degree to which different genders, 'otherwise social equals, are unequal in their access to the scarce and valued resources of their society.'[10] In addition, while seed-and-soil probably pertains to some non-Western cultures, I have confined my study to the West because space is limited. Finally, on a minor technical point, all emphases within quotation marks throughout the book are as per the original works.

THE SEED AND THE SOIL: HISTORICAL CONDITIONS

The prehistoric origins of seed-and-soil are lost in unrecorded time. For millions of years our human forager ancestors would have observed seeds growing in the soil and, for thousands more, our horticultural and agrarian ancestors grew and lived by seed. What, then, were the historical conditions that produced the notion that only men can engender life? Immediately, the perennial issue arises as to the existence of archaic matriarchal societies and patriarchal takeovers. Suffice to say here, all earlier claims about the existence of matriarchies have since been overturned. Yes, there were more or less egalitarian, matrifocal, matrilineal, or matrilocal cultures, but there is no evidence for matriarchy, that is, rule by women over everyone, including all men.[11] I return to this issue in chapter 1. Yet, even if matriarchy in the strict sense of the word has never existed, early humans clearly held Woman's life-giving and nurturing capacities in high esteem. Forager communities associated Woman with Earth as parthenogenic creator of new life and perpetual all-giving nourisher of humankind. Why did this dynamic, high-status construction give way to the notion of Woman as passive soil awaiting insemination by the life-giving male—the idea of male supremacy in procreation?[12]

Presumably, ideas about paternity came first, probably via observation of domesticated animals.[13] Indeed, according to Gabrielle Rubin, although the construction of paternity cannot be dated, archaeological evidence in the form of cave art and figurines from around 17,000 BCE suggests that phallic symbolism was overtaking goddess iconography.[14] However, ideas about paternity—which might simply have offset prior beliefs about women's primacy—do not equate to a theory of male monogenesis. In Delaney's view, male monogenetic theory correlates with Abrahamic monotheism. As Genesis has it, says Delaney, God invests Adam with His divine seed to be transmitted down the patriline from father to son forever.[15] David Bakan, too, examines the way Genesis uses the term 'seed' for the male exudate—a way of thinking that attributes the embryo's entire genetic endowment to the father and none to the mother.[16] Not only is seed-and-soil implicit in Hebrew texts that were compiled early in the first millennium BCE, the trope also appears in Greek poetry of the eighth century BCE.[17] Some scholars contest Delaney's argument that male monogenetic theory correlates with Abrahamic monotheism, their point being that many ancient polytheistic creation myths reference male monogenesis.[18] I examine this debate in chapter 3.

In line with the biological and theological oneness of ancient perception, seed-and-soil is simultaneously biological and theological: Man is to Woman as God is to his creations. While God creates man in his image and invests him with the divine and ensouling seed, he fashions Woman, man's helpmate, from a piece of bone and a handful of earth. Even worse than her bodily inferiority, however, is Woman's moral inferiority, in particular, her inability to resist Satan's seductions.

If that is the view from Genesis, what of the ancient Greeks? Classicist Page duBois maintains that a remarkable shift occurred in the Greek conceptualization of the feminine principle from its primeval origins as the all-giving Gaia, to—after the Neolithic revolution—the enclosed field.[19] duBois's thesis, however, is largely technoeconomic and pays less attention to the cultural factors in which Delaney, for example, is particularly interested. Moreover, extreme forms of male domination, even patriarchy, have existed in nonagrarian communities, among the ancient Hebrews, for example. A contemporary case is the Papua New Guinean highland subsistence community, the Sambia, whose cosmology has it that semen is divinely imbued with the capacity to enhance growth and strength. Sambia husbands believe that holding back from inseminating their spouses prevents the wives from becoming dominant. Conversely, the men ritually inseminate the boys of the tribe in order to grow them into warriors.[20] I examine the particular case of the Sambia in chapter 10.

To return to ancient Greece, the Homeric landscape has Gaia providing abundantly for all,[21] and Hesiod has mortal men living like gods, 'free from toil and grief,' and having 'all the good things; for the fruitful earth unforced bore them fruit abundantly and without stint.'[22] With the shift to agriculture, however, men began to appropriate, fence, plough, and seed the once-communal lands. According to Hesiod, it is Pandora, crafted by Zeus to punish men, who brings evil into men's once-idyllic lives, including forcing them into harsh, back-breaking labor in those fields. From the 'beautiful evil,' Pandora, springs an entirely new species—Woman—created for the express purpose of causing trouble for men. Hesiod has Woman always taking and never reciprocating; her desire to injure men is both cosmologically and biologically determined. It follows that Woman's malevolence, as divinely produced, is beyond her control. Hesiod constructs her as trapped in misandry, able neither to escape nor be cured. Is it this particular aspect of the Woman myth that explains the sheer virulence of institutionalized misogyny? Such venom has never remotely been matched in any kind of organized misandry.

Genesis, too, has it that Woman and her evil-doing force Man to live by the sweat of his brow. Both traditions agree that Woman is to blame for the perpetual toil men must endure to put bread on her table. As to the beautiful evil, the dualistic tradition accords Woman-as-evil her sacred, redemptive complement in Woman-as-angel, as in the Virgin Mary and Sarah, long-suffering wife of Abraham and Mother of Israel. The message to women is that, by following the example of the Virgin, they may partly atone for their diabolical nature. This damned-whores-versus-God's-police stereotype about women has been the subject of much feminist critique, particularly since the 1970s.[23] The good woman/bad woman binarism is embedded in each of the constructions of Woman examined in this book.

The ancient Hebrew and Greek traditions constructed and canonized Woman as not merely inferior to Man, but also, by her very nature, programmed with the desire to manipulate, deceive, use, betray, impoverish, and otherwise harm him. Furthermore, they said, Woman's malevolent desire is inseparable from her sexuality. While debate continues about the degree to which Hesiod's construction, for example, permeated the wider Greek culture,[24] the brute fact of the canon remains. Regardless of the complex mix of women's agency, victimhood, and resistance in Greek society, the ancient patriarchal texts, Aristotle's in particular, became some of the most influential in the Western natural-philosophical canon and, hence, central to the construction of powerful male elites throughout millennia. The archaic notion that Woman's innate sexuality programs her to harm men was pivotal to the early modern central-European witch hunts and is still embedded in some exceptionally misogynous contemporary traditions. History, says historian Gerda Lerner, has always mattered to those in power. Narratives of the powerful both legitimize and maintain power by establishing the official version of events as the dominant version. For women in the past, *not* having a history has truly mattered.[25]

Greece's classical period saw the reinscription of seed-and-soil in Aristotelian and Hippocratic theories of procreation.[26] In Aristotle's view, the man contributes form, life, and soul to his offspring, while the woman contributes matter—'catamenia,' or menses.[27] God created the male in His image, says Aristotle, as active, spiritual, dry, warm, pure, and superior to the female, who is the inverse—passive, profane, damp, cold, and polluted. And only the male, he says, is the product of normal fetal development. The non-male fetus is a result of deformity due to one or more inauspicious variables—a chilly sou-wester blowing at the time of conception, for example. 'The female,' says Aristotle, 'is, as it were, a mutilated male.'[28]

Debate continues about the meaning and appropriateness of the categorization 'monogenetic' or 'single-seed' as applied to Aristotle's theory.[29] Of critical importance, however, is that, as a result of the revival of Aristotelian thought in thirteenth-century Europe, this dangerously wrong notion about procreation—that women merely provide base matter to be ensouled by men—entered the European natural-philosophical canon and remained there for five hundred years. Of course it is not that simple, and debate continues about how Aristotelian and Galeno-Hippocratic ideas ebbed, flowed, and intermingled over centuries.[30] I examine this issue in chapter 5.

Although scientific accounts of mammalian fertilization began in 1875,[31] only with the rediscovery of Mendelian genetics in 1900 was it understood that the ovum contributes 50 percent of the genetic material to the child.[32] However, says Delaney, that knowledge and its ramifications have not been widely propagated; nor have Westerners assimilated it symbolically, metaphorically, and linguistically.[33] Naturalized millennia ago, the Western analogy of semen with seed, while inaccurate, cannot properly be described as irrational, as there was no way of observing how sperm and ovum contribute equally to the formation of the embryo.[34] To use the idea to assign inferior status to women, however, and hence to deny women full recognition as human beings and as parents, was ideological, not rational. Only by elaborating difference into a marker of inferiority, says Lerner, did proto-patriarchal cultures construct the handful of biological differences between women and men as socially significant, reinventing, for example, the hitherto sacred capacity of Woman to bear children as signifying weakness and deficiency. What makes Woman, that is, Woman under patriarchy, is not her biological difference but her historically created genderedness.[35]

Under what kinds of political conditions did seed-and-soil survive and, at times, thrive? Did it emerge and flourish at times when societal pressures, for example, to educate or enfranchise women, threatened to destabilize gender stratification? These are questions of history.

DOES HISTORY MATTER FOR FEMINISM?

Historical insight is critical to building a feminist future.[36] Yet, among other things, the myth that feminism's goals have been met has seen the pursuit of historical continuities become an endangered species within feminist theory.[37] More specifically, that pursuit stands accused of privileging continuity over change, similarity over difference. But history-as-continuity, says historian Judith Bennett, does not exclude history-as-change.[38] Moreover, the achievement of a feminist world relies on well-informed knowledge of the past. Since the productive partnership of history and feminist theory in the 1970s, says Bennett, the two disciplines have grown so far apart that feminist theory has become remarkably uninformed by the historical insight on which a feminist future depends.[39]

What happened to all that historically informed activity? What brought about the myth that feminism had accomplished its goals? Why did some theorists label the scholarly investigation of patriarchy an essentializing and universalizing pursuit, which led to its near abandonment? Why does 75 percent or more of recent historical enquiry focus only on the last two hundred years?[40] Why, often, does social change gather momentum only to retreat? In Bennett's analyses of transformations in women's experiences over time, she notes a curious pattern of change and reversion, namely, that women's status compared to men's always reverts to a certain balance, which Bennett calls 'patriarchal equilibrium.'[41] Such equilibrium means, for example, that, proportionately, the wage gap between men and women is roughly the same today as it was in the Middle Ages. For children born into the same social circumstances today, girls, compared to boys, will face more constraints and restrictions.[42] Female disadvantage remains. Women as a group are never more advantaged than men as a group, says Joan Huber; what varies is the extent of women's disadvantage. This pattern, she says, justifies attention to possible causes.[43]

Taking another example of continuity, this time in ancient Western ideas about sex and sexuality, within the constellation of similarities and differences among sexed bodies, as well as among philosophical approaches, the theme persists of the female body as inferior to that of the male.[44] Our current relations of domination and our ideology about gender hierarchy, says duBois, have been embedded in culture since the initial sexual division of labor. Produced historically, they endure through the 'recoding of difference in different terms in different historical moments.'[45]

As I track how debates about gender unfold in a variety of fields, the question of whether feminist scholars ought to privilege similarity or difference, continuity or change, women as agents or women as victims is a constant in itself. So, too, are assertions that this or that scholar is universalizing, totalizing, naturalizing, essentializing, anachronizing, and so forth. While such criticism sometimes hits the mark (and certainly *did* in targeting old trait-based theories), at other times it is so clearly inapt as to be puzzling. Its charges are familiar: to generalize about certain trends in a data set, no matter how tentatively, is inevitably to universalize; to track continuities is to dishonor change and difference; to use terms such as 'patriarchy' is to treat as monolithic what is in fact multifaceted, fragmented, and partial; to attend—in addition to discursive and other cultural factors—to biological factors in human behavior is inevitably to essentialize. To these transgressions—at least the first parts thereof—I plead guilty and here briefly state my case, which chapter 1 examines in more depth.

First, to generalize is not inevitably to universalize. Moreover, in addition to making fine distinctions, constructing careful and circumscribed generalizations is fundamental to scholarly practice. Second, tracking continuities and searching for differences are vital and complementary processes. Although scholars may find it impractical to pursue both at once, they can still acknowledge the mutually beneficial tension between big-picture studies and detailed, locally focused analyses. My project has an eye for continuity, a psychological slant, and a scope across millennia, and, as such, of necessity omits myriad details and debates of significance and interest. But just as to generalize is not inevitably to universalize, nor ought it to be taken as dishonoring change and difference. Nor is the associated charge of anachronization always justified. Again, while the criticism is sometimes warranted, at other times it targets the legitimate deconstruction of antecedents of contemporary problems, which, far from suggesting that past societies are ignorant or blameworthy, aims to understand how past events—including accidents of history—have contributed to shaping the present. In chapter 7, apropos the early modern European witch hunts, I discuss claims such as John Bossy's that it is nonsense to say 'a whole culture or population was mistaken for what it customarily did.'[46]

Third is the question of treating as monolithic what is in fact multifaceted. Over the past few decades, many scholars have claimed that the very use of the term 'patriarchy' assumes that the relationship between women and men is unchanging and universal.[47] So vexed is the issue of patriarchy-as-continuity that they reject the use of the term itself because of its allegedly totalizing baggage. But why criticize the term per se, thereby ignoring the distinction between the term and its usage? If the label 'patriarchy' signifies a single monolithic entity against which women have no agency and are purely victims, it follows that the label 'Church' signifies a single monolithic entity of which atheists are merely victims. Or might such criticism amount to a conflation of the category with its membership, namely, *actually existing* patriarchal systems?

Taking a parallel case, the term 'monarchy' does not imply a monolith entity; to see one (Tongan, Bhutanese, English, and so forth) is not to have seen them all. The purpose of categorization is not to deny difference but to abstract what is common among members. Like 'oligarchy,' 'monarchy,' and other '-archies,' 'patriarchy' as category abstracts what is common to all patriarchies. While some scholars may read the category 'patriarchy' as positing 'an implausibly universal human nature,' says Michael McKeon, it can be historicized as 'a general continuity complicated by specific and divergent discontinuities.'[48] In the Indian context, for example, specifically brahminical, tribal, and dalit patriarchies, as well as regional and religious variations, are nonetheless patriarchies.[49] The conflation of the different ontological levels, the category and its membership, together with criticism of the term itself rather than of its inapt usage, has caused much confusion in the literature as well as establishing a form of academic 'groupthink.'[50]

Fourth, to attend to the biological factors in human behavior as well as to the cultural is not inevitably to essentialize. Take again the example of patriarchy, which, while entirely culturally constructed, is unlikely to have taken hold so implacably without the assistance of evolutionary factors such as women's capacity to bear children, the human infant's prolonged period of dependency, people's desire to win their group's acceptance, and their fear of outsiders.[51] Of course, the desire for belonging and the wariness of strangers, while biologically adaptive, are also culturally shaped, but that is my very point: the cultural and the biological are inextricable. Similarly, aspects of subjects' agency and victimhood are in constant dialectical tension, not in dualistic opposition.

In summary, patriarchy was and is temporally socially constructed and reconstructed in numerous specific forms, locations, and circumstances. There are multiple patriarchies with similarities to and differences from each other just as there are multiple resistances, agencies, and dialectical relationships among all of these.[52] Local, regional, and global patriarchal institutions invite or compel everyone to comply with their rules and norms and occasionally they co-opt the odd patriarchally inclined woman to help out at the top. In whichever of many changing forms it assumes, contemporary patriarchy is embedded globally.[53] Nowadays, to the old-style 'fathers' of the tribe add the global old boys, exemplars of, in R. W. Connell and Julian Wood's terms, 'transnational business masculinity.'[54]

What are the ramifications for feminist scholarship? What does it mean to privilege continuity or change, similarity or difference, women as agents or women as victims? What would it mean to move beyond these dualities to engage with their dialectics in everyday realities? In their 2005 review of the notion of hegemonic masculinity, for example, Connell and Messerschmidt conclude that gender dominance involves numerous factors including the 'interplay of costs and benefits'; the 'protest masculinities' of marginalized groups that challenge hegemonic masculinity; the appropriation of some aspects of hegemonic masculinity by bourgeois women in the construction of corporate careers; and the 'mutual conditioning' of men and women in gender and other social dynamics.[55] And doing masculinity (or femininity) in different relational situations involves forming compromises among conflicting demands and making complex cost-benefit analyses among different gender strategies. In addition, the authors convincingly refute a list of charges including that their concept of masculinity essentializes the character of men, imposes a false unity on reality, is heteronormative, and marginalizes and naturalizes the body.[56]

From a biocultural-dialectical perspective, it is essential to document the highly contextualized, complicated, transactional, and dialectical interweaving of agency and victimhood under patriarchy. Some scholars, however, press the self-evident proposition that 'Women are not merely victims' to the point where it pushes other complexities aside. Concomitantly, as broader complexities disappear from view, so, too, does the fact that not only women, but also the vast majority of men, albeit in different ways, suffer under patriarchy's Othering: its sexism, classism, racism, genderism, imperialism, and so forth. If we are to develop a potent political agenda, we must hold in mind simultaneously the extent of women's exclusion from public life, the ingrained misogyny and genderism of phallocentric cultures, their attendant harms, and women's and others' agency and resistance. Hence, my stance is to read similarities and differences within human projects and relationships as in constant dialectical tension. For feminist scholars to keep track of hegemonic interests is not to deny or neglect women's agency and resistance within and outside fortress walls.

For women under patriarchy, as for gender-nonconforming men, safety has always meant appearing to conform. Invalidating environments enculturate a conscious or unconscious sense of personal and social inadequacy and shame even in the children of the privileged (no child can meet or escape from the impossible demands of the Father). The mechanism by which people internalize external invalidation, however, is far from what Freud called masochism.[57] On the contrary, women's participation in punitive power relations, for example, springs from a constellation of self-protective mechanisms, cost-benefit choices, lack of consciousness, and so forth, and such participation cannot properly be attributed to free and informed choice.[58] Contrary to Freud's assumption, women are not 'naturally' masochistic; patriarchal culture, however, tells them they must accept their subordination on pain of social condemnation, even death.[59]

THE PSYCHOLOGY OF GENDER STRATIFICATION

Certain prejudices about Woman came to the fore during particular historical periods, beliefs that separately or together cast her as enclosed field awaiting the life-giving seed of the male; unwitting and helpless tool of the Devil; largely incapable of intellectual, moral, or spiritual development; polluted and polluting; and legitimate target of men's anxieties, envy, and hostility. Of course, these constructions overlap and, to reiterate, embedded in them is the good woman/bad woman binarism that defines how women ought and ought not to behave. No doubt there also were more reasoned and even appreciative counter-narratives. The rest of this section focuses on the psychosocial underpinnings of the objectifications outlined above.

Woman's Power as Sacred and Taboo

What are the psychosocial roots of anti-woman prejudice? The scenario begins with the prehistoric awe of women's reproductive capacities and apparent control over life and death. It proceeds through constellations of anxiety, love, envy, and ambivalence to the psychological defense mechanism of projecting disturbing affect onto the Other. This process of objectification constructs Woman's power as sacred and taboo, extraordinarily dangerous to men, and with her sexuality as its fountainhead.[60] Patriarchy, says Julia Kristeva, abjects women, femininity, and the maternal and, hence, oppresses, denigrates, and subordinates women.[61] The purpose of religious ritual, says Kristeva, is to ward off the fear of one's identity sinking irretrievably into the abjected mother.[62] And psychological displacement renders anxiety about an unspecific, potentially limitless phenomenon such as 'the feminine' less disturbing by attaching it to a specific, more manageable object such as women's sexual behavior.

Thus, the sacred and the taboo, purity and impurity, come together in the figure of Woman. The ancient Greeks perceived women as holding the power to give or to withdraw life. Women allegedly monitored passage into and out of mortal existence and, hence, presided over the transitional experiences of birth and dying, that is, the passages out of and into darkness.[63] In the archaic age, says Ruth Padel, the Greeks feared death with an increasing sharpness that spread to the very rites of burial and mourning, and they linked women's knowledge of darkness to that of making contact with what is polluting.[64] Greek society used and controlled women's supposed kinship with the dark side of divinity, setting women up to mediate between men and the potentially contaminating forces that attend, for example, childbirth, menstruation, babies, children, the old, the ill, and the dead.

What comes out of darkness—whether from the underworld or the female body—said the Greeks, was polluting to men, even a drop of nursing milk on a manly chest.[65] Ancient practices of barring women from entering sacred areas or touching sacred items often related to men's fear of contamination by menstrual fluid.[66] The early Christian fathers held that the mere presence of a woman acting as priest pollutes the sanctuary.[67] 'Nothing is more filthy, unclean than a menstruant; whatever she will have touched, she makes it unclean,' said Jerome.[68] And powerful menstrual taboos are still with us, a theme to which I return in the concluding chapter.

The dialectics of veil societies, including those of ancient Greece, have it that women embody a voracious sexuality that threatens not only men but the entire social order.[69] The Greek concept of *miasma* or pollution acts as a social divider, a science of division that marks higher from lower, and better from worse,[70] and it is the veil that confines the female *miasma*, particularly the pollution inherent in female sexuality.[71]

As a result of Woman's role as mediator between men and the forces of impurity, she became an even more potent source of contamination. To complicate matters, traditional male ideology enculturates men to experience surrender to the pleasure and intimacy associated with becoming close to another as a kind of psychological and moral weakening. In this, the psychological and the corporeal collapse into a pleasure-anxiety complex shot through with an ambivalence that can be felt as intolerable. Objectifying Woman (by idealizing, denigrating, or oscillating between the two) comprises an unconscious escape mechanism.[72] In the process of denigration, corporeal and psychological aspects of purity/impurity merge in framing the Other as physically and psychologically dangerous, confound the corporeal with the moral, the thought with the action, and the idea of wanting-to-harm with the actual infliction of injury.

Woman, of course, is not the only Other constructed as potentially polluting and incapable of self-control. Throughout history, cultures have constructed other genders, ethnicities, classes, and so forth as dirty, lazy, and disease-ridden, that is, as dangerous sources of physical, mental, and moral pollution.[73] They have also characterized the Other as weak and womanly; Woman, it appears, is the prototype for all Othering.[74]

Associating the Other with disease is a particularly potent weapon, given that human beings have evolved to fear polluting themselves by coming into contact with disease-related phenomena such as putrifying food, bodily excreta, illness and its products, dead bodies, and rotting corpses.[75] Primarily, patriarchy makes abject and, hence, subject to taboo, all threats that cannot be excluded.[76]

Over time, the invalidation of marginalized groups often worsens as subsequent generations conflate the effects of the discrimination practiced on the targeted group with the cause or reason for instituting the discriminatory practice in the first place. And prior explicit justificatory narratives for unjust restrictions become implicit social norms;[77] no one remembers the original story; people simply ascribe the targeted group's behavior to its so-called innate inferiority, laziness, lack of intelligence, and so forth. In this way, societies' taken-for-granted assumptions contribute to the maintenance of discriminatory cycles that can last indefinitely.[78] Throughout Western history, for example, virtually all women were excluded from education and concomitant employment, and the intellectual accomplishments of family- or nunnery-educated women were not necessarily documented.

Woman as Enclosed Field

Turning to the work of a much documented scholar, Aristotle's version of seed-and-soil was deeply implicated in the creation of powerful male elites throughout the European medieval and modern periods. How did these elites view the Woman question? In other words, what were men to do—particularly those who wished to beget progeny—in the face of the purportedly divinely constructed evil: female sexuality? If only Woman's sexuality could be curtailed, she may gain some hold on herself. What were the methods of choice? Various forms of 'female enclosure' included clitoridectomy, infibulation, veiling, seclusion, early marriage, psychological conditioning, harem and eunuch guards, and even murder.[79] It goes without saying that much of this required constant surveillance. From the ancient Egyptians, Arabs, Romans, and dozens of African, Middle Eastern, Asian, and Indigenous Australian peoples, including to the present day, the procedure of choice for permanently curbing woman's sexual desire was and is genital mutilation.[80] Herodotus reported the existence of female genital mutilation in ancient Egypt in the fifth century BCE.[81]

This is not to suggest that most cultures felt the need to control women's sexuality by covering or cutting them; after all, modern Western cultures are highly successful at constructing self-disciplining and self-invalidating subjects. Around the globe, however, 2 million women each year undergo forced genital surgery.[82] Read in a Hesiodian light, perhaps its key purpose is to obliterate the capacity of women's sexuality to harm men. In some countries, more than 90 percent of the women have been cut and in the contemporary Egyptian countryside, says David Gollaher, for a woman to remain uncut 'is to risk becoming an outcast.'[83]

Sections of nineteenth- and twentieth-century Western medicine, too, reiterated the need for men to control women's dangerous sexuality. On the eve of the twentieth century, David Gilliam, later honorary fellow of the American Gynecological and Obstetrical Society, called for the increased use of oophorectomy (ovary removal) for moral purposes: 'Why do we alter our colts and calves? That we may make them tractable and trustworthy, that we may convert them into faithful, well disposed servants.' Those physicians who have acted on their convictions, says Gilliam, 'tell us that castration pays; that the moral sense of the patient is elevated, that she becomes tractable, orderly, industrious, and cleanly.'[84] Husbands who thought their wives oversexed brought them to gynecological surgeons for clitoral surgery, which, decades into the twentieth century, was still publicly advocated and practiced in the United States and beyond.[85] 'Many women need circumcision,' said S. I. Kistler,[86] following the infamous advocate-obstetrician Isaac Baker Brown. Of course, while clitoridectomy may enable a woman to perform as an adequately orderly wife and mother, there was no bringing her allegedly inferior mental and moral capacities up to the level required by civic duty. Chapter 9 takes up these issues in more detail.

The trope of seed-and-soil associates Woman with the less appealing aspects of soil, that is, as dark, damp, cold, passive, and polluting, and, hence, perpetuates the Aristotelian illusion embodied as scientific theory more than two millennia ago. The polysemous nature of the term 'passive' (receptive, inert, latent, enduring without resistance, and so forth) has inevitably made for some confused ideas about what it means to be female. The quality of receptivity, for example, whether in a sexual partner or a party host, is neither passive nor inert. And close to a century of scientific research has shown that the so-called passive ovum is active in several senses during conception, including actively engulfing and drawing in the spermatozoon to initiate fertilization: 'The first visible response to sperm contact is the engulfment of the sperm by a protrusion of cytoplasm, the *fertilization cone*.'[87] Since the late nineteenth century, scientists have recognized the ovum's active, participatory role in fertilization. The ovum, says Paul Gross, cannot be described as 'passive' 'in any scientifically meaningful way.'[88]

With its construction of women as innately morally inferior, the legacy of seed-and-soil still haunts contemporary debates about sexual ethics, particularly apropos sexual harassment and assault. 'What can we do?' chorus leaders of church and state. 'It's just human nature.'[89] This is no surprise given that several still-prominent relationships-guidance institutions originated in the cross-over between clergy and eugenicists, and the Church is still a remarkably influential lobbyist in such secular debates.[90] As an aside, eugenicists, in spite of their awareness of existing evidence that all sex hormones are present in all sexes, promoted sex-hormone experimentation and misused it to promulgate the essentialist case.[91]

SEED-AND-SOIL IN HISTORY

Allusions to seed-and-soil proliferate throughout Judaic, Greek, Roman, and Christian texts in which procreative, pedagogical, and proselytizing references merge.

In the Jewish work commonly known as Pseudo-Philo, God tells the people he 'will bless their seed and the earth will quickly yield its fruit.'[92] As in Deuteronomy, the reference is not only to agriculture but also to 'abundant human progeny.' 'Blessed shall be the fruit of thy body and the fruit of thy ground.'[93] And the *Testament of Solomon* is laced with generic references to sexual transgression as 'sowing seed other than in one's own field.'[94]

The first-century Hellenistic Jewish philosopher Philo Judaeus describes the man sowing seed into the womb as into a field. During the woman's menses, Philo says, the man behaves 'like a good husbandman . . . while the field is still inundated he will keep back the seed.'[95]

The Quran, too, employs seed-and-soil, describing those 'seeking bounty from Allah' as 'sown corn that sendeth forth its shoot and strengtheneth it and riseth firm upon its stalk, delighting the sowers.'[96] It speaks of 'the seed of Adam and of those whom We carried (in the ship) with Noah, and of the seed of Abraham and Israel.'[97]

The poets and playwrights of ancient Greece took seed-and-soil for granted as procreative metaphor:[98]

Hesiod: 'No younger man will be better than he at scattering the seed and avoiding double-sowing; for a man less staid gets disturbed, hankering after his fellows.'

Pindar: 'On that eventful day or in the nights of love the seed of your greatness fell in foreign furrows: for then it was that Euphamos' race was sown to endure forever.'

Aeschylus has Apollo proclaim in *Eumenides*: 'The mother of what is called her child is not the parent, but the nurse of the newly-sown embryo. The one who mounts is the parent.'

Sophocles has Antigone's sister ask of Kreon: 'But will you kill your own son's promised bride?' and Kreon's response: 'Oh, there are other furrows for his plough.'

And Oedipus: 'Give me a sword, I say, to find this wife no wife, this mother's womb, this field of double sowing whence I sprang and where I sowed my children!'

Euripides: 'None other was your father: you are the seed of that divine spirit of Heracles' and 'Telamon was my father / He planted my seed in Salamis / the land that reared me.'

Scholars favored pedagogical as well as procreative allusions:[99]

Ezra: 'Give us a seed for our heart and cultivation of our understanding so that fruit may be produced.'

Hippocrates: 'For our natural disposition is, as it were, the soil; the tenets of our teacher are, as it were, the seed; instruction in youth is like the planting of the seed in the ground at the proper season.'

Aristotle: 'The soil must have been previously tilled if it is to foster the seed; the mind of the pupil must have been prepared.'

Third-century Greek biographer Diogenes Laertes describes the Stoic view of philosophy as 'a fertile field: Logic being the encircling fence, Ethics the crop.'

Aristotle calls on men to attend to the training of their wives, which 'should be the object of a man's unstinting care': 'For the tiller of the soil spares no pains to sow his seed in the most fertile and best cultivated land. . . . Surely every care should be taken on behalf of our own children's mother and nurse, in whom is implanted the seed from which there springs a living soul.'

Good doctrines are sown as seed, says Plutarch, and women's minds must not be left unsown:

> It is said that no woman ever produced a child without the cooperation of a man, yet there are misshapen, fleshlike uterine growths. . . . Great care must be taken that this sort of thing does not take place in women's minds. For if they do not receive the seed of good doctrines and share with their husbands' intellectual advancement, they, left to themselves, conceive many untoward ideas and low designs and emotions.[100]

Even when the passive soil-partner is male (the sower is *always* male), the analogy still holds. Like Philo and Plato, second-century Clement of Alexandria depicts intercourse in terms of the penetration model, that is, as asymmetrical pairings in which one partner is active (the plowing and sowing farmer) and the other is passive, regardless of the gender of the passive partner.[101] Within the seed-soil binarism, passive male partners can be figured as barren ground as well as traitors to their kind and their seed line. The fathers of Christianity, following in the Greco-Roman tradition, had it that passivity implies femininity and, hence, inferiority. The Quran, too, records the divinely ordained inferiority of women: 'Men are in charge of women, because Allah hath made the one of them to excel the other. . . . So good women are the obedient. . . . As for those from whom ye fear rebellion, admonish them and banish them to beds apart, and scourge them.'[102]

For early Christianity, which was outlawed under Roman imperial rule until 313, metaphors of procreation and kinship, says Denise Buell, function polemically in its multiple competing discourses of self-identity. Even the procreative term 'begat'—as in 'Clement begat Origen and Origen begat Gregory'—is used to establish the 'intellectual patrilines' stretching back to the disciples and apostles of Jesus.[103] Seed-and-soil imagery abounds in Christian teachings; always the male farmer sows seed into the passively waiting ground:

John 1.35: 'What is said settles in the mind with difficulty, because the ground is very hard . . . and there are many which lay plots, and carry away the seed.'

Clement of Rome: 'As land neglected . . . produces thorns and thistles, . . . there is need now of much care in cultivating the field of your mind . . . lest a noxious growth choke the good seed of the word.'

'The earth seems to take the place of the womb, into which the seed being cast is both formed and nourished.'

'The body takes its beginning from the seed of a man . . . which is conveyed into the womb as into soil.'[104]

St. Leo: 'The seed of the Word . . . ought to grow in the soil of your heart.'

Theophila: 'He who sows in a strange soil by unlawful embraces.'[105]

Seed-and-soil has been and remains an organizing metaphor for Judaism, Christianity, and Islam and, hence, modern Western culture. How did it come so far? As outlined in this chapter, from naturalized trope in the literature of the ancient Hebrews and Greeks, Aristotle elevated seed-and-soil to the status of science. Then, in the thirteenth century, Europe's love affair with Aristotle carried the Aristotelian version of seed-and-soil into the natural-philosophical canon. And the Church backed its demand for priestly celibacy by driving home Aristotle's construction of Woman as damp and cold: women seduced men, said the Church, not only to harm them but also to 'steal' their heat. Political and religious elites used seed-and-soil as a tool of oppression not only during the early modern witch hunts but also in subsequent reactionary movements against women's suffrage. And, as Irigaray meticulously demonstrates, there is little, if any, doubt that Aristotelian assumptions underpinned Freud's construction of Woman.[106] The gender politics of orthodox psychoanalytic theory (currently, the Anna-Freudian school) may be differentiated from the culturally oriented and gender-inclusive work of neo-Freudians such as Karen Horney and Erich Fromm. Orthodox Freudian theory has yet to adequately deconstruct its notions of women and their mental health. That deconstruction would entail an acknowledgement of the power dynamics that operate both within and outside the consulting room. And with thousands of Freud's papers still under orthodox embargo, some for decades to come, I question the possibility of an adequate overhaul in the near future.[107]

Part I of the book lays the philosophical and conceptual groundwork for the rest. Chapter 1 argues for a feminist re-recognition of the importance of women's history and of the category 'patriarchy' as signifying a complex, omnidirectional, and multifaceted phenomenon by which to focus feminist strategy and study. The chapter examines the historical conditions of the construction of patriarchy and enquires into the debate around identity and identity politics. It also tackles some key debates in feminist politics, namely, biology/culture, essentialism/anti-essentialism, difference-based versus equality-based feminism, and the notion of woman-as-agent versus that of woman-as-victim, as well as the typically dualistic style of these debates. The chapter highlights Lerner's crucial point that difference does not necessitate hierarchy; it is the cultural elaboration of biological difference into a marker of inferiority that constructs gender and social stratification.[108] Within the parameters of the book, chapter 2 theorizes the construction of gender and genderism and shows how seed-and-soil engenders a taboo against male femininity, which in turn produces femiphobia—the fear of being feminine or feminized.

Part II of the book considers the origins and import of the doctrine of male monogenesis and its primary metaphor—seed-and-soil—during certain periods of history from Genesis to Freud. Chapter 3 (on Genesis) and chapter 4 (on honor codes) reference Delaney's work in particular. At the heart of chapter 3 is the story of Abraham. Who begat whom—the patriline that stems from God—is central to Genesis, no more so than to the patriarchal narratives, which celebrate the founding of Abrahamic monotheism with its assumption of male monogenesis, institutionalization of Father right, and requirement for conquest. Chapter 4 examines seed-and-soil in the context of honor and shame codes in some Middle-Eastern and Mediterranean cultures. As among the ancient Hebrews and Greeks, seed-and-soil is embedded in culture-specific societal codes for the bestowal and withdrawal of honor. Male honor depends, among other things, on the ability to control the sexual behavior of female relatives. Under honor codes, men have reproductive control, Woman is a factory for making men, and men lose honor if they cannot protect the boundaries of their women. Hence, central to honor codes are chastity codes that govern the way in which women must demonstrate sexual unavailability so as to avoid bringing shame on themselves and their families; the mere perception that a woman has behaved in a way that 'dishonors' her family or community can trigger an attack on her life. The chapter draws on anthropological and sociological studies of ancient and modern honor societies.

Chapter 5 investigates how Aristotle constructed seed-and-soil as science, which, with the reintroduction of the Aristotelian corpus to Europe in the thirteenth century, entered the natural-philosophical canon and remained there for five hundred years. The chapter draws out some of the continuities in Aristotle's theory of procreation, the notion of gender hierarchy, for example.

Chapter 6 considers the legacy of seed-and-soil in the medieval Church. In order to forge bonds among the brotherhood and to warn it off women, the Christian fathers amplified existing constructions of Woman as cold-blooded, evil temptress. In addition, the chapter compares and contrasts Galeno-Hippocratic and Aristotelian theories of procreation, and examines the medieval Church's connection to a particularly misogynous clerical manual, *Women's Secrets*, on how priests were to advise women about their 'infirmities' and devise appropriate penances.

Chapter 7 examines how seed-and-soil underpinned the early modern witch hunts. Whereas the medieval witch hunts had been sporadic rather than methodical, the accusations more likely to be of heresy than witchcraft, and the prosecuted just as likely to be men as women, all that changed in the early modern period. After several hundred years of sporadic activity, why did witch hunting become widespread, systematized, and gendered? Why did the Church, from the thirteenth century onward, begin to take seriously a phenomenon they had previously decreed nonexistent? Why did the case types (heresy versus witchcraft) reverse in prevalence and the witch become, in the main, female? Was a certain type of woman targeted and, if so, what role did the Christian canon play in that targeting? The chapter concludes that the inquisitors did hunt down a certain type of so-called deviant woman and that the canon was crucial in the cultural production of that deviance. The early modern construction of the witch as female opened up a new era that criminalized gender nonconformity in both women and men as so-called perversion.

Chapter 8 argues that the nineteenth- and twentieth-century construction of Woman as ineducable, hysterical, and deviant rested on Aristotelian assumptions about Woman as the inferior inverse of Man. The Academy held that too much learning would masculinize women and endanger their wellbeing to the point of producing 'a puny, enfeebled, and sickly race.'[109] The same Aristotelian assumptions underpinned the Academy's case for the medicalization (as Freud's hysteria) and criminalization (as perversion) of gender nonconformity. As the fathers of modern medicine and social science saw it, the most dangerous threat of their time was the feminization of men (through absence of war) and the masculinization of women (through education). Hence these men's focus was more about preventing women's suffrage than advancing scientific knowledge.

Chapter nine enquires into Freud's theories in the light of femiphobia. Women, says Freud, are highly prone to neurosis, particularly hysteria; hence, for the good of their mental health they must transfer their sexual pleasure from the 'masculine' clitoris to the 'feminine' vagina and convert their penis-envy into the wish for a baby and a husband (as possessor of a penis). Such Freudian notions are pivotal to the transmission of Aristotelian notions of gender well into the twentieth century. Historically, Freud's construction of Woman—as caught between hysteria and accepting her 'natural' inferiority and subordination to men—conferred legitimacy on the nineteenth- and twentieth-century campaigns against women's emancipation and also on the medicalization of women's responses to their subjugation. In addition, Freud's idealization of the Father skewed psychoanalysis, further legitimizing and propagating the Abrahamic myth and its values.

Part III of the book examines seed-and-soil's contemporary legacy. Chapter 10 focuses on the effects of femiphobia on love and family relationships and sexual violence. The chapter argues that boys' enculturation into manhood puts them at risk of psychological and physical harm and, thereby, so, too, the women around them and the societies in which they live. As femiphobia is readily transformed into anger at women and the desire to punish them, some femiphobic men become perpetrators of sexual violence. The chapter explores the link between femiphobia and sexual violence through a survey of recent findings from several major international studies, contemporary literature on traditional masculine ideology, anthropological material on an extraordinarily femiphobic culture, and findings from ongoing reviews into a self-admitted culture of abuse in the Australian Defense Forces. Cross-culturally, the prevalence of sexual violence correlates highly with gender inequality.[110] Clearly, then, at least part of the solution to the problem lies in designing, implementing, and, above all, enforcing measures to achieve gender equality.

Chapter 11 concludes the book and summarizes the central characteristics of seed-and-soil and its legacy. What will it take to shift an institutionalized and invisible power network that was constructed prehistorically over no less than 2,500 years and further culturally entrenched over an additional three millennia? Most take the system as given. While we must address the psychosocial aspects of traditional gender enculturation, a broad political agenda is also vital. Cross-cultural research shows that relatively few perpetrators of sexual violence suffer legal consequences for their crimes. In the absence of adequate prosecution, there is no reason to expect that sexual violence will decline. In addition, to understand that the way we see the world is the outcome of a chain of historical, sometimes accidental, events, is to have more freedom in constructing a feminist future. The chapter concludes with one or two examples of change in the making.

NOTES

1. Aeschylus (525-456 BCE), *Eumenides,* in *Aeschylus with an English Translation by Herbert Weir Smyth,* vol. 2 (Cambridge, MA: Harvard University Press, 1926), 605-778. 'CE' stands for 'Common Era,' and 'BCE' for 'Before Common Era.'

2. Carol Delaney, "The Legacy of Abraham," in *Beyond Androcentrism: New Essays on Women and Religion,* ed. Rita Gross (Missoula, MT: Scholars Press, 1977), 227ff; Carol Delaney, *The Seed and the Soil: Gender and Cosmology in Turkish Village Society* (Berkeley: University Of California Press, 1991).

3. Delaney, *Seed and Soil,* 27; Mary Daly, *Gyn/Ecology* (Boston: Beacon Press, 1978).

4. Delaney, *Seed and Soil.*

5. Delaney, *Abraham on Trial.* Abrahamic religions have a common origin in Abraham; the three largest are Judaism, Christianity, and Islam.

6. Francis Fukuyama, *End of History and the Last Man* (New York: Free Press, 1992). Fukuyama has since distanced himself from Bush-style neoconservatism.

7. Mary Caputi, *Feminism and Power: The Need for Critical Theory* (Lanham, MD: Lexington Books, 2013), xviii; Jacques Derrida, *Specters of Marx: The State of the Debt, the Work of Mourning, and the New International,* trans. Peggy Kamuf (New York: Routledge, 1994), 64-65; Susan Douglas, *Enlightened Sexism: The Seductive Message That Feminism's Work Is Done* (New York: Times Books, 2010); Susan Douglas, "The Rise of Enlightened Sexism," *On The Issues* (Fall 2012); Susan Faludi, *Backlash: The Undeclared War Against American Women* (New York: Three Rivers Press, 2006).

8. E.g., Catalyst, "No Change for Women in Top Leadership," in 2012 Catalyst Census: Fortune 500 (Catalyst, 2012); EOWA, "2012 Census of Women in Leadership," (Equal Opportunity for Women in the Workforce Agency, 2012), 9; CEDA, *Women in Leadership: Understanding the Gender Gap* (Committee for Economic Development of Australia [CEDA], 2013), 14; Catherine Fox, "The Higher You Go, the Wider the Gap," in *Women in Leadership: Understanding the Gender Gap* (Committee for Economic Development of Australia [CEDA], 2013), 22.

9. Luce Irigaray, *Speculum of the Other Woman,* trans. Gillian Gill (Ithaca, NY: Cornell University Press, 1985).

10. Janet Saltzman Chafetz, ed., introduction to *Handbook of the Sociology of Gender* (New York: Springer, 2006), 10.

11. Ian Hodder, "New Finds and New Interpretations at Çatalhöyük," in *Çatalhöyük 2005 Archive Report* (Catalhoyuk Research Project, Institute of Archaeology, 2005); Brooke Holmes, *Gender: Antiquity and its Legacy* (London: I.B. Tauris, 2012); Gerda Lerner, *The Creation of Patriarchy* (New York: Oxford University Press, 1986); Douglass Bailey, "Figurines, Corporeality, and the Origins of the Gendered Body" in *Companion to Gender Prehistory* ed D. Bolger (Oxford: John Wiley, 2012), 246.

12. Delaney, *Seed and Soil.*

13. Gabrielle Rubin, *Les Sources Inconsciente de la Misogynie* (Paris: R. Laffont, 1977); Elizabeth Fisher, *Women's Creation: Sexual Evolution and the Shaping of Society* (New York: McGraw-Hill, 1979), 192. According to Rubin, animal domestication in the Middle East began around 17,000 BCE, whereas Fisher estimates 11,000 BCE.

14. Rubin, *Les Sources Inconsciente* in David Greenberg, *The Construction of Homosexuality* (Chicago: University of Chicago Press, 1990), 34.

15. Carol Delaney, "The Abrahamic Theory of Procreation and Spiritual Kinship" (paper presented at Wenner-Gren Workshop on "The Sacred Social: Investigations of Spiritual Kinship among Abrahamic Faiths," University of Virginia, March 27-29, 2014), 4; Delaney, *Abraham on Trial.* Delaney is not, of course, claiming a causal relationship.

16. David Bakan, *And They Took Themselves Wives: The Emergence of Patriarchy in Western Civilisation* (New York: Harper Row, 1979), 27; Delaney, *Abraham on Trial.*

17. Delaney, *Abraham on Trial.*

18. Kenneth Stone, *Queer Commentary and the Hebrew Bible* (Cleveland, OH: Pilgrim Press, 2001), 133; Marcia Inhorn, *Local Babies, Global Science: Gender, Religion, and In Vitro Fertilization* (New York: Routledge, 2003), 64.

19. Page duBois, *Sowing the Body: Psychoanalysis and Ancient Constructions of Woman* (Chicago: University of Chicago Press, 1991), 39.

20. Greenberg, *The Construction of Homosexuality.*

21. duBois, *Sowing the Body*, 42.

22. Hesiod (c.750–650 BCE), *Works and Days*, in *Theogony, The Homeric Hymns and Homerica,* trans. Hugh G. Evelyn-White (Cambridge, MA: Harvard University Press, 1914), 117–18.

23. Anne Summers, *Damned Whores and God's Police* (New York: Penguin, 1975).

24. Holmes, *Gender Antiquity.*

25. Gerda Lerner, *Why History Matters* (New York: Oxford University Press, 1997), 202, 208.

26. Delaney, *Abraham on Trial*, 157; Holmes, *Gender Antiquity*. Respectively denoted 'single-seed' and 'two-seed' theories, although the labels mask considerable complexity, including the so-called inferiority of the female seed.

27. Aristotle, *On the Generation of Animals,* trans. Arthur Platt (Whitefish, MT: Kessinger Publishing, 2004).

28. Aristotle, *Generation of Animals*, (2004 edition) 43.

29. Holmes, *Gender Antiquity.*

30. Holmes, *Gender Antiquity.*

31. C. R. Austin, *The Mammalian Egg* (Oxford: Blackwell Scientific Publications, 2011).

32. Jonathon Harwood, *Styles of Scientific Thought: The German Genetics Community: 1900–1930* (Chicago: University of Chicago Press, 1993), 36.

33. Delaney, "The Abrahamic Theory of Procreation," 12.

34. A seed contains, in a protected and resting state, a plant egg fertilized by pollen. The planting of seed into soil is somewhat akin to the implantation of a fertilized egg into the uterine wall.

35. Lerner, *Why History Matters*, 209.

36. duBois, *Sowing the Body*; Holmes, *Gender Antiquity*; Judith Bennett, *History Matters: Patriarchy and the Challenge of Feminism* (Philadelphia: University of Pennsylvania Press, 2006).

37. Bennett, *History Matters*, 2.

38. Bennett, *History Matters*, 80.

39. Bennett, *History Matters*, 2.

40. Bennett, *History Matters*, 35-37.

41. Bennett, *History Matters*, 55.

42. Bennett, *History Matters.*

43. Joan Huber, "Comparative Gender Stratification," in *Handbook of the Sociology of Gender,* ed. Janet Saltzman Chafetz (New York: Springer, 2006), 66.

44. Holmes, *Gender Antiquity*, 35.

45. duBois, *Sowing the Body*, 10.

46. John Bossy, "Thinking with Clark," *Past and Present* 166, (2000): 242.

47. Sheila Rowbotham, "The Trouble with Patriarchy," in *No Turning Back: Writings from the Women's Liberation Movement 1975–80* (London: The Women's Press, 1981); Michèlle Barrett and Mary McIntosh, "The 'Family Wage': Some Problems for Socialists and Feminists," *Capital and Class* 4, 2, (1980); Judith Butler, *Gender Trouble: Feminism and the Subversion of Identity* (New York: Routledge, 1990), ch. 2, 1-2.

48. Michael McKeon, "Historicizing Patriarchy: The Emergence of Gender Difference in England 1660–1760," *Eighteenth-Century Studies* 28, 3, (1995), 295.

49. Suranjita Ray, "Understanding Patriarchy," (online paper, University of Delhi, 2006), 1–2.

50. Kazem Sadegh-Zadah, *Handbook of Analytic Philosophy of Medicine* (New York: Springer, 2011), 155; Irving Janis, *Groupthink: Psychological Studies of Policy Decisions and Fiascos*, 2nd ed. (New York: Houghton Mifflin, 1982). 'Groupthink' is Irving Janis's term for the way a cohesive group sees things when its desire for unity overrides its capacity to think critically.

51. Peter Singer, *A Darwinian Left: Politics, Evolution and Cooperation* (New Haven, CT: Yale University Press, 1999).

52. R. W. Connell and J. W. Messerschmidt, "Hegemonic Masculinity: Rethinking the Concept," *Gender and Society* 19, (2005): 849.

53. Greg Forter, *Gender, Race, and Mourning in American Modernism* (Cambridge: Cambridge University Press, 2011); Yuval Noah Harari, *Sapiens: A Brief History of Humankind* (London: Random House, forthcoming September 2014).

54. R. W. Connell and Julian Wood, "Globalization and Business Masculinity," *Men and Masculinities* 7, 4, (2005).

55. Connell and Messerschmidt, "Hegemonic Masculinity," 847-48.

56. Connell and Messerschmidt, "Hegemonic Masculinity," 852, 836-37.

57. Paula Caplan, *The Myth of Women's Masochism* (Lincoln, NE: iUniverse, 1993), 74-77.

58. Christiane Olivier, *Jocasta's Children: The Imprint of the Mother* (London: Taylor and Francis, 1989).

59. Caplan, *Myth of Women's Masochism*, 74-77.

60. Julia Kristeva, *Powers of Horror: An Essay on Abjection,* trans. Leon Roudiez (New York: Columbia University Press, 1982).

61. Julia Kristeva, *Tales of Love,* trans. Leon Roudiez (New York: Columbia University Press, 1989), 374.

62. Kristeva, *Powers of Horror*, 64.

63. Ruth Padel, "Women: Model for Possession by Greek Daemons," in *Images of Women in Antiquity,* eds. Averil Cameron and Amélie Kuhrt (Oxon: Routledge, 1993), 5.

64. Padel, "Women: Model for Possession," 5.

65. Padel, "Women: Model for Possession," 5.

66. Grietje, "The Better Blood," 144. 'Fluid' is the more accurate term, as menses contains endometrial and other matter apart from blood, but I refer to menstrual 'blood' as well as 'fluid.'

67. Rosemary Ruether, "Women's Body and Blood: The Sacred and the Impure," in *Through the Devil's Gateway: Women, Religion and Taboo,* ed. Alison Joseph (SPCK Publishing and Channel Four Television Company, 1990), 17, 16.

68. Jerome, "Commentary on Zachariah, 3:13:1," in *Voices of Early Christianity: Documents from the Origins of Christianity*, ed. Kevin Kaatz (Santa Barbara, CA: Greenwood, ABC-CLIO, LLC., 2013), 117.

69. Lloyd Llewellyn-Jones, *Aphrodite's Tortoise* (Swansea, UK: The Classical Press of Wales, 2003), 259.

70. Llewellyn-Jones, *Aphrodite's Tortoise,* 260.

71. Llewellyn-Jones, *Aphrodite's Tortoise*, 18.

72. Olivier, *Jocasta's Children*; Karen Horney, *Feminine Psychology* (New York: W. W. Norton, 1993).

73. Harari, *Sapiens.*

74. Hélène Cixous, "Sorties: Out and Out: Attacks/Ways Out/Forays," in *The Newly Born Woman*, eds. Hélène Cixous and Catherine Clément, trans. Betsy Wing (London: I. B. Tauris, 1996), 79.

75. Harari, *Sapiens.*

76. Georges Bataille, *Death and Sensuality: A Study of Eroticism and the Taboo* (New York: Walker and Co., 1962); Kristeva, *Powers of Horror.*

77. Harari, *Sapiens*; Judith Butler, *Gender Trouble*; Pierre Bourdieu, *Masculine Domination* (Cambridge: Polity Press, 2001).

78. Harari, *Sapiens.*

79. Delaney, *Seed and Soil*, 40.

80. David Gollaher, *Circumcision: A History of the World's Most Controversial Surgery* (New York: Basic Books, 2000), 193.

81. Hanny Lightfoot-Klein, "Prisoners of Ritual: Some Contemporary Developments in the History of Female Genital Mutilation," (Paper Presented at the Second International Symposium on Circumcision, San Francisco, CA, May 1991).

82. Gollaher, *Circumcision*, 189.

83. Gollaher, *Circumcision*, 193.

84. David Gilliam, "Oophorectomy (Removal of the Uterine Appendages) for the Insanity and Epilepsy of the Female: A Plea for its More General Adoption," *Transactions of the American Association of Obstetricians, Gynecologists, and Abdominal Surgeons* 9, (1896): 559.

85. W. G. Rathmann, "Female Circumcision: Indications and a New Technique," *General Practitioner* 20, (1959); Gollaher, *Circumcision*, 201.

86. S. I. Kistler, "Rapid Bloodless Circumcision," *Journal of the American Medical Association* 54, (1910): 1782.

87. Norman John Berrill and Gerald Karp, *Development* (New York: McGraw-Hill, 1976), 118; Gerald Schatten and Daniel Mazia, "The Penetration of the Spermatozoon through the Sea Urchin Egg Surface at Fertilization," *Experimental Cell Research* 98, (1976), 333; Michael

Eisenbach and Ilan Tur-Kaspa, "Do Human Eggs Attract Spermatozoa?" *BioEssays* 21.3, (1999): 203–210; Colin Russell Austin, *Ultrastructure of Fertilization* (New York: Holt, Rinehart and Winston, 1968), 89.

88. Paul Gross, "Bashful Eggs, Macho Sperm, and Tonypandy," in *A House Built on Sand,* ed. Noretta Koertge (New York: Oxford University Press, 1998), 65.

89. ABC, "Facebook Expose Prompts Further Questions," 7.30 program (Australian Broadcasting Corporation, March 1, 2012). Army Chief David Morrison on the ADF's plan to combat sexual harassment and assault in the ADF and recruit more women.

90. Angus McLaren, *Impotence: A Cultural History* (Chicago: University of Chicago Press, 2007), 216. The founder of the American Association of Marriage Counsellors in the 1930s was a Quaker and 'similar religious preoccupations coloured the views of the [1938] National Marriage Guidance Council of Great Britain.' Two clergymen founded the Australian Marriage Guidance Councils in 1948: Relationships Australia (Relationships Australia, S. A., 2012).

91. Robyn Ryle, *Questioning Gender: A Sociological Exploration* (London: Sage, 2012); Robert Nye, "Sexuality," in *A Companion to Gender History,* eds. Teresa Meade and Merry Wiesner-Hanks (Hoboken, NJ: Wiley, 2008), 20.

92. William Loader, *The Pseudepigrapha on Sexuality: Attitudes toward Sexuality in Apocalypses, Testaments, Legends, Wisdom and Related Literature* (Cambridge: Wm. B. Eerdmans Publishing, 2011), 13.10.

93. Loader, *Pseudepigrapha on Sexuality*, 272, Deut 28:4.

94. Loader, *Pseudepigrapha on Sexuality*, 509.

95. Philo, *Special Laws III*, IV: 32–33, in Maren Niehoff, *Philo on Jewish Identity and Culture.* Tübingen: J. C. B. Mohr, 2001, 229.

96. Mohammed Pickthall, trans. *The Meaning of the Glorious Koran* (Hyderabad-Deccan: Government Central Press, 1938), Sura 48: 29.

97. Pickthall, Quran, Sura 19:58.

98. Hesiod, *Works and days*, Card 405; Pindar, "Pindar's Fourth Pythian Ode (254-57)," in *Pindar's Victory Songs,* trans. F. J. Nisetich (Baltimore, MD: Johns Hopkins Press, 1980), 174-88; Aeschylus, *Eumenides,* 770; Sophocles, *Antigone,* in *The Complete Greek Tragedies,* eds. D. Grene and R. Lattimore (Chicago: University of Chicago Press, 1959), 566–69; Sophocles, *Antigone*, 1255-58; Sophocles, *The Complete Sophocles,* vol. 1, *The Theban Plays,* eds. Peter Burian and Alan Shapiro (New York: Oxford University Press, 2010), 345; Euripides, *Heraclidae*, in *Euripidea Altera,* vol. 2, ed. David Kovacs (Leiden: E. J. Brill, 1996), Card 520; Euripides, *The Essential Euripides: Dancing in Dark Times,* ed. Robert Meagher (Wauconda, IL: Bolchaxy-Carducci Publishers, 2002), 229.

99. Theodore Bergren, "2 Esdras," in *The New Oxford Annotated Bible,* eds. Michael Coogan, et al. (New York: Oxford University Press, 2007), 8:6; Hippocrates, *Lex*, in *The Genuine Works of Hippocrates,* ed. Charles Darwin Adams (New York: Dover Books, 1868), sec. 3; Aristotle, *Nicomachean Ethics,* ed. and trans. H. Rackham (Cambridge, MA: Harvard University Press, 1926), 1179b, 6; Diogenes Laertius, *Lives of Eminent Philosophers,* ed. R. D. Hicks (Cambridge, MA: Harvard University Press, 1972), bk. 7, ch. 1, 40; Aristotle, *Economics*, in *Aristotle in 23 Volumes,* vol. 18, trans. G. C. Armstrong (Cambridge, MA: Harvard University Press, 1935), bk. 3, sec. 2, 70-80.

100. Plutarch, *Conjugalia Praecepta,* trans. Frank Cole Babbitt (London: William Heinemann, 1928), 2, sec. 48.

101. Denise Buell, *Making Christians: Clement of Alexandria and the Rhetoric of Legitimacy* (Princeton, NJ: Princeton University Press, 1999), 42.

102. Pickthall, Quran, Sura 4: 34.

103. Buell, *Making Christians,* 4.

104. Thomas Smith, trans. "The Recognitions of Clement," in *Ante-Nicene Fathers,* vol. 8, ed. Philip Schaff (Grand Rapids, MI: Christian Classics Ethereal Library, 1886), bk. 6, ch. 2; bk. 8, ch. 26; bk. 8, ch. 28.

105. Leo the Great, "Sermons of Leo the Great," in *Nicene and Post-Nicene Fathers: Second Series,* vol. 12, *Leo the Great,* eds. Philip Schaff and Henry Wallace (New York: Cosimo, 2007), Sermon 72, 184; Theophila, "Banquet of the Ten Virgins," in *Ante-Nicene Fathers,* vol.

6, *Fathers of the 3rd Century,* ed. Philip Schaff (Grand Rapids, MI: Christian Classics Ethereal Library, 2004), ch. 4, disc. 2.

106. Irigaray, *Speculum.*

107. Library of Congress, "A finding aid to the papers in the Sigmund Freud Collection in the Library of Congress" (Library of Congress, Washington, DC, April 2012), http://lcweb2.loc.gov/service/mss/eadxmlmss/eadpdfmss/2004/ms004017.pdf. Accessed January 3, 2013,

108. Lerner, *Why History Matters*, 209.

109. Maudsley, "Sex in Mind and in Education," 206.

110. Emma Fulu, Xian Warner, et al. "Why Do Some Men Use Violence Against Women And How Can We Prevent It? Quantitative Findings from the United Nations Multi-Country Study on Men and Violence in Asia and the Pacific" (Bangkok: UNDP, UNFPA, UN Women and UNV, 2013).

Part I

Seed and Soil: The Groundwork

Chapter One

History Matters

This chapter argues for a re-reading, vis-à-vis feminism, of the biology/culture debate, the historical construction of patriarchy, and notions of identity. In the introductory chapter, I raised several questions: What happened to the historically informed feminist activity of the 1970s? What prompted the myth that feminism had accomplished its goals? Why did some feminist theorists label the term 'patriarchy' out of date, resulting in the near abandonment of associated studies? Why is the majority of current historical enquiry focused only on the last two hundred years?[1] Why is it often the case that social change gathers momentum only to retreat? Without studying history, how well-equipped are we to judge the reliability of our current sources of information and the wisdom of current trends, and to answer the repeated sociopolitical-economic question, *Cui bono* (Who benefits)? The question of whether history matters—or, more accurately, *which* histories matter—is a political one. Often, histories of great and glorious military victories matter, for example, whereas so-called 'black-armband' accounts do not.[2] As to the issue of neoliberalism's mooted appropriation of second-wave-feminism's agenda, I refer readers to analyses by, for example, Rebecca Stringer, Susan Douglas, Mary Caputi, and Jacques Derrida. As the latter explain, neoliberalism and neoconservatism fail to discern the tragic realities they often produce.[3]

While analyzing the above questions in-depth is beyond the book's brief, this chapter examines some problematic underlying debates: biology versus culture, essentialism versus anti-essentialism, difference-based versus equality-based feminism, and the notion of woman-as-agent versus that of woman-as-victim, as well as the typically dualistic style of these debates.[4] Key to this examination is Lerner's point that there is no reasonable need to construct difference as hierarchical; it is the politically motivated elaboration of difference as inferiority that constructs gender and social stratification.[5]

SEED AND SOIL AND OTHER BINARISMS

Seed-and-soil is perhaps the most ancient construction of gender elevated into the Western natural-philosophical canon as science. In its Aristotelian form, seed-and-soil is associated with the binary opposition of activity versus passivity, which, says Hélène Cixous, is an absolute constant in the history of philosophy.[6] Binarisms are always about power relations, always hierarchical; always one is superior to the other. Victory is always associated with activity, and, in philosophy, Woman is always associated with passivity.[7]

While Aristotelian binary oppositions can appear obvious—it is either raining or it is not raining (if you ignore fog)—they rarely apply to even relatively simple situations involving human consciousness. Yet philosophy is drenched in either-ors—mind/body, culture/biology, active/passive, male/female, and so forth. Applied inappropriately, either-or thinking is capable of transforming the complex yet comprehensible into the simplistic and incomprehensible. Throughout the following discussion, I use the terms 'Aristotelian,' 'binary,' 'either-or,' and 'dualistic' more or less interchangeably.

A vexed question in contemporary feminist scholarship is whether it is possible to move beyond dualistic approaches such as difference-based versus equality-based feminism, essentialism versus anti-essentialism, and biology versus culture. The aim of such moving beyond need not be to simulate synthesis at the expense of difference and complexity, but to find ways of acknowledging those differences and that complexity that also enable the conception of and movement toward political goals.

While it is beyond the scope of this book to detail specific theoretical perspectives, biocultural dialectics and critical biocultural anthropology, among others, incorporate critical, political-economic, ecological, and human adaptability approaches to understanding biological conditions that arise from political-economic processes and local conditions, for example, of poverty and inequality. Importantly, biocultural dialectics is not to be confused with genetically deterministic biosocial perspectives.[8]

This chapter's examination of particular politico-psychological aspects of patriarchal dynamics relies broadly on biocultural-dialectical principles. These may be read as differing in some ways from competing paradigms such as social constructionism and classical psychoanalytic theory, but there are also commonalities. For example, although this book does not accept the biological determinism of orthodox Freudianism, it embraces certain psychoanalytic notions such as the unconscious and the pleasure principle. Hence, it follows recent schools of thought that integrate selected psychoanalytic notions into a conceptually coherent biocultural framework, in particular, Dialectical Behavior Therapy (DBT).[9]

As well as being a therapeutic method, DBT is a separate school of psychology that presents its own theory of pathogenesis, pathotopology, and psychotherapeutic intervention. DBT accepts traditional psychoanalytic constructs such as compromise formation (among competing, contradictory, and irreconcilable demands on the person); cathexis and anticathexis (redefined as attachment and aversion); emotional trauma as entailing extreme psychological defense mechanisms; and the role of transference in psychological healing; but it also embraces Adlerian and similar emphases on conflict between people and their environment and issues of power.[10] Depth psychology in general theorizes that human behavior serves several, often competing purposes simultaneously; that motivations have multiple sources, including the needs, desires, and goals of the individual and those of the cultural and social groups with whom the individual identifies; and that only the actor knows his or her own motivations and then often only unconsciously.[11] Why dialectics? Dialectics, says Thomas Marra, are the continua that define life's conflicting, sometimes incompatible differences in demands, expectations, and desires on the person. Unlike psychoanalytic theory, DBT constructs the dialectic among competing demands as being between self and society rather than between consciousness and the unconscious. The person who is in critical emotional pain, says Marra, may be living out the dialectic of wanting to embrace life yet fearing they cannot tolerate the pain that life brings to them: 'They want both to live and to die. These are incompatible desires, but they are experienced simultaneously in the moment by the person who suffers such a dialectical conflict.' In another situation, a person might choose to endure the pain associated with an abusive relationship rather than the perceived costs of aloneness or abandonment.[12]

Just as DBT applies the principle of dialectics in the intrapersonal, interpersonal, and social domains, so I apply it to the phenomena I describe in this book. For example, DBT's explicit theorizing of processes of invalidation (intrapsychic as well as external) speaks to how societies marginalize and punish gender nonconformity. An environment or culture that denies, trivializes, or punishes the person's inner experiences invalidates those experiences by suggesting that the person is not only wrong (both cognitively and morally) for having such experiences but also has socially undesirable personality characteristics.[13] Those subject to habitual invalidation tend to identify (at least partly) with their persecutors and become self-invalidating and unwilling or unable to tolerate unpleasant affect.[14] Attempts to block out feelings of sadness, anger, anxiety, and so forth can range from mildly risk-taking behavior and emotional rigidity to severe mental illness and suicide.[15]

DBT's explicit denaturalizing of invalidation is a useful tool for theorizing patriarchal dynamics. As bureaucratized, invisible, and omnidirectional networks of power relations, patriarchies invalidate in myriad naturalized ways not just girls and women but also—even less visibly but, arguably, more brutally—boys and men. Two associated theoretical constructs, among others, speak to institutionalized invalidation: the notion of 'groupthink' and that of allostatic load, or being 'stressed out.'[16] Groupthink, as discussed earlier, is the way a cohesive group sees things when its desire for unity overrides its capacity to think critically.[17] To preserve group cohesion, members suppress personal doubts, follow the leader, and silence or exclude dissenters.[18] How do silencing, social exclusion, and similar forms of invalidation relate to allostatic load and stress-related illness? In evolutionary terms, humankind's hypersociality attaches extreme survival importance to being accepted by one's social group.[19] Given a mere hint of rejection or belittlement, the brain's emotional system reacts involuntarily—just as it did millions of years ago—with intense fear. On the one hand, given that the prehistoric forager abandoned by their group faced certain death, this automatic fear response is adaptive. In the contemporary context of ubiquitous social evaluation, however, an involuntary fear response—particularly one unable to distinguish between impending death and a cold shoulder—is highly problematic.[20] It also speaks to why groupthink is so effective in bringing about conformity.

As for the ill effects of allostatic load, while stress under acute conditions is protective, chronic activation causes damage and accelerates disease. While stress associated with challenging events is inevitable, being 'stressed out' is not.[21] People suffering from intense or ongoing stress may develop cardiovascular problems, including heart attacks, hardening of the arteries, and stroke. The immune system can be compromised, making people more susceptible to colds and infections, for example, or—in ratcheting up its response—bring on allergies, asthma, or autoimmune conditions. Other conditions include clinical anxiety and depression, diabetes, colitis, chronic fatigue syndrome, fibromyalgia, eczema, and ulcers.[22] In short, chronic stress from social sources can make people ill and shorten their lives. And the stress that most reliably raises levels of cortisol (the central stress hormone) is social-evaluative threat, that is, threats to self-esteem or social status in which others can negatively judge one's performance.[23]

For women under patriarchal groupthink, as for gender nonconforming men, as for all who live under punitive regimes, safety has meant appearing to conform. During the early modern witch hunts, for example, women gained security by conforming to the 'good woman' stereotype and being seen to criticize those who deviated from it.[24] Via internalizing and displacing external invalidation, traumatized people may turn against nonconforming neighbors in the hope of averting further terror.[25] Similarly, unconscious self-protective strategies may contribute to women's participation in punitive power relations.[26] That gender as metaphor mystifies and renders power relations invisible hinders women's consciousness of their situation and makes consciousness-raising an ongoing requirement. The metaphoric constructs of the Bible's fallen Eve and Aristotle's Woman-as-mutilated-male, says Lerner, 'assert and assume the existence of two kinds of human beings—the male and the female—different in their essence, their function, and their potential.' Hence, the metaphor of the inferior and not quite completed female is invisibly embedded in every major explanatory system in Western philosophy.[27]

In reading patriarchy, can we do better than the parodied version in some feminist literature: the bunch of nasty, white-haired, long-bearded, black-robed Moses-figures with their stone tablets?[28] While 'patriarchy,' literally the rule of the fathers, names a category, the membership of that category is irreducibly complex, ranging in style from bearded Taliban to Fortune 500. Here it is important to note that while all patriarchies are male dominated, not all male domination is patriarchal. Patriarchy, says Delaney, is about the glorification of the male as 'father,' a term that derives from specific, non-universal beliefs about procreation. When people projected creative power onto God, that power became invisible and omnipresent. In constructing an alliance between God and men, men partake of this power. Henceforth, men's dominance appears to be God-given and natural. The Father, in fact, stands in for God.[29] Robert Filmer, in the 1680 treatise *Patriarcha*, explains how the patriarch inherits the absolute authority of Adam granted by God:

> And indeed not only Adam, but the succeeding patriarchs had, by right of fatherhood, royal authority over their children. . . . And this subjection of children being the fountain of all regal authority, by the ordination of God himself; it follows that civil power not only in general is by divine institution. . . . This lordship which Adam by command had over the whole world, and by right descending from him the patriarchs did enjoy, was as large and ample as the absolutest dominion of any monarch which hath been since the creation. . . . If we compare the natural rights of a father with those of a king, we find them all one, without any difference at all but only in the latitude or extent of them: as the father over one family, so the king, as father over many families, extends his care to preserve, feed, clothe, instruct, and defend the whole commonwealth.[30] Not only until the Flood, but after it, this patriarchal power did continue, as the very name patriarch doth in part prove.[31]

Patriarchal power in some ancient cultures extended to the father's right to kill his children, even as adults. In Roman law, for example, the male head of the family held the power of life and death over all his patrilineal descendents.[32]

Perhaps the most contested question about male domination is whether it is rooted in a natural, biologically based order. Yes, say some scholars; given, for example, the brief lifespan of prehistoric women, the majority of it spent bearing and suckling children, wouldn't women willingly exchange independence for protection?[33] It is a plausible theory, says Lerner, but there is no archaeological evidence to support it.[34] In addition, recent archaeological evidence suggests that, even twenty thousand years ago, individual foragers who survived childhood could live into their sixties, and that the more varied and nutritious diet of foragers kept them healthier than their agrarian grain-fed peasant descendents, despite the rise in fertility rates after the agricultural revolution.[35] Typically, says Janet Chafetz, forager societies are among the most gender-egalitarian economies known, particularly those in which women contribute highly to subsistence.[36] Some Stone Age sites, for example, the burial locations of the 'Sunghir children' in Russia, show signs of male chiefdom. However, just as goddess iconography does not necessarily signify matriarchy, chiefdom does not equate to patriarchy.[37] Generally, it appears that gender equality declined as the technoeconomic base of societies shifted from horticultural to agrarian plow cultivation;[38] the most extreme male domination occurs in groups where men carry out herding and agrarian plow cultivation.[39]

Apropos the rise of male domination, Delaney argues that a pivotal change occurred when men mistakenly interpreted their role in conception as primary. This appropriation of women's reproductive capacity, says Lerner, occurred prior to the creation of private property; in fact, women were the first private property. While theories differ on the relative importance of technoeconomic factors versus cultural-religious factors,[40] those that best fit the historical evidence, says Lerner, recognize as a factor men's control of women's sexuality and procreation.[41] Friedrich Engels emphasized the connection between socioeconomic change and gender relations, says Lerner, but while he eschewed the biological determinism of the traditionalists, he constructed the sex wars as class antagonism and, hence, impeded the theorizing of the differences between the two.[42] The 'original state,' says Engels, was egalitarian, matrilineal, and socially harmonious. However, the construction of private property and the overthrow of the mother right brought about 'the *world-historic defeat of the female sex*':

> The man took command in the home also; the woman was degraded and reduced to servitude; she became the slave of his lust and a mere instrument for the production of children. . . . In order to make certain of the wife's fidelity and therefore of the paternity of the children, she is delivered over unconditionally to the power of the husband; if he kills her, he is only exercising his rights.[43]

Unlike Engels, Freud famously applauded the ancient overthrow of Mother right as establishing the law as the basis for civilization. In *Moses and Monotheism*, Freud characterizes the turning from the mother to the father as signifying 'above all a victory of spirituality over the senses, that is to say, a step forward in culture.'[44] Previously, in "Totem and Taboo," Freud had described patriarchy as the result of a long process sparked by men's revolt against the elemental power of the father. All we find in 'Darwin's primal horde,' he says, 'is a violent and jealous father who keeps all the females for himself and drives away his sons as they grow up.'[45] The question, says Freud, is how did later (including still existing) egalitarian bands of males—subject to matrilineality—evolve out of the primal horde? The answer, he says, is that 'the brothers who had been driven out came together, killed and devoured their father and so made an end to the patriarchal horde.' Brotherly co-existence required renouncing the women they desired (their chief reason for dispatching their father) and, hence, a law against incest. The brothers' fatherless society 'may have been based on homosexual feeling and acts,' says Freud, as well as on the germ of matriarchy, which in turn is 'replaced by the patriarchal organization of the family.'[46]

Delaney, among others, subjects Freud's story—his 'origin myth'—to rigorous anthropological critique in which she lays bare Freud's unexamined assumptions regarding primal sexuality and violence, in particular those about innate male aggression and female passivity. Delaney points to a contradiction at the heart of Freud's story about the origins of patriarchy—which Freud assumed took place over centuries—namely, that it presupposes the 'patriarchal horde' as the original society. This error suggests that in order for the story to make sense, Freud needed the primal male to be the father-patriarch.[47] I return to Freud's myth and Delaney's critique in chapter nine.

What of contemporary readings of patriarchy? Lerner, for example, describes patriarchy as 'the manifestation and institutionalization of male dominance over women and children in the family and the extension of male dominance in society in general.' The term, says Lerner, 'does *not* imply that women are either totally powerless or totally deprived of rights, influence, and resources.'[48] Deniz Kandiyoti's term 'classic patriarchy' refers to the structure of the extended family that 'gives the senior man authority over everyone else, including younger men' and 'is bound up in the incorporation and control of the family by the state.'[49] For Marcia Inhorn, 'generalized patriarchy in Middle-Eastern gender relations is seated in familial patriarchy; the generalized valorization of father figures of all types, be they teachers, religious leaders, politicians, or God himself is seated in the glorification of men as fathers to their children; and such glorification of fathers is supported by monogenetic procreative ideologies giving men exclusive life-giving powers.'[50] For Adrienne Rich, patriarchy is the power of the fathers, which works, by direct force or pressure, influence, law, tradition, customs, etiquette, education, and the division of labor to determine what part women shall or shall not play. However, says Rich, this is not to say that all men have power or that no women do.[51] Characterized by the ancient binarism of active-penetrating-superior versus passive-penetrated-inferior,[52] patriarchy involves multiple layers of Othering. Gender stratification, says Bennett, is imbricated with imperialism, racism, feudalism, capitalism, classism, and heterosexism.[53]

Acknowledged as nothing short of a feminist theory of history, Lerner's project is both ground-breaking and, inevitably, over-reaching; it is criticized specifically for neglecting some significant historical continuities, particularly in regard to patriarchy in ancient Greece.[54] In order to identify some of the historical conditions that produced patriarchy, I briefly summarize Lerner's thesis. The available archaeological evidence, says Lerner, suggests that the process of constructing patriarchy had begun by late in the fourth millennium BCE.[55] During the agrarian revolution, many relatively egalitarian societies adopted private property, the exchange of women (because of taboos on endogamy) and, consequently, patriliny and patrilocality.[56] The latter 'had devastating effects on women,' says Hunter, in forcing relocation from their own kinship group to that of their husband.[57] The Code of Hammurabi, says Lerner, 'marks the beginning of the institutionalization of the patriarchal family as an aspect of state power.' It reflects a class society in which women's status depended on the social status and property of the male head of household.[58]

Precursors to organized slavery appeared in typical practices, for example, in which conquerors killed their male captives and raped and enslaved the women. Such slavery occurred, for example, in ninth- and tenth-century BCE Greece, prior, it seems, to the existence of private property.[59] Slave women produced slave children, whose labor was used to increase production and amass surpluses.

Plausibly, women under constant threat of rape and enslavement by enemy tribes would seek the protection of the most powerful members of their own communities. Intertribal exchange of women became a way of honoring taboos on incest and endogamy and of cementing marriage alliances, thus avoiding incessant warfare. At the same time, the practice reified women's reproductive capacity, says Lévi-Strauss, and dehumanized women.[60] Men-as-a-group had rights in women that women-as-a-group did not have in men. Women themselves became a resource acquired by men, commodified via bride price and sale price.[61] The agglomeration of agrarian communities into city states saw the construction of hierarchies, kingship, propertied and warrior classes, organized trade, institutionalized slavery, and patriarchal family structure. By 3500 BCE, the first large-scale agrarian societies in the ancient Middle East appeared in Sumer.[62] Their cosmoganies had subordinated female deities to chief male gods and their origin myths legitimated male ascendency. Henceforth, it was as if gender-based slavery was how things had always been. By the time men were beginning to theorize the universe, centuries of patriarchal culture had naturalized the subordination of women.[63] And by the fifth century BCE, Greece's slave system was on par with later forms in Rome and the Americas.[64]

There are archaeological indications of relatively egalitarian societies moving toward proto-patriarchal organization. In Cyprus, for example, says Stephanie Budin, women appear to have lost status during the Secondary Products Revolution after the Chalcolithic period.[65] Communities in which women had been producers later saw them processing the products produced by men. Other archaeological evidence from the period shows women withdrawing to the inner spaces of their homes (spaces devoid of openings onto the outside world) and the curtailment of their activities both within and beyond the domestic sphere (for the protection of their virginity, etc.). The iconography suggests the formation of a system of social elites, says Budin, in which women appeared in the same category as cows.[66]

'Would you have us return to matriarchy?' is a common response when patriarchy comes under fire. And it is true that Paleolithic goddess iconography had scholars speculating about the nature of pre-patriarchal societies. But their consensus is that the Woman of Willandorf evidences matriarchy about as much as statues of the Virgin prove that women run the Catholic Church. Neolithic and Chalcolithic evidence of more egalitarian systems than our own, such as at Çatalhöyük in Turkey, is exactly that—evidence that more egalitarian communities existed prior to the rise of the city-state.[67] As the state matured, however, bureaucratic (patriarchal) power trumped kinship systems and influence.[68] There is no evidence for the existence of matriarchal societies in the sense where women-as-a-group has decision-making power over men, defines the rules of sexual conduct, or controls marriage exchanges.[69] However, even if earlier societies were not truly matriarchal, says Delaney, they accorded women the religious power accruing from the miracle of birth.[70] That power balance began to shift when humans—probably as a result of observing domesticated animals—constructed notions of a male role in conception, notions that later transmogrified into the mistaken and discriminatory assumption that it was a primary role.[71] Chapters 3 and 5 examine this issue in more detail.

In summary, according to the available evidence, patriarchy is a human invention of a specific historic period. Western civilization, says Lerner, is founded on systems of theology, philosophy, and science that take the preexisting subordination of women for granted, which is why it was invisible for so long.[72] In comparison to other oppressed groups—peasants, slaves, colonials, ethnic minorities—who knew they were subordinated and who developed theories and strategies of liberation, she says, women have often been framed as passive subjects of patriarchy (which is why some feminists avoid speaking of patriarchy as an institution). However, the reason women did not know they were subordinated is not because of a failing on their part, but because, by the time Western civilization was invented, the 2.5-millennia process of constructing patriarchy (from approximately 3100 to 600 BCE) had long been completed and had gone unrecorded.[73]

Male dominated societies of all kinds construct the few biological differences between women and men—women's capacity to bear children, for example—as signifying women's weakness and inferiority. Societies that are more egalitarian, says Lerner, tend to express those differences in the form of slightly different economic activities, but without the connotation of inferiority. Hence, to reiterate a crucial point, differences between women and men neither logically nor inevitably lead to gender stratification.[74] Regardless of whether women and men are more alike than different (or whether they are opposed to one another in any sense), the direction for societal change is the same, specifically, toward abolishing sexism, genderism, racism, and so forth.[75]

The purpose of the foregoing discussion is to move beyond dualistic biology-versus-culture approaches to reading patriarchy. The next section asks what it may mean to move beyond essentialism-versus-anti-essentialism arguments in the debate about identity.

IDENTITY AND IDENTITY POLITICS

What is meant by the term 'women'? Over the past two decades, says Allison Weir, the notion of the collective social identity 'women' has been criticized as a claim to sameness or essence—and thus as 'traps that inhibit our freedom' or 'sources of retreat and withdrawal into safe homes that shelter us from public life and from recognition of difference.'[76] Identity politics, according to this view, involves 'misguided affirmations of the very identities that colonize us,' a 'self-defeating politics of balkanization, closure and resistance to change.'[77]

What makes for this kind of hyperbole? What fuels the essentialism-anti-essentialism debate? Is this another example of category-membership conflation?[78] Judith Butler and Elizabeth Weed's collection on Joan W. Scott's critical feminism points to the paradoxes thrown up by Scott's questioning, not just about how women are treated differently in certain contexts, but about how gendered meanings are produced as part of the operation of power, or, in Scott's words, as 'a primary way of signifying power.'[79] Following Scott, say the editors, feminist scholarship began to question not merely how women achieved rights but 'how the very notion of rights was gendered from the start' and how women's struggles reshaped the very meaning of rights.[80] Undecidable issues arose and opinions split.

Scott reminds us about the origins of French feminism (until 1944, women could not be citizens) and the contemporaneous questions about gender: 'Did biology determine the capacity for reason? . . . Did reproduction conflict with intelligence?' 'Precisely because these questions were undecidable,' says Scott, law was substituted for truth, 'but this substitution was not acknowledged as such.'[81] Debates about gender typically invoked 'nature' as authority to explain the differences between the sexes. Such differences became the justification for laws and policies when in fact they were largely the effect of those laws and policies.[82]

The resulting feminist arguments were powerful, says Scott, but also paradoxical: 'in order to protest women's exclusion, they had to act on behalf of women and so invoked the very difference they sought to deny.'[83] This dilemma has come to us in the form of debates about equality or difference: 'Are women the same as men? And is this sameness the only basis upon which equality can be claimed? Or are they different, and, because or in spite of their difference, entitled to equal treatment?' In fact, says Scott, the meaning of sexual difference, like race and ethnicity, is 'always in dispute,' and 'the intensity of feminist politics' follows from the 'undecidability of sex differences.'[84]

Scott's account is plausible but it raises as many questions as it answers. Why were the questions she describes articulated in this way? Why ask the patently paradoxical (and, as it stands, relatively meaningless) question—'Are women the same as men?'—which reduces a complex constellation of differences and similarities to a simplistic binary opposition—sameness or difference? Not least among the differences is women's capacity to bear children, and not least among the similarities is membership of the same species. Rather than asking 'Are women the same as men?' why not ask, for example, 'Are women human?' as does Catharine MacKinnon.[85]

Here is Scott again:

> Feminism was a protest against women's political exclusion; its goal was to eliminate 'sexual differences' in politics, but it had to make its claims on behalf of 'women' (who were discursively produced through 'sexual difference'). To the extent that it acted for 'women,' feminism produced the 'sexual difference' it sought to eliminate. This paradox—the need both to accept *and* to refuse 'sexual difference'—was the constitutive condition of feminism as a political movement.[86]

However, feminism acts for the membership of the category 'women,' not for the category itself. And, yes, all human conflict is fraught with paradox, but here Scott is demanding the impossible. Whereas, for example, 'to eliminate *discrimination* based on sexual difference' or 'to gain the same rights that men enjoy' are plausible goals, 'to eliminate "sexual differences"' is not. Might the historic-political circumstances, however, have precluded even the notion of discrimination? Did feminism find itself caught up in the undecidable questions levelled by its opponents? Did it have any choice? I don't know the answers to these questions, but the degree to which the debate may have been undermined by dualistic thinking deserves attention. Alongside the differences among categories such as gender, race, class, and religion, there are similarities in terms of human rights. The question of whether Muslims are the same as Christians (or the blind as the sighted, Asiatics as Caucasians, the working class as nobility, and so forth), does not bear on their human rights. Moreover, to refute a common misapprehension, to pursue equality in human rights is not to pursue the impossible—for example, genomic, constitutional, or demographic equality.

One problem with contemporary criticism of identity and identity politics, says Weir, is that while identities have been constituted in part through systems of domination and subordination, they also have enabled solidarity in resistance to those systems. To further complicate matters, notwithstanding the extensive criticism of the idea of identity, the problem remains that identities are inescapable. This is acknowledged, says Weir, by even the fiercest critics.[87] Of course the old confidence that the category 'women' can be clearly identified by bodily characteristics no longer stands, nor that 'women' constitute a transcendent sisterhood that bridges differences in class, race, religion, sexuality, and world region.[88]

Although in theoretical terms the category 'women' is unstable, says Bennett, in practice it usually acts as a 'stable category—for its time and place—that can critically determine a person's life chances.'[89] Similarly, one can hold on to the theory that, as a category, 'women' does not exist, says Denise Riley, 'while maintaining a politics "as if they existed"—since the world behaves as if they unambiguously did.'[90] This paradox has been problematic in feminist history since the 1790s.

For Butler, there is no choice but to acknowledge the paradox as central to the notion of identities as the subjections that enable.[91] And then what? As we can neither escape nor radically change the structures of power, resistance is located in symbolic gestures, transgressions in speech, and parodic performances that subvert the dominant categories. Weir accepts Butler's critique as legitimate but criticizes it as based on the same understanding of identity as paradox as is central to the liberal conception of the individual. And, while admiring Butler's 'brilliant and compelling' insight into the constitution of identity through the law that enables, Weir argues that Butler is 'blind to the possibility of an identity that is not, or not only, produced by the law, or that is more or other than a subjugation that enables.' The problem is that Butler extends the model of the paradox of the subject to *every* form of self, agency, and identity.[92]

While Butler's formulation of identities as paradoxical is useful, some see it as impeding the construction of collective political agendas. In contrast, Linda Alcoff interrogates the broader question: What brought about these notions that 'social identity itself is an a priori problem,' that identities 'under *any* description pose dangers,' and that identities commit us to 'mistaken assumptions' when they are reified or acted on politically?[93] Why criticize the notion of identity per se, says Alcoff, when you can simply question 'overly homogenizing, essentialist, reductive, or simplistic' constructions of identity?[94]

And if identity has become suspect, Alcoff says, identity politics has been prosecuted, tried, and sentenced to death. By whom? Not just by conservatives, she says, but also by liberals, Rorty and Schlesinger, for example, who offer neither definition nor elaboration of the historical genesis of identity politics even in its most basic thesis, namely, that identity in all its heterogeneity is relevant to politics. 'My question,' says Alcoff, 'is how did the legitimate concern with specific instances of problems in identity-based movements become a generalized attack on identity and identity politics in any form?'[95]

Schlesinger's attack on multiculturalism, she says, offers a clue. In his brief best-seller, *The Disuniting of America*, Schlesinger interprets pre-1960s America as happily progressing as a melting-pot nation in which individuals lost their old identities and left behind their 'ancient prejudices and manners.'[96] He omits, says Alcoff, to mention the discriminatory treatment of non-Europeans. According to Schlesinger, the 1960s 'cult of ethnicity' derailed this steady progress and, moreover, was driven not by ethnic groups themselves but by leaderships with vested interests.[97] The cult of ethnicity, says Schlesinger, 'exaggerates differences and intensifies resentments and antagonisms,' the end-game being 'self-pity and self-ghettoization.'[98]

However, Schlesinger's main concern, says Alcoff, is not to eradicate all identities but to prevent non-European identities from dominating.[99] The danger from a wholesale criticism of identity is that minority agendas are constructed as a threat to progressive politics. The real challenge for identity politics, Alcoff says, is to articulate its relation to class and ensure universal rights, while imagining justice across cultural difference.[100]

Weir, too, argues that we can understand identities as sources, not only of subjugation, but also of freedom. Identity as *category* must be differentiated from identity as *connection to* and *identification with* ideals, each other, and defining communities. In other words, identities are formed on multiple levels through multiple relations, not all of them chiefly subjugatory.[101]

The key point here is that Weir's analysis takes us from a metaphysical to an ethical and political conception of identities. Appropriative identifications are distinct from transformative identifications. Identities constituted through denial of difference are distinct from complex heterogeneous identities constituted through difference and conflict. Identities are produced through multiple contesting relations.[102]

More so than being paradoxical, says Weir, identities are better understood as complex webs of interactions among diverse relations, including of power, meaning, and solidarity.[103] Weir brings together theories of identity as produced by subjugation with relational theories of identity as produced through intersubjectivity and defining communities.

Alcoff adopts a similar position, holding that when one is identified what is actually identified is one's 'horizon of agency.'[104] Identities imposed on people from the outside are more a brand than an identity. Identity can be understood as a way of inhabiting, interpreting, and working through, collectively and individually, a social location and group history. Identities are not lived as discrete and stable sets of interests but as sites from which to engage in meaning-making.[105]

Alcoff offers an alternative feminist conception of identity in which gender is a position one occupies and can act from politically (as in the game of chess). The position of women is relative to a cultural and social network, says Alcoff, and yet it is not, as Derrida would have it, undecidable. When women become feminists, it is not that they have learned new facts but that they read those facts from a different position, from their own position as subjects.[106] As I see it, to be raised under patriarchal tradition is to think dualistically *by default*, that is, unconsciously; patriarchal systems cannot survive without binaries. Without the kind of conscious-raising that shatters binaries, the so-called sisterhood, for example, has no conscious model for dealing with internal dissent other than that of the brotherhood-in-self-protection mode. Just as patriarchy needs dualistic thinking, democracy needs dissent.

While we ought to avoid defining women in terms of essential internal characteristics, says Alcoff, we also ought to avoid being driven by our fear of being essentialist into a mere politics of negation. We can make many accurate claims about women as they exist here and now in particular locations, she says, and thus we can make demands that reflect women's needs. What it means to be a woman varies but that does not mean we can say nothing about women, only that we must refrain from universal pronouncements: feminist theory must locate and limit the scope of its presumed applicability.[107]

I agree with Alcoff and Weir that critics of notions of identity often have applied the label 'essentialist' prematurely, reductively, or simplistically, usually by setting up straw opponents and framing them as incapable of resisting entrapment by classification or differentiating a limited but useful abstraction from a decree. And it is not labels per se that ensnare us but how we relate to them. That is not to dismiss the problem; it is to suggest that we must develop more nuanced ways of understanding it. Neither essentialist nor anti-essentialist positions, says Alcoff, emphasize enough the temporal dimension and the changeable nature of human characteristics and political contexts. Positionality emerges from constantly changing relational circumstances.[108] It also accords with Iris Young's conceptualization of 'women' as a serial collectivity, a reworking of the Sartrean distinction between 'seriality' and 'group.'[109] While a group recognizes itself as unified in a common undertaking, a series (e.g., a queue of people) is passively unified by circumstances. Moving beyond essentialism-versus-anti-essentialism is a prerequisite for moving toward positional and relational notions of identity along the lines suggested by Young, Weir, and Alcoff that will enable a collective political agenda. Toward what end? In the terms articulated by the World Health Organization and the United Nations: toward securing freedom from violence and gaining access to education, meaningful work, fair remuneration, and a rightful share in our societies' political self-determination.[110] Perhaps a time will come when factors such as gender and race carry as much weight in a job interview as do eye color and blood type. Then again, perhaps genomics will trump all.

NOTES

1. Bennett, *History Matters*, 35–37.
2. Australian examples of military victories that matter include those tragic losses amenable to reconstruction as noble, for example, Australia's part in the World War I Gallipoli campaign. Examples of so-called 'black-armband' views of history include accounts of repeated massacres of Australian Indigenous peoples.
3. Rebecca Stringer, *Knowing Victims: Feminism, Agency, and Victim Politics in Neoliberal Times* (New York: Routledge, 2014), ch. 1; Caputi, *Feminism and Power*, xviii; Derrida, *Specters of Marx*, 64–65; Douglas, *Enlightened Sexism*; Douglas, "The Rise of Enlightened Sexism"; Faludi, *Backlash*.
4. Caputi, *Feminism and Power*, 4–5.
5. Lerner, *Why History Matters*, 209.
6. Hélène Cixous, "Sorties: Out and Out: Attacks/Ways Out/Forays," in *The Newly Born Woman*, eds. Hélène Cixous and Catherine Clément, trans. Betsy Wing (London: I. B. Tauris, 1996), 68.
7. Toril Moi, *Sexual/Textual Politics: Feminist Literary Theory* (London: Methuan, 1985), 104–105.
8. E.g., Tom Leatherman and Alan Goodman, "Critical Biocultural Approaches in Medical Anthropology," in *A Companion to Medical Anthropology*, eds. Merrill Singer and Pamela Erickson (Chichester, West Sussex: Blackwell, 2011), 29; Alan Goodman and Thomas Leatherman, "Traversing the Chasm between Biology and Culture: An Introduction," in *Building a New Sociocultural Synthesis: Political Economic Perspectives on Human Biology*, eds. Alan Goodman and Thomas Leatherman (Michigan: University of Michigan Press, 1998), 5.
9. Linehan, *CBT for Borderline Personality Disorder*; Thomas Marra, *Dialectical Behaviour Therapy in Private Practice* (Oakland, CA: New Harbinger, 2005).
10. Marra, *Dialectical Behavior Therapy*, 9.
11. Anne Dean, *Teenage Pregnancy: The Interaction of Psyche and Culture* (Hillsdale, NJ: The Analytic Press, 1997), viii–ix.
12. Marra, *Dialectical Behavior Therapy*, 9–11.
13. Linehan, *CBT for Borderline Personality Disorder*.
14. Connell and Messerschmidt, "Hegemonic Masculinity."
15. Linehan, *CBT for Borderline Personality Disorder*.
16. Bruce McEwan, *The End of Stress as We Know It* (Washington, DC: Joseph Henry Press, 2002), 7.
17. Irving Janis, *Groupthink: Psychological Studies of Policy Decisions and Fiascos*, 2nd ed. (New York: Houghton Mifflin, 1982).

18. Paul 't Hart, "Irving L. Janis' Victims of Groupthink," *Political Psychology* 12, 2, (1991): 247.

19. Michelle Wirth and Allison Gaffey, "Hormones and Emotion: Stress and Beyond," in *Handbook of Cognition and Emotion*, eds. Michael Robinson et al. (New York: The Guilford Press, 2013), 74.

20. Wirth and Gaffey, "Hormones and Emotion"; Julian Short, *An Intelligent Life* (Sydney: Random House, 2005).

21. McEwan, *The End of Stress,* 4.

22. McEwan, *The End of Stress,* 4.

23. Richard Wilkinson, "How Economic Inequality Harms Societies," (TEDGlobal transcript October, 2011), 3–4, accessed July 9, 2013, http://www.ted.com/talks/richard_wilkinson.html.

24. Merry Wiesner-Hanks, *Women and Gender* (Cambridge University Press, 2000); Christina Larner, "The Crime of Witchcraft in Early Modern Europe," in *The Witchcraft Reader,* ed. Darren Oldridge (London: Routledge, 2002).

25. Barbara Whitmer, *The Violence Mythos* (Albany: State University of New York Press, 1997).

26. Olivier, *Jocasta's Children.*

27. Lerner, *Creation of Patriarchy*, 211.

28. Leila Rupp, "Revisiting Patriarchy," *Journal of Women's History* 20, 2, (2008): 140.

29. Delaney, *Seed and Soil*, 35.

30. Robert Filmer, *Patriarcha*, (1680), ch. 1.

31. Filmer, *Patriarcha*, ch. 2.

32. Bruce Frier and Thomas McGinn, *A Casebook on Roman Family Law* (New York: Oxford University Press, 2004), 191; Judith Evans Grubbs, Tim Parkin, and Roslynne Bell, eds., *The Oxford Handbook of Childhood and Education in the Classical World* (New York: Oxford University Press, 2013), 85–86.

33. Barbara Smuts, "The Evolutionary Origins of Patriarchy," *Human Nature* 6, 1, (1995); Malcolm Potts and Martha Campbell, "The Origins and Future of Patriarchy: The Biological Background of Gender Politics," *Journal of Family Planning and Reproductive Health Care* 34, 2, (2008).

34. Lerner *Creation of Patriarchy.*

35. P. M. Lambert, "Health Versus Fitness: Competing Themes in the Origins and Spread of Agriculture," *Current Anthropology* 50, 5, (2009), 603; Jay Stock and Ron Pinhasi eds., *Introduction to Human Bioarchaeology of the Transition to Agriculture* (Chichester, UK: Wiley-Blackwell, 2011), 2001.

36. Chafetz, *Handbook of the Sociology of Gender.*

37. E. Trinkaus and A. Buzhilova, "The Death and Burial of Sunghir 1," *International Journal of Osteoarchaeology* 22, 6, (2012).

38. Margaret Gonsoulin, "Women's Rights and Women's Rites: Religion at the Historical Root of Gender Stratification," *Electronic Journal of Sociology* (2005); Rae Lesser Blumberg, "Toward a Feminist Theory of Development," in *Feminism and Sociological Theory,* ed. Ruth Wallace, 161–199 (Beverley Hills, CA: Sage, 1989).

39. Randall Collins et al., "Toward an Integrated Theory of Gender Stratification," *Sociological Perspectives* 36, 3, (1993), 189.

40. Gonsoulin, "Women's Rights and Women's Rites"; Chafetz, *Handbook of the Sociology of Gender*; Huber, "Comparative Gender Stratification."

41. Virginia Hunter, "The Origins of Patriarchy: Gender and Class in the Ancient World," *Labour/Le Travail* 22, (1988), 244.

42. Lerner, *Creation of Patriarchy,* 23.

43. Friedrich Engels, *Origin of the Family, Private Property and the State* (London: Penguin, 2010), note 21.

44. Sigmund Freud, "Moses and Monotheism," trans. Katherine Jones (*SE 23,* 1939), 180.

45. Freud, *Totem and Taboo,* 164.

46. Freud, *Totem and Taboo,* 167.

47. Delaney, *Abraham on Trial*, 193, 195.

48. Lerner, *Creation of Patriarchy*, 239. The Merriam-Webster Online Dictionary defines patriarchy as: 'social organization marked by the supremacy of the father in the clan or family, the legal dependence of wives and children, and the reckoning of descent and inheritance in the male line; *broadly*: control by men of a disproportionately large share of power.'

49. Deniz Kandiyoti, "Bargaining with Patriarchy," *Gender and Society* 2, 3, (1988): 278.

50. Marcia Inhorn, *Infertility and Patriarchy: The Cultural Politics of Gender and Family Life in Egypt* (Philadelphia: University of Pennsylvania Press, 1996), 25.

51. Adrienne Rich, *Of Woman Born: Motherhood as Experience and Institution* (New York: W. W. Norton, 1995).

52. Carden, "Endangered Ancestress Revisited"; Foucault, *Use of Pleasure*; Dover, *Greek Homosexuality*.

53. Bennett, *History Matters*, 58; Rich, *Of Woman Born*, xiv.

54. Hunter, "The Origins of Patriarchy," 246.

55. Lerner, *Creation of Patriarchy*, 212.

56. Claude Lévi-Strauss, *The Elementary Structures of Kinship* (Boston: Beacon Press, 1969), 115.

57. Hunter, "The Origins of Patriarchy," 241.

58. Lerner, *Creation of Patriarchy*, 140.

59. Hunter, "The Origins of Patriarchy," 244.

60. Lévi-Strauss, *Elementary Structures of Kinship*, 115.

61. Lerner, *Creation of Patriarchy*, 112, 213.

62. Michael Mann, *The Sources of Social Power*, vol. 1, *A History of Power from the Beginning to AD 1760* (New York: Cambridge University Press, 2012), 38.

63. Lerner *Creation of Patriarchy*, 53, 211.

64. Hunter, "Origins of Patriarchy," 245.

65. Stephanie Budin, *Images of Woman and Child from the Bronze Age: Reconsidering Fertility, Maternity and Gender in the Ancient World* (New York: Cambridge University Press, 2011), 245.

66. Budin, *Images of Woman and Child*, 245, 255, 257.

67. Hodder, "New Interpretations at Çatalhöyük."

68. Richard Lee and Richard Daly, "Man's Dominance and Women's Oppression: The Question of Origins," In *Community Power and Grassroots Democracy*, eds. Michael Kaufman and Haroldo Alfonso (London: Zed Books and Ottawa: International Development Research Centre, 1997), 36–37.

69. Holmes, *Gender Antiquity;* Lerner, *Creation of Patriarchy*, 30.

70. Carol Delaney, "The Legacy of Abraham," in *Anti-Covenant: Counter-reading Women's Lives in the Hebrew Bible*, ed. Mieke Bal (Sheffield: Sheffield Academic Press, 1989), 36.

71. Delaney, "The Legacy of Abraham," in *Anti-Covenant*, 28.

72. Gerda Lerner, "Women and History Part I: The Creation of Patriarchy," (Interview with Gerda Lerner by Jeffrey Mishlove, 1999), 6.

73. Lerner, "Women and History," 8.

74. Lerner, *Why History Matters*, 209.

75. World Health Organization, "Global and Regional Estimates of Violence against Women," (Geneva: World Health Organization, Department of Reproductive Health and Research, 2013).

76. Allison Weir, *Identities and Freedom: Feminist Theory between Power and Connection* (New York: Oxford University Press, 2013), 2.

77. Weir, *Identities and Freedom*, 2.

78. Sadegh-Zadah, *Handbook of Analytic Philosophy of Medicine*, 155.

79. Judith Butler and Elizabeth Weed, introduction to *The Question of Gender: Joan W. Scott's Critical Feminism*, eds. Judith Butler and Elizabeth Weed (Bloomington: Indiana University Press, 2011), 3.

80. Butler and Weed, *Question of Gender*, 3.

81. Joan Scott, *Only Paradoxes to Offer: French Feminists and the Rights of Man* (Cambridge, MA: Harvard University Press, 1996), ix.

82. Scott, *Only Paradoxes to Offer*, ix.

83. Scott, *Only Paradoxes to Offer*, x.
84. Scott, *Only Paradoxes to Offer*, x.
85. Catharine MacKinnon, *Are Women Human? And Other International Dialogues* (Cambridge: Belknap Press, 2006).
86. Scott, *Only Paradoxes to Offer*, 3–4.
87. Weir, *Identities and Freedom*, 2–3.
88. Bennett, *History Matters*, 9.
89. Bennett, *History Matters*, 9.
90. Denise Riley, *Am I That Name? Feminism and the Category of "Women" in History* (Minneapolis: University of Minnesota Press, 1988), 112.
91. Butler, *Gender Trouble.*
92. Weir, *Identities and Freedom*, 8.
93. Linda Martín Alcoff, *Visible Identities: Race, Gender and the Self* (New York: Oxford University Press, 2006), 14.
94. Alcoff, *Visible Identities*, 14.
95. Alcoff, *Visible Identities*, 14–16.
96. Arthur Schlesinger, *The Disuniting of America: Reflections on a Multicultural America* (New York: W. W. Norton, 1991), 1.
97. Schlesinger, *Disuniting of America*, 34.
98. Schlesinger, *Disuniting of America*, 102.
99. Alcoff, *Visible Identities*, 18.
100. Alcoff, *Visible Identities*, 18.
101. Weir, *Identities and Freedom.*
102. Weir, *Identities and Freedom*, 3.
103. Weir, *Identities and Freedom*, 3.
104. Alcoff, *Visible Identities*, 42.
105. Alcoff, *Visible Identities*, 42.
106. Alcoff, *Visible Identities*, 148. This is similar, says Alcoff, to how the colonized come to identify with their own group rather than with the colonizer.
107. Alcoff, *Visible Identities*, 152.
108. Alcoff, *Visible Identities*, 151–52.
109. Iris Young, "Gender as Seriality: Thinking about Women as a Social Collective," *Signs* 19, 3, (1994): 713.
110. E. g., Fulu, "Why Do Some Men Use Violence."

Chapter Two

The Psychology of Selling Gender

'Am not'! yells the toddler—his eyes filling—to the big boys on the see saw. 'Am not, am not, am NOT'! And his little body is wracked with sobs as he turns toward his mother.

Australian playground, 2013.

How does the legacy of seed-and-soil contribute to the taboo against male femininity? How in turn does this taboo engender femiphobia, the culturally constructed fear of being perceived as feminine or indeed of being feminine or feminized ?[1] The aim of this short chapter is to lay out, within the parameters of this book, the psychosocial roots of gender, genderism, and femiphobia. A major premise of the chapter is that male privilege comes at enormous cost, not only to women and girls but also, albeit in different ways, to boys and men themselves. 'Male privilege is also a trap,' says Bourdieu, 'and it has its negative side in the permanent tension and contention, sometimes verging on the absurd, imposed on every man by the duty to assert his manliness in all circumstances.'[2]

Phallocentric cultures protect their interests by denigrating, shaming, and attacking boys and men who exhibit so-called 'feminine' traits or pursuits. The pressure on men to conform to traditional male ideology is relentless and destructive. Chapter 10 demonstrates its deleterious effects on love and family life and examines the link between femiphobia and gendered violence.

GENDER AND GENDERISM

Why is it a boy's worst nightmare to be called a 'girl'? Why do most men dread being like women? Why does the phallocratic state portray its external enemies as 'weak' and 'womanly' and women as the internal enemy? Why, asks psychologist Stephen Ducat, do many men perceive strong independent women, gay men, social welfare, and corporate regulation as threatening? Is their masculinity so tenuous that men must constantly prove it?[3] All these characteristics relate to femiphobia, the fear of being feminine or feminized.[4] In the 1980s, psychologists used the term 'femiphobia' to help explain men's reluctance to admit illness, to seek medical advice (that's what women do), and consequently to die prematurely.[5]

In the earlier psychoanalytic literature, Karen Horney had theorized that men suffer from sexual-performance anxiety in a way that women do not,[6] Bruno Bettelheim claimed that men's denial of their feminine tendencies lay at the root of their hypermasculine behaviors,[7] and Erich Neumann suggested that patriarchal normality is a type of fear of the feminine.[8]

As a culturally created phenomenon—albeit one that cannot be dissociated from biology and embodiment—femiphobia is part of and produced by the taboo against male femininity. Male-dominant cultures subordinate women and despise feminine wishes or behavior in boys and men.[9] In some contexts, appearing to be 'womanly' can put men's mental and physical wellbeing—even their lives—at risk. Male-on-male violence almost invariably opens with the accusation 'Cunt!' as if classifying a man as Woman justifies violating him.

What of other gender-related fears and prejudice? Transphobia, or trans-prejudice, is a set of antagonistic beliefs about and attitudes toward people who are gender-identity variant, whereas the broader notion of genderism refers to a set of unsympathetic judgments of and beliefs about people who do not present as stereotypical men or women.[10] Arguably, genderists perceive all variations from stereotypical gender-role behaviors as threatening.

Why is gender nonconformity often constructed as menacing? Psychologists posit interrelated factors such as religious and essentialist views, psychiatric pathologization, and societal ignorance, but, according to Sam Winter and colleagues, precise reasons are not known.[11] In postcolonial nations, additional factors may include militarization of the society, institutionalized homophobia, and a sense of emasculation by colonialism, neo-colonialism, and structural inequality.[12] Depth psychology would add that the genderist man's hatred of his own internal feminine aspects (which he projects onto others and then attacks) is aroused when he sees a feminine man. In contrast, he may perceive the masculine woman as mocking him or encroaching on his territory, threatening to usurp his privilege as a 'real man.' Genderist cultures frame female masculinities as rejected scraps of dominant masculinity, says Judith Halberstam, so that they can label male masculinity 'the real thing.'[13] Perhaps this is one reason why conservative critics in the United States are complaining that allowing women into the military is undermining the integrity of the nation.[14]

This book has much to say about how the taboo against male femininity affects masculine gender enculturation; it has little to say about other groups targeted by the generic taboo on gender nonconformity. That is not to devalue the experiences of these groups but rather to say that the book is concerned with the effects of institutionalized femiphobia (both men and women can be femiphobic). Furthermore, because people's inner, subjective fears and prejudices about gender do not necessarily correlate with their gender performance (although there will be links in some cases), so-called feminine men and women are not necessarily prejudiced against masculinity, and so-called masculine women and men are not necessarily femiphobic. Stemming as it does from gender hierarchy, femiphobia may be involved in many forms of genderism.

If femiphobia is the fear of being feminine or feminized, what is the meaning of the term 'femininity' in this context and in biocultural theory generally? Essentialist notions of femininity and masculinity stretch back at least to Aristotle. Recall that, for Aristotle, only the male is the product of normal fetal development. The non-male fetus is 'misbegotten' due to one or more inauspicious variables.[15] While Aristotle was by no means the first genderist, from his undisguised anti-woman hierarchy, two millennia of essentialist theories followed.

A twentieth-century exemplar is Lewis Terman and Catherine Miles's sex-role identity theory,[16] which posits the existence of invariant masculine and feminine essences, normative constellations of traits, attitudes, and interests shown by members of the given sex.[17] Essentialist psychoanalytic theories have dwelt on the developmental ramifications of the differing nature and level of the infant boy's—compared with the girl's—identification with his mother and of his putatively stronger need to separate or disidentify from her,[18] but more gender-inclusive dialectical models depict this as a culturally mandated psychic separation from maternal and paternal caregivers.[19]

From the 1980s onward, the essentialist paradigm largely gave way to social constructionist and dialectical models, which envisaged gender as plural, co-constructed, relational, and situational.[20] Femininities and masculinities, these models say, are co-constructed in the act of relating to others within a given cultural context, particularly its power relations including class and race.[21] They are culturally constructed norms for male and female behaviors as seen by the dominant culture. By such norms of masculinity and femininity, societies decide which individuals are worthy members of their sex as well as worthy moral actors in general.[22] Throughout history, societies have marginalized and punished gender-nonconforming women and men via such judgments.

Rejecting the idea of fixed trait- or character-based masculinity, Connell proposed the notion of 'hegemonic masculinity' to encapsulate a type of masculinity that occupies the current hegemonic position in a given pattern of gender relations.[23] Hegemonic masculinity relates to historically flexible and locally based hierarchies of masculinities. At any given time, says Connell, 'one form of masculinity rather than others is culturally exalted'; it legitimizes the current form of patriarchy and thereby guarantees the continuing dominance of men and the subordination of women.[24] Hegemonic masculinity, then, is a pattern of practice distinguishable from subordinate masculinities. Although not always the most commonally observed or applied, hegemonic masculinity nevertheless presents itself as exemplar. The term 'complicit masculinity,' on the other hand, describes a non-dominating practice that nevertheless receives the benefits of patriarchy.[25]

WHAT COST ENTITLEMENT?

> *'I divorce my son!' she said to her friend. 'I go there and he has an apron on and he's doing the dishes! I didn't bring up my son to do this! He's a woman! I divorce him!!'*
>
> Greek-Australian woman, 2013.

Contemporary scholars differ in their emphasis on cultural-political versus biological drivers of femiphobia, particularly regarding the theorized need for boys to disidentify from the feminine. A survey of contested models of gender socialization is beyond the scope of this book; suffice to say that, while the broader cultural turn spawned models that de-hierarchized gender, others simply seemed to invert the hierarchy, particularly in regard to counseling unhappy men. What became known as 'deficit models' of men's relational capacities generated intense criticism and some posed a series of alternative factors that may inhibit men's expressivity. Others argued that 'feminist' ideas about men needing to learn skills from women merely reinforced notions of men's 'defectiveness.' Some scholars criticized leading feminists' focus on gender as difference, while others applauded its explicit respect for the feminine. Minimizing difference, on the other hand, potentially opens the way to equal opportunity for women. On numerous psychological measures, Janet Shibley Hyde's meta-analysis showed that men and women are very similar, along with moderate differences in physical aggression and large differences in some aspects of sexuality.[26]

Some theorists focus less on macro-structural factors such as institutionalized patriarchy and more on how gender-role enculturation is more developmentally challenging for boys than for girls.[27] While the boy's struggle to separate from his mother can be a feature of healthy development, says Michael Diamond, repudiation or disidentification from the feminine can bring about a brittle masculine identity and unhealthy narcissism.[28]

And whereas little boys go through a premature and abrupt separation from their mothers, girls enjoy a prolonged symbiotic maternal attachment.[29]

As a result, in comparison to girls, boys may be vulnerable to traumatic actual separations—disruptions later experienced as loss or abandonment. Here, potentially, says William Pollack, is a developmental basis for gender-specific vulnerability to the 'traumatic abrogation of the early holding environment.'[30] While some scholars place the mother-son duo center stage, others include the entire cast, claiming that the disconnection is not from the mother per se but from an empathic relationship that happens to be with the mother. Everything in the culture forces the boy in the name of 'growth' to disconnect in order to achieve maleness. The boy's disconnection is not just from his mother but also from the very process of growth in relationship. He learns to turn away from the whole relational mode and hence 'never really learns *how* to do it,' namely, be in relationship with another and grow.[31] As the boy matures, he may denigrate the relational mode and even discount it as possible. As the theory goes, men may experience developmental deficits in their capacity for intimacy, empathy, and relational commitment and adopt defensive, pseudo-self-sufficient character styles. Once thus characterized, they are easily shamed into stifling the expression of a range of needs and feelings.[32]

This male experience of shame is a primary self-affect embedded in the man's gender identity, with shame-based community sanctions enforcing gender-role conformity.[33] Do men and women experience shame similarly? Brené Brown and colleagues interviewed seven hundred men and women about their experiences of shame and concluded that shame is a deeply gendered psycho-social-cultural construct. They defined shame as 'the intensely painful feeling or experience of believing we are flawed and therefore unworthy of love and belonging.'[34] Although the way in which men and women experience shame is similar, the reasons they experience it are highly gendered. Whereas the critical social message to men is, 'Don't be weak,' to women it is, 'Do it all, do it perfectly, and take care of everyone while you're doing it.'[35]

Compared to men, women have more direct shame experiences; they learn how to deal with shame in relationship rather than allowing it to destabilize them psychologically. The cost, however, is that women suffer more shame-oriented depression and are twice as likely as men to take anti-depressant medication.[36] Men, on the other hand, are enculturated to experience shame states as threats to their gender identity and psychological wellbeing and are more likely to act out impulsively. On penalty of exclusion and denigration, phallocentric cultures force men to disidentify with the feminine and to adopt rigid stereotypical attitudes that require the suppression of vulnerability, tolerance, intimacy, and empathy. Under such circumstances, shame feels intolerable and typically induces impulsive, fight-or-flight reactions. Such unmanaged shame experiences can result in men becoming isolated, both socially and emotionally.[37]

In this view, manhood is more psychologically problematic and difficult to achieve than womanhood, and boys must accomplish it through a variety of ongoing tests.[38] In a study of U.S. college students, Joseph Vandello and colleagues found that students perceived manhood (in contrast to womanhood) as an elusive and tenuous state that men must prove constantly and publicly.[39] In response to negative feedback about their gender status, participating men experienced heightened feelings of threat and physically aggressive thoughts, seemingly anxiety-based responses to the destabilizing of their gender status.[40]

Theories of masculinity argue that striving for manhood in the face of threat is a near ubiquitous experience for men. The hierarchical nature of masculinity, its narrowly defined ideal traits, and the esteem accorded to it place men's gender behavior under constant surveillance.[41] Men typically measure themselves and other men against a monolithic standard of hegemonic masculinity that prescribes socially desirable behaviors including competitiveness, assertiveness, physical strength, aggression, risk-taking, courage, heterosexuality, and lack of feminine traits.[42]

What men dread most, says Michael Kimmel, is being shamed or humiliated in front of other men. Boys learn that their peers are gender police, ever alert and threatening to unmask them as 'sissies.' A boy surrounded by a group chorusing 'He's a sissy!' will either burst into tears and run home in disgrace or fight the group to prove he's not a sissy.[43]

Homophobia, then, Kimmel says, is men's fear of being unmasked by other men, emasculated, and revealed as not measuring up, as not real men. This fear is a source of shame and must be covered up (along with any feminine traits) by displays of hypermasculinity. And because masculine status is hierarchical, one man's gain is another's loss: insulting another man's masculinity is a way of establishing one's own. Because men who fear being perceived as unmanly often deny manhood to other men,[44] masculine insecurity is contagious.

A long-held thesis of masculine overcompensation has it that when men who doubt their masculinity perceive it as under threat, they tend to react with hypermasculinist displays.[45] Testing this thesis experimentally for the first time, Willer and colleagues explored the effects on men and women of receiving random feedback about their gender characteristics. Women who received a 'masculine' rating showed no effects. Compared to men rated as 'masculine,' men rated as marginally feminine went on to express higher scores on factors pretested as culturally associated with masculinity (e.g., homophobia, support for war, and interest in purchasing an SUV).[46]

Furthermore, related results indicated that 'threatened' men showed greater homophobic and pro-dominance attitudes and belief in male superiority. Societal change perceived as threatening to male status appeared to have a similar effect to individual-level threat. The authors suggested that these men's response was a reaction formation: when the men were told that they possessed an unacceptable trait, they overcompensated by enacting its extreme opposite.[47]

What are the psychosocial roots and power dynamics of femiphobia? According to object-relations theory, psychosocially, children come to see themselves as being like those who nurture them, but phallocentric cultures will not abide the idea of boys identifying with their mothers.[48] Back in the 1950s, researchers asked children if they wanted to be mothers or fathers when they grew up. Boys aged four and over identified with their same-sex parent, but a majority of three-year-old boys said they wanted to be mothers.[49] Perhaps this finding reflects the relative physical and psychological absence of men within the traditional nuclear family, particularly in the 1950s. The little boy who identifies with his mother suffers socially for expressing it. As the theory goes, the boy may experience the thwarting of his maternal identification as a sense of shame, inadequacy, and lack—feelings that he denies and represses.[50] Out of this psychosocial conflict, the boy develops an abiding but unconscious envy of the mother for possessing feminine attributes that, biologically, socially, or culturally, he cannot own or express. This envy eventually extends to all women. However, envy entails hatred for the desired object because it reminds one of one's lack and dependency.[51] The boy puts these conflicts out of his mind via defensive strategies such as idealizing, appropriating, provoking envy in others, devaluing women, and transforming love and longing into hate and fear.[52] For what society has done to him, the boy blames women.

Patriarchal gender norms, says Ducat, harm girls and boys in different ways. Boys, for example, learn that cross-gender behavior is taboo. In Freud's view, taboo is an ancient prohibition originally imposed on a group (probably by a previous generation) concerning actions for which a strong desire existed. Taboo cultures, he says, assume 'an ambivalent attitude toward their taboo prohibitions; in their unconscious they would like nothing better than to transgress them but they are also afraid to do it.' Just as in the neurotic, the basis of taboo is 'a forbidden action for which there exists a strong inclination in the unconscious.'[53] While Freud's interpretation lacks political nuance, it does shed light on femiphobia as a psychosocial and political phenomenon. I return to Freud's theory in chapter 9. Men's participation in the enforcement of the taboo, says Ducat, helps them keep hidden from themselves as well as others their mainly unconscious inkling that they are in fact or want to be like women.[54] Here, however, the psychological and the political merge; such a taboo is necessary only in anti-women cultures.

The purpose of this short chapter has been to outline the theoretical foundation of femiphobia. Part II of this book examines femiphobia, male monogenetic theory, and seed-and-soil during certain historical periods from Genesis to Freud.

NOTES

1. O'Donovan, "Health and Femiphobia"; Bailey, "Gender Identity."
2. Bourdieu, *Masculine Domination*, 50.
3. Ducat, *Wimp Factor*, 24.
4. O'Donovan, "Health and Femiphobia"; Bailey, "Gender Identity."
5. O'Donovan, "Health and Femiphobia"; "Femiphobia: Unseen Enemy of Intellectual Freedom."
6. Karen Horney, *Feminine Psychology* (New York: Norton, 1993), 145.
7. Bruno Bettelheim, *Symbolic Wounds: Puberty Rites and the Envious Male* (New York: Collier Books, 1962), 148–149.
8. Erich Neumann, *The Fear of the Feminine and Other Essays on Feminine Psychology* (Princeton, NJ: Princeton University Press, 1994), 261.
9. Ducat, *Wimp Factor*.
10. Mark King, Sam Winter, and Beverley Webster, "Contact Reduces Transprejudice: A Study on Attitudes towards Transgenderism and Transgender Civil Rights in Hong Kong," *International Journal of Sexual Health* 21, 1, (2009); Darryl Hill and Brian Willoughby, "The Development and Validation of the Genderism and Transphobia Scale," *Sex Roles* 53, 7–8, (2005), 534; Sam Winter, et al., "Transpeople, Transprejudice and Pathologisation: A Seven-Country Factor Analytic Study," *International Journal of Sexual Health* 21, (2009), 97.
11. Winter et al., "Transpeople, Transprejudice," 98.
12. Sarah Coughtry, "Patriarchy and the Trap of Masculinity" (MA thesis, Smith College, 2011), 100.
13. Judith Halberstam, *Female Masculinity* (Durham, NC: Duke University Press, 1998), 1.
14. Claire Sjolander and Kathryn Trevenen, "One of the Boys? Gender Disorder in Times of Crisis," *International Feminist Journal of Politics* 12, 2, (2010): 163.
15. Aristotle, *Generation of Animals*, 46.
16. Lewis Terman and Catherine Miles, *Sex and Personality: Studies in Masculinity and Femininity* (New York: McGraw-Hill Book Company, 1936); Sigmund Freud, "Some Psychological Consequences of the Anatomical Distinction between the Sexes," (*SE*, vol. 19, 1925); Sigmund Freud, "Female Sexuality," (*SE*, vol. 21, 1931); Sigmund Freud, "Femininity," in *New Introductory Lectures on Psycho-Analysis* (*SE*, vol. 22, 1932).
17. Ronald Levant and William Pollack, introduction to *A New Psychology of Men*, eds. Ronald Levant and William Pollack (New York: Basic Books, 1995), 3.
18. Nancy Chodorow, *The Reproduction of Mothering* (Berkeley: University of California Press, 1978); Nancy Chodorow, *Feminism and Psychoanalytic Theory* (New Haven, CT: Yale University Press, 1989); Ralph Greenson, "Dis-identifying from Mother: Its Special Importance for the Boy," *International Journal of Psychoanalysis* 49, 3, (1968).
19. William Pollack, "No Man is an Island: Toward a New Psychoanalytic Psychology of Men," in *A New Psychology of Men*, eds. Ronald Levant and William Pollack (New York: Basic Books, 1995), 35.
20. Joseph Pleck, *The Myth of Masculinity* (Cambridge, MA: MIT Press, 1981); Michael Kimmel, *Changing Men: New Directions in Research on Men and Masculinity* (New York: Sage Publications, 1987); Connell, *Masculinities*; M. Zinn, P. Hondagneu-Sotelo, and M. Messner, *Gender through the Prism of Difference* 3rd ed. (New York: Oxford University Press, 2005).
21. Ann McKinley, "Identities Cubed: Perspectives of Multidimensional Masculinity Theory," Scholarly Works, Paper 761.
22. Gilmore, *Manhood in the Making*, 10.
23. Connell, *Masculinities*, 76.
24. Connell, *Masculinities*, 77
25. Connell and Messerschmidt, "Hegemonic Masculinity."
26. Pollack, "No Man is an Island," 57; Martin Heesacker and Shawn Prichard, "In a Different Voice Revisited: Men, Women and Emotion," *Journal of Mental Health Counseling* 14, (1992); Joel Wong and Aaron Rochlen, "Demystifying Men's Emotional Behavior: New Directions and Implications for Counselling and Research," *Psychology of Men and Masculin-*

ity 6, 1, (2005); Rachel Hare-Mustin and Jeanne Merecek, "The Meaning of Difference: Gender Theory, Postmodernism and Psychology," *American Psychologist* 43, (1988); Janet Shibley Hyde, "The Gender Similarities Hypothesis," *American Psychologist* 60, 6, (2005): 586. Large differences exist in aspects of sexuality such as incidence of masturbation and attitudes about casual sex.

27. Levant, "Reconstruction of Masculinity."
28. Diamond, "Masculinity Unravelled," 1.
29. Levant, "Reconstruction of Masculinity," 242.
30. Pollack, "No Man is an Island," 41.
31. Stephen Bergman, "Men's Psychological Development: A Relational Perspective," (Lecture, Wellesley Center for Research on Women: Work in Progress, 1991), 4–5.
32. Levant and Pollack, *New Psychology of Men*, 5.
33. Steven Krugman, "Male Development and the Transformation of Shame," in *A New Psychology of Men* eds. Ronald Levant and William Pollack (New York: Basic Books, 1995), 121.
34. Brené Brown, Virginia Rondero Hernandez, and Yolanda Villarreal, "Connections: A 12-Session Psychoeducational Shame Resilience Curriculum," in *Shame in the Therapy Hour,* eds. Ronda Dearing and June Tangney (Washington, DC: American Psychological Association, 2011), 358.
35. Brown et al., "Connections: A 12-Session Psychoeducational Shame Resilience Curriculum," 358.
36. Krugman, "Male Development," 121.
37. Krugman, "Male Development," 122.
38. Gilmore, *Manhood in the Making*, 11.
39. Joseph Vandello et al., "Precarious Manhood," *Journal of Personality and Social Psychology* 95, 6, (2008): 1326.
40. Vandello et al., "Precarious Manhood," 1325.
41. Robb Willer et al., "Overdoing Gender: A Test of the Masculine Overcompensation Thesis," *American Journal of Sociology* 118, (2013): 1013.
42. Willer et al., "Overdoing Gender."
43. Kimmel, "Masculinity as Homophobia," 131–32.
44. Kimmel, "Masculinity as Homophobia," 131, 135.
45. Alfred Adler, "Inferiority Feeling and Masculine Protest," in *The Individual Psychology of Alfred Adler: A Systematic Presentation in Selections from his Writings,* ed. Heinz Ansbacher and Rowena Ansbacher (New York: Basic Books, 1956); Bourdieu, *Masculine Domination*; Michael Kimmel, *Manhood in America: A Cultural History* (New York: The Free Press, 1996); Ducat, *Wimp Factor.*
46. Willer et al. "Overdoing Gender," 991.
47. Willer et al. "Overdoing Gender," 982.
48. Ducat, *Wimp Factor.*
49. Ducat, *Wimp Factor*, 32.
50. Ducat, *Wimp Factor*, 33.
51. Ducat, *Wimp Factor*, 33–34.
52. Ducat, *Wimp Factor*; Melanie Klein, *Envy and Gratitude: A Study of Unconscious Sources* (London: Tavistock Publications, 1957), 62–67.
53. Sigmund Freud, *Totem and Taboo: Resemblances between the Psychic Lives of Savages and Neurotics* (London: George Routledge and Sons, 1919), 53–55.
54. Ducat, *Wimp Factor*, 2; Gary Brooks, "Masculinity and the U.S. Presidency," *Psyccritiques* 50, 10, (2005); David Lotto, "Review of *The Wimp Factor* by Stephen Ducat" *Journal of Psychohistory* 34, 2, (2006): 171–73; Sjolander and Trevenen, "One of the Boys?" There appears to be little scholarly criticism of Ducat's thesis beyond the typical and justifiable rebuke of the psychoanalytic habit of couching hypothesis as fact. While it is entirely plausible that men might desire the procreative power of Woman and the freedom from always having to 'act like a man,' Ducat's claim must stand as hypothesis rather than fact.

Part II

Seed and Soil: From Abraham to Freud

Chapter Three

Abraham's Seed

Side by side, father and son walk the trail to the mountain to worship. The father carries the fire and the knife; the boy carries the firewood.
'My father.'
'Here I am, my son.'
'Behold the fire and the wood: but where is the lamb for the burnt offering?'[1]

This chapter examines the roots of seed-and-soil in Genesis, particularly in the story of Abraham. Who begat whom—the patriline that stems from God—is central to Genesis, no more so than to the patriarchal narratives, which celebrate the founding of Abrahamic monotheism with its assumption of male monogenesis, institutionalization of Father right, and requirement for conquest. Male monogenetic theory, says Delaney, glorifies Man as God's chosen vehicle for creating new life and, concomitantly, demeans Woman as little more than a medium for that divine creation.[2] Indeed, Genesis has it that males contain females within themselves and hence do not need women to reproduce:[3] 'And the rib, which the Lord God had taken from man, made he a woman, and brought her unto the man. And Adam said, "This is now bone of my bones and flesh of my flesh: she shall be called Woman, because she was taken out of man."'[4] Male monogenetic theory is at the heart of the three largest Abrahamic religions: Judaism, Christianity, and Islam.[5] Although the theory is obvious, says Delaney, as well as critical to notions of gender, family, kinship, and state, it is usually dismissed as no more than metaphor, despite the fact that metaphor conditions how we imagine things to be.[6]

The ancient Greek canon, too, institutionalizes male monogenetic theory, for example, in Hesiod's *Theogony*, which celebrates Zeus's rise to the top of the Greek pantheon, and in Aeschylus's *Eumenides*, which commemorates

Zeus's decisive overthrow of Mother right and institution of patriarchal law. While *Theogony* is a product of the eighth century BCE and the patriarchal narratives a century prior, both sources refer back to a time when Abraham and Zeus were establishing themselves as fathers of new faiths.[7] Chapter 5 examines seed-and-soil in ancient Greece.

MALE MONOGENETIC THEORY

According to male monogenetic theory, only the male has seed and, hence, only the male is the true (begetting) parent. In the biblical version, Adam receives God's divine seed, which carries both identity and soul, to transmit to his sons, to be transmitted to their sons, and so on throughout eternity. The theory constructs women as either fertile or barren, says Delaney; it symbolically allies women with the earth, the field, and that which God created. And although fathers endow their daughters with soul, daughters are unable to transmit it.[8]

Male monogenetic theory was by no means universal in antiquity. Alternative theories of procreation existed then and still do. In some Australian Aboriginal societies, for example, there is no notion of blood relationships or physiological connection between child and mother or father; rather, the embryo is constructed as a reincarnated spirit or matrilineal ancestor. The same spirit-child could have entered any number of different women.[9]

In addition, to say that seed-and-soil stretches back into prehistory and is still extant is not to say it is timeless or immutable. As one of many persistent naturalizers, however, of patriarchal status quo across Western cultures, the trope has been singularly effective. The assumption that the male is the only generative agent in procreation, says Delaney, is implicit in the Abrahamic myth, which marks the beginning of a new order in which gender, creativity, and divinity are inseparable.[10] Like Delaney, I adopt a secular view of the myth, namely, that it has been constructed and reconstructed over millennia and is culturally significant whether or not Abraham ever lived or if the events described actually occurred.[11]

Whereas the stories of Adam and Eve, Cain and Abel, Noah and the Flood, and the Tower of Babel have parallels in the ancient Sumerian, Akkadian, and Babylonian literature, says Delaney; the myth of Abraham, the first Western patriarch, is found only in Genesis.[12] It features in the patriarchal narratives (typically those dealing with Abraham, Isaac, Jacob, and Joseph in Genesis 12–50)[13] and contains the story known as the Binding of Isaac ('Akedah' in Judaism), which glorifies Abraham's willingness to sacrifice his son on God's command. In Delaney's view, the story underpins the institutionalization of patriarchy in Western culture; it establishes the model of faith for Hebrew monotheism and the model for how far this faith must be taken to avoid obliteration by a jealous deity.[14]

The Abrahamic myth is set in the second millennium BCE. As no extra-biblical sources establish Abraham as a historical figure, scholars cannot trace the story to a period prior to its appearance in the Hebrew Scriptures, the dating of which is contested. Some archaeologists say there are no surviving fragments dating back to before the third century BCE, whereas some Hebrew scholars maintain a provenance going back to the beginnings of Israelitic culture.[15] Delaney and others place the source of the story at around the ninth century BCE.[16] The setting is the Mesopotamian city of Ur (south of modern-day Baghdad), during the reign of Nimrod of Babylon.[17] According to Genesis, God calls Abraham to leave his home and travel to a new land:

> I am the Almighty God; walk before me, and be thou perfect. . . . As for me, behold, my covenant is with thee, and thou shalt be a father of many nations. . . . And I will give unto thee, and to thy seed after thee, the land wherein thou art a stranger, all the land of Canaan, for an everlasting possession; and I will be their God. . . . And ye shall circumcise the flesh of your foreskin; and it shall be a token of the covenant betwixt me and you. And the uncircumcised man child . . . shall be cut off from his people; he hath broken my covenant.[18]

We can read Genesis, says Delaney, as the legitimating document for male monogenetic theory and the basis of patriarchy: 'symbolically, the monogenetic theory is the human analogue of divine, monotheistic creation,' and Abraham is its symbol.[19] In return for God's gift of the Promised Land and his vow to multiply Abraham's seed 'as the stars in heaven,' Abraham and his tribesmen must undergo circumcision. The sign of God's covenant with Abraham, says Delaney, was carved on the penis, the organ understood as 'the fountain of generativity, the vehicle for the transmission of seed.'[20] Clearly, she says, women are not part of the Covenant.[21]

In exchange for Abraham's willing obedience, God promises the lands of Palestine-Canaan to Abraham's seed in perpetuity. Exodus describes the nature of the settlement:

> The Lord is a man of war. . . . Thy right hand, O Lord, is become glorious in power; thy right hand, O Lord, hath dashed in pieces the enemy. . . . The people shall hear and be afraid: sorrow shall take hold on the inhabitants of Palestina . . . all the inhabitants of Canaan shall melt away.[22] . . . For I will cast out the nations before thee and enlarge thy borders.[23]

The patriarchal narratives glorify conquest, appropriation, and enslavement of the Other. Such expansionary aggression requires rapid deployment of resources; men own and enclose their assets—land, livestock, slaves, wives, and children—and mobilize these for the war effort; women (and those men who would) must not interfere with the conscription of sons; and sons must obey. For such a system to work, fathers must be held as the primary (begetting) parents and must willingly sacrifice their children on the altar—religious or secular—of patriarchal authority.[24] In other words, the doctrine of male monogenesis fits and serves the subjugator mindset.

Just as seed-and-soil depicts the male as active and the female as passive, so it constructs male sacrifice in tribal and religious conquest as active, intentional, controlled, noble, and pure. Female sacrifice, on the other hand, comprises menstruating and giving birth, processes constructed as passive, unintentional, uncontrolled, profane, and polluting, but which nonetheless replenish the stock of warriors.[25]

Given that in all likelihood humankind in the ancient Middle East had been aware of men's role in procreation since around 17,000 BCE,[26] when was the idea inflated into a theory of male monogenesis? When did men appropriate to the male alone the divine capacity to generate life? Does male monogenetic theory have a special link to Abrahamic monotheism, as Delaney claims, or were versions of the theory integral to monotheism's polytheistic past and peers? As the debate is highly specialized as well as contested, I summarize just one point. There appears to be little scholarly doubt that male monogenetic theory existed in, and gradually came to dominate, many ancient polytheistic cultures. For this reason, Kenneth Stone queries Delaney's emphasis on the correlation between male monogenetic theory and Abrahamic monotheism.[27] The crucial dimension of monogenesis is not its link with monotheism, Stone concludes, but rather, 'the fact that it attributes to male gods (and male humans) the principal role in the generation of life.'[28] And in polytheistic Egypt, says Inhorn, written versions of male monogenetic theory persisted throughout three millennia and across myriad versions of creation myths and deities.[29] Not only myth but also most of the literate medical traditions that have existed in Egypt, she says, supported monogenetic theories of procreation. When European colonial biomedicine (first French and then British) entered Egypt in the early part of the nineteenth century, the European penchant for preformation models of procreation lent considerable support to the established monogenetic models under pharaonic and Greek medical systems.[30]

Among numerous versions of ancient Egyptian creation myths, one prehistoric line of narrative has the primordial-waters Goddess generating the sun-god Atum, who then creates the rest of the universe.[31] The Pyramid Texts contain two accounts of how Atum brought other gods into existence by masturbating: 'All manifestations came into being after I developed. . . . I created on my own every being. . . . My fist became my spouse. . . . I copulated with my hand.'[32] By the time of the Pharaonic Texts, says Leeming, 'the male force had achieved dominance.'[33] Other scholars agree:

'Atum's act of masturbation not only shows that the male factor in generation was already understood: it indicates that the male factor was regarded as decisive.'[34] Amid massive cross-cultural diversity, many narratives redacted over centuries contain particular continuities:

> First, the demotion of the Mother-Goddess figure and the ascendance and later dominance of her male consort/son; then his merging with a storm-god into a male Creator-God, who heads the pantheon of gods and goddesses. Wherever such changes occur, the power of creation and fertility is transferred from the Goddess to the God.'[35]

The same goes for the polytheistic origins of Abrahamic monotheism. The Genesis compilers, ignoring history and legend as they knew it, constructed God as preceding the universe itself:[36] there is only one (male) God who transmits his creative capacity to mortal men. 'Only men produce seed and only men have the power to transmit the line.'[37]

Given the evidence of male monogenetic theory in ancient polytheistic cultures, is Delaney actually claiming that the theory originated with monotheism? Or is she saying, less contentiously, that Genesis makes pivotal its association with Abraham's God? The doctrine of male monogenesis, Delaney says, imagines procreation 'as the vehicle of channeling divine creativity to earth. And ever since Abraham, it is men who embody the power to do so':

> Seed is the essence of the father, and his child is of the same essence. Because the seed 'belongs' to the man, so too does the child belong to him in a way that it does not belong to the mother. In addition to being of the same essence as the father, a male child additionally has the ability to reproduce that essence. Father and son *are* one. This is the Christian definition of God.[38]

As Genesis has it, from the very beginning, God constructed Woman as inherently inferior and 'naturally' subservient to Man. And not only did Abrahamic monotheism systematize, institutionalize, and canonize male privilege and the subordination of women as supposedly God given, it similarly legitimized conquest and enslavement of all those defined as Other. By historical accident, says Huber, an ancient tribe of Hebrew herders uniquely influenced the modern world when their gender norms became embedded in laws and customs across much of Eurasia and North Africa owing to Christian and Muslim political and military victories.[39] According to those norms, a woman is owned by a master, first her father, then her brothers, her husband and, as she could not inherit, her eldest son. Judaism, Christianity, and Islam, as well as secular Western thinking, says Blumenthal, all have their roots in the patriarchal-rural life of ancient Israel.[40]

Genesis is acutely concerned with the interconnected issues of paternity, legitimacy, inheritance, and succession: 'who shall be chosen to carry on the divinely ordained line?'[41] These issues are about the protection of the divine seed and its procession down the patriline. Whereas the woman who is a virginal bride and an unfailingly faithful wife bolsters her husband's confidence in his paternity, the lack of such trust can mean death for the wife. According to Deuteronomy, the penalty enforced on a bride unable to produce evidence of her premarital virginity is death by stoning:

> If any man take a wife . . . and say, I took this woman, and when I came to her, I found her not a maid . . . if this thing be true, and the tokens of virginity be not found for the damsel: Then they shall bring out the damsel to the door of her father's house, and the men of her city shall stone her with stones that she die: because she hath wrought folly in Israel, to play the whore in her father's house.[42]

Furthermore, as portrayed in Genesis, efforts to protect the patriline extended to passing over the firstborn son for blessing and inheritance in favor of the second son. This pattern, says Delaney, began with the first children of Adam and Eve. The firstborn son, Cain, was a tiller of the ground; his brother, Abel, kept sheep. Both brought offerings to God:[43]

> Cain brought of the fruit of the ground an offering unto the Lord. And Abel, he also brought of the firstlings of his flock and of the fat thereof. And the Lord had respect unto Abel and to his offering: But unto Cain and to his offering he had not respect. And Cain was very wroth . . . and it came to pass, when they were in the field, that Cain rose up against Abel his brother, and slew him.[44]

Why did the writers of Genesis want to show that God passes over the firstborn? Even if primogeniture later became the norm in Israelite society, says Delaney, there is no reason to assume it was so in Abraham's society.[45] While sociological analyses of the Cain and Abel story are often class based,[46] Delaney enquires into the gendered nature of the story and the theological motives of the compilers. Whereas we can read Cain's offering—the fruits of the soil—as associated with the feminine, Abel's offering of the firstlings of the flock is associated with blood sacrifice and the patriline. The pattern in Genesis in which firstborn children are passed over for inheritance, suggests Delaney, relates to men feeling more secure about their paternity of a second child than of a firstborn.[47]

THE BINDING OF ISAAC

The pattern of anxiety about seed, paternity, God's covenant, and the purity of the patriline is repeated in the story of Abraham. Childless in their old age,

Abraham and his wife Sarah travel to Canaan, the land promised by God to Abraham's seed, a province of the Babylonian Empire on the eastern Mediterranean seaboard (the Levant).[48] In Canaan, Abraham fathers two sons, Ishmael, who Abraham banishes, and Isaac:[49]

> And it came to pass after these things, that God did tempt Abraham, and said unto him . . . Take now thy son, thine only son Isaac, whom thou lovest, and get thee into the land of Moriah; and offer him there for a burnt offering upon one of the mountains which I will tell thee of. And Abraham rose up early in the morning . . . and took the wood of the burnt offering, and laid it upon Isaac his son; and he took the fire in his hand, and a knife; and they went both of them together. And Isaac spake unto Abraham his father, and said, My father: and he said, Here am I, my son. And he said, Behold the fire and the wood: but where is the lamb for a burnt offering? And Abraham said, My son, God will provide himself a lamb for a burnt offering: so they went both of them together. And they came to the place which God had told him of; and Abraham built an altar there, and laid the wood in order, and bound Isaac his son, and laid him on the altar upon the wood. And Abraham stretched forth his hand, and took the knife to slay his son.[50]

As Abraham raises his knife to slit Isaac's throat, God sends an angel to stay Abraham's hand. The angel points to a thicket in which a ram is trapped, a ram intended for sacrifice in Isaac's stead. Isaac, it seems, is saved and God promises to reward Abraham's willing obedience: 'I will multiply thy seed as the stars of the heaven and as the sand which is upon the sea shore; and thy seed shall possess the gate of his enemies. And in thy seed shall all the nations of the earth be blessed; because thou hast obeyed my voice.'[51] According to Genesis, Abraham returns home without Isaac, and Sarah dies shortly thereafter, hence prompting scholarly speculation that Isaac may actually have been sacrificed.[52] Jewish Midrashim offer countless variations on the story but these do not bear substantively on my arguments.

According to religious scripture, Abraham's line, through Isaac, leads directly to the twelve tribes of Israel and eventually to Jesus of Nazareth and, through Ishmael, to the twelve tribal chiefs of the Arab peoples and to the prophet Muhammad, seventh-century founder of Islam.[53] Early Christian exegeses of the Akedah depict it as prefiguring the death of Jesus. By this interpretation, the figure of the ram represents Jesus—whose crucifixion completes the sacrifice interrupted by the angel on Moriah—thereby 'proving' Christianity's 'superiority' over Judaism.[54]

The Quranic version also contains substantive differences from the Hebrew. In the Quran, for example, Abraham tells the boy: 'O my dear son, I have seen in a dream that I must sacrifice thee. So look, what thinkest thou?' The Quran does not name this son but most Muslims assume he is Ishmael.[55]

According to the *Encyclopedia Judaica*, the myth has a predominantly ethical intention, namely, to symbolize the end of the practice of child sacrifice in the Middle East.[56] But is that the end of the story?

Parents in some parts of the Middle East still use the Akedah much as those elsewhere use tales of witches, wolves, and trolls. Children in Delaney's Turkish village told her the Akedah meant that, had God not provided the ram for the sacrifice, 'fathers would still be slaughtering their children.' And it was common for the villagers to threaten misbehaving children with: 'You'd better be good or I'll make a *kurban* [sacrifice] out of you.'[57] On the holiest day of the Muslim calendar, every male head of household in the village sacrifices a sheep and reminds every male child that, but for the grace of God, he might have been sacrificed. Hence, says Delaney, while the ritual appears to mitigate patriarchal power, it actually 'reinforces it on a cosmic scale.'[58]

What is the myth saying and to whom? Did the children get it right? Would fathers still be slaughtering children were it not for the Akedah, or is the story about the power of the father (God and his stand-ins) and his demand for absolute obedience? Laissez-faire sacrifice of youth for imperial gain, in other words? Delaney believes so. Prevailing exegeses, however, focus on everything *but* power. Søren Kierkegaard, for example, insists that the story is about the preeminence of faith.[59] Others say it is a tale of love: Abraham loved his son but he loved God more.[60] Then again, it could be an account of God's punishment of Abraham for casting out Ishmael, his first son.[61]

There is disagreement among Muslim scholars about certain details of the Akedah—whether the sacrifice was of Isaac or Ishmael, for example, and when the story entered the tradition—but little or no contestation that it is about faith.[62] The Akedah stresses the obedience of both Abraham and Ishmael, who symbolize what it means to be Muslim.[63] Abraham's great faith is proven by his obedience/submission to God. His son is a willing and obedient sacrifice: 'O my father! Do that which thou art commanded. Allah willing, thou shalt find me of the steadfast.'[64] The Quran establishes Abraham as faithful friend of God and unparalleled father of the faith: 'Who is better in religion than he who surrendereth his purpose to Allah while doing good (to men) and followeth the tradition of Abraham, the upright? Allah (Himself) chose Abraham for friend.'[65]

Why is there so little argument among Muslim scholars about the essential meaning of the Akedah? When it comes to worldview, says Delaney, despite local and regional variations, the Muslim ideal is 'a return to the pristine pattern of society and model of behavior' as conveyed by God through Muhammad.[66] Islam focuses not on societal transformation but on recuperation: 'the struggle to recapture and immobilize an eternal experience.'[67] Consequently, Muslim commentaries on Abraham are about absolute submission to God, with the Akedah encapsulating what Islam understands as: 'the utter appropriateness of its Arabic tradition. The dominant school of Sunni theology . . . denied that values might exist prior to God's commands. Infanticide is wrong because God forbids it; God does not forbid it because it is wrong.'[68]

What, then, of the claim by the Turkish village children and the *Encyclopedia Judaica* that the myth is telling the faithful that God no longer wants children slaughtered in His name? While the idea is plausible, Delaney says, it has little evidential basis; child sacrifice does not appear to have been common practice in the ancient Middle East; nor is it mentioned again in Genesis. Furthermore, she says, claims about its likely occurrence are based on highly contested archaeological evidence of much later provenance, as well as outdated assumptions that the earlier the time, the more barbaric the people. In actuality, human sacrifice, when it is found, is in societies with relatively complex forms of social organization including priests and temples.[69] In addition, if the purpose of the narrative was to end the practice of child sacrifice, says Delaney, wouldn't the authors have underlined that message? The Akedah itself suggests that animal—not human—sacrifice was the custom: 'Where is the lamb for the burnt offering?' Isaac asks. Moreover, a merely negative meaning—prohibition against a current practice—would not lend appropriate theological status to a foundational narrative for a new faith.[70]

Does the Akedah, then, symbolize the need to sacrifice *all*, including children, for faith in God, as Kierkegaard claims? Not only the life of a beloved child, suggests Kierkegaard, but also ethics itself, is less important than faith. Abraham resists the temptation of the ethical and makes the leap of faith, a paradox 'capable of transforming a murder into a holy act, well-pleasing to God.'[71] Abraham's act, says Delaney, can be distinguished from murder only if it can be seen as somehow more than human and therefore beyond human categories.[72] For Kierkegaard, Abraham is above the law: 'Abraham is great by reason of a personal [not moral] virtue. . . . Why then did Abraham do it? He did it for God's sake because God required this proof of his faith; for his own sake he did it in order that he might furnish the proof.'[73]

In the fifth century, Augustine argued similarly: 'Abraham was not only free from the guilt of criminal cruelty, but even commended for his piety, when he consented to sacrifice his son, not, indeed, with criminal intent but in obedience to God.'[74]

Many scholars agree. Abraham, they say, was asked to do what all subsequent generations of Jews would be asked to do: 'We have to really pay a price for our ideals and for our faith even if it means sacrificing our children.'[75] But this model of detaching oneself from affective ties is not just about loving God, says Delaney; it is also about defining masculinity: Abraham was heroic *because* he was able to cut the tie. Central to the Akedah is the idea of *willing* obedience. Just as Abraham shows his love for God by his obedience, that is how Isaac shows his love for his father.[76] On learning that he was to be sacrificed, Isaac, says first-century historian Josephus, 'was of such a generous disposition as became the son of such a father, and was pleased with this discourse; and said, that he was not worthy to be born at first, if he should reject the determination of God and of his father . . . even if his father alone had so resolved. So he went immediately to the altar to be sacrificed.'[77]

Isaac becomes the willing victim, says Delaney; his and his father's motives coincide. Abraham obeys God, Isaac obeys Abraham; the story establishes a hierarchy of command. The son's interests are irrelevant, or, rather, assumed to be identical with the father's, as the needs of Jesus were identical with God's.[78] The Akedah, says Delaney, constructs not only a model of faith but also a model of how far that faith is to be taken. Furthermore, the myth makes it seem sacrilegious to question these issues. Following Abraham, fathers channel God's will and stand in for God. And they can always say, 'I was only obeying a higher order.'[79]

Fifty years before Kierkegaard's exegesis, Immanuel Kant castigated Abraham:

> We can use, as an example, the myth of the sacrifice that Abraham was going to make by butchering and burning his only son at God's command (the poor child, without knowing it, even brought the wood for the fire). Abraham should have replied to this supposedly divine voice: 'That I ought not to kill my good son is quite certain but that you, this apparition, are God—of that I am not certain, and never can be, not even if this voice rings down to me from (visible) heaven.'[80]

For Kant, it is impossible to know that the voice we hear is the voice of God; thus, if the command is to go against moral law, we must consider it an illusion. For Derrida, however, the true nature of Abraham's sacrifice is of ethics to obligation. Responsibility and faith go together. By commanding Abraham to sacrifice his son, to offer a gift of death to God, God leaves him free to refuse. That, according to Derrida, is the test.[81] God's command to Abraham is to turn toward God as the 'unique one' and, in order to do that, make a gift of death. In essence, God says to Abraham: I see you understand what absolute duty toward the unique one means—responding when there is no reason to be asked—and furthermore that you have acted on it. At this point, Abraham is at the same time the most responsible and the most irresponsible of men, irresponsible in the face of the ethical because he responds to absolute duty.[82]

Carol Gilligan questions the ethics of those who would stand with Kierkegaard as dangerously abstracted from life. She contrasts the biblical Abraham with the biblical woman who appears before King Solomon—the woman whose motherhood is verified by her willingness to relinquish the truth, and thus her child, in order to save his life.[83] Will we choose to sacrifice our children for our faith or will we say, with Kant, that we can never be sure that our faith is not based on illusion? Will we put love of God before our love for children and each other? Or will we choose responsibility—to and for each other—over authority and obedience?[84] Genesis is first and foremost about obedience, says Delaney, and the Akedah is about power—the power of the fathers.[85] The priestly voice in the Hebrew Scriptures is obsessed with order and obedience, says David Janzen: God can dwell in Israel's midst only if Israel shows absolute obedience to His will.[86]

The Akedah, says Magda Romanska, prefigures the patriarchal nature of the sacred and the Western construction of masculinity as intimately connected with death. It provides the performative establishment of patriarchy as a vehicle for the divine and creates a 'fraternity of faith that demands and gives death as a price of belonging.'[87] Notions of belonging and death are key to identity narratives, which are chiefly directed at insiders who may threaten the nascent hierarchy. The message is that 'the punishment of otherness is death, and that insiders can easily become outsiders (others) by failure to submit.'[88] From this perspective, the main danger for ancient Israel is from within, namely, insubordination by 'individuals who do not strictly adhere to the moral code and the worldview that undergirds it.'[89] The biblical narrator, says Grace Jantzen, glories in the never-ending cycle of invasion, overthrow, slaughter, and occupation, orchestrated by God and into which the patriarchs throw their sons with relish—the constant spilling of enemy blood over centuries.[90] On the death of Moses, Joshua inherits the leadership of Israel and the Lord declares 'Whosoever he be that doth rebel against thy commandment . . . he shall be put to death'[91]

Israelite society, says Jantzen, represents masculinity as the attribute of engaging in violence against other men, and the biblical stories of 'violence, covenant, and conquest were accepted as historically true,' as representing God and his relationship to his people.[92] Because many of these stories were written or compiled during the Persian-Hellenistic period, says Lori Rowlett, they were intended as a narrative of self-identity via the past. After the period of subjugation by the Babylonians, the Israelites are unconsciously in the grip of the very structures they seek to reverse. The resulting violence remains hidden in plain sight in the texts because the adherents deny its centrality in the construction of their religion.[93] And the more conciliatory counter-narratives within the Bible are largely pushed aside by the realities of violence and segregation and the myth that feeds them: the Israelites are God's chosen people and God intends them to have the land of Canaan. The violence inflicted is thus mandated by God; it is holy war. In spite of significant dissent, says Rowlett, this ideology still shapes the contemporary Middle East.[94]

Dissenters include scholars who have tracked the tragic legacy of the alignment of Israel's national identity with the Abrahamic myth: 'It is precisely the endurance of "noble death", both martyric and military, within national narratives,' says Yael Feldman, 'that has rendered them both "theological" and, alas, violent.'[95] Modern Jewish literature, too, is undergoing a reorientation. Oft represented as 'faith beautiful,' says Michael Brown, the Akedah is now just as likely to stand for faith misguided and destructive: 'Abraham and Isaac no longer serve as role models to be emulated, as they did in pre-modern sources, but as object lessons to be avoided.'[96] For some contemporary Jewish writers, the Akedah has become a metaphor for 'the unfeeling, self-aggrandizing sacrifice of children by their actual or communal fathers on the altars of purblind commitment to Zionism and to Israel's wars.'[97] And Hebrew protest poets are revealing the ideological function of the myth, exploiting its persuasive power and surrounding sacredness to create an anti-myth to 'vanquish the original.'[98]

What, then, of power, gender, and violence in the Abrahamic myth? A critical purpose of blood sacrifice, for example, is to maintain kinship lines between men in patriarchal society.[99] And men's involvement with sacrifice stands in structural tension with women and childbirth;[100] indeed, says Jay, sacrifice can be seen 'as remedy for having been born of woman.' In the kinship forged among men by means of sacrifice, the blood of the sacrifice neutralizes the blood of kinship via birth and provides a rebirth (being 'born again') into the brotherhood. The motivation is to control both life and death.[101] Arguably, sacrifice is the most intentional form of life-taking,[102] and the Abrahamic model 'genders life-taking as a male enterprise.'[103] Life-making, too, is gendered as male, with the penis constructed as the instrument of procreation.[104] As portrayed in the Hebrew Scriptures, says Jantzen, the system of sacrifice that maintains kinship and purity among the Israelites is the counterpart to holy war against the external Other.[105]

The charge that Genesis is patriarchal does not go uncontested. Whereas some critics frame Delaney's argument as self-evident, others read it as a recipe for despair. Male domination in the scriptures, says critic Jenny Diski, 'has been noted before, and it is clear that neither God nor his chosen ones were signatories to the International Convention on Human Rights. That's a shame.'[106] In addition, says Diski, Delaney's call for a new moral vision, a new myth to live by, is 'to accept that we are helpless victims rather than interpreters of myth,' a 'bleak view of humanity's capacity for analytical thought.'[107] However, given Delaney's consistent emphasis on people's agency in interpreting, deconstructing, and reconstructing their myths and their world, her call for 'a new myth to live by' is hardly intended literally. In addition, Diski's don't-call-*us*-victims argument falsely pits the practice of criticism against the recognition of human agency.[108]

At the opposite pole, Harvey Goldberg sees nothing self-evident about Delaney's feminist assault on Genesis. Delaney, he says, 'insists on the gendered nature of the story,' in particular the seed-and-soil notion of procreation.[109] Goldberg counters Delaney's analysis with selected biblical examples: In the story, he says, 'Sarah appears immediately' and 'is also prominent' at the end. 'When Isaac marries, he brings Rebecca to the tent of "Sarah his mother".' 'Rather than the passive receptacle of Abraham's creative seed, Sarah is the impresario of Isaac's success.' And thus, 'women are not undifferentiated "soil."' Moreover, 'an alternative reading of the Abraham story is that it establishes the importance of the conjugal family within an overall "patriarchal" framework.'[110] Goldberg's reading, however, reduces the complex metaphorical import of seed-and-soil to dualistic literality and overlooks how violence and obedience, for example, bear heavily in shaping boys and men under patriarchy.

Patriarchy is not kind to men. Some die for their obedience, some for their disobedience, and some suffer the death by a thousand slights to their sense of manhood. That is the seed's side of the seed-soil bargain. In order to maintain superiority over the Other (whether soil-woman, soil-men, or outsiders) and remain inside the brotherhood, men must be willing to kill and be killed on the command of the father.

Abraham's obedience to God, says Delaney, 'is a model for the son's obedience to the father.'[111] As well as establishing this ethic of obedience, the Akedah legitimizes the doctrine of male monogenesis and paternal ownership of children.[112] Abraham's willingness to sacrifice his son was meant to convey the recognition that his love of God must come before his son and that everyone is dependant on God from whom 'all blessings flow.'[113]

The meaning of the story, says Delaney, is not in the ending of child sacrifice but in the establishment and canonization of Father right, the foundation of patriarchy. The idea of paternity became the essence of God.[114]

> Isaac said to his father, how pleasant is the altar you built me, my father, stretch out quickly, and take your knife, while I pray before my Lord.
>
> Uncover your arm and gird your loins, like a man who prepares a meal for his lord, this is the day about which they will say: a father did not pity, a son did not tarry.
>
> How will you go and tell my mother Sarah, how will you leave me and go home? And Isaac kissed his father Abraham, and commanded him, told him this:
>
> Sprinkle my blood over the altar, assemble my dust and bring it to my mother, my life and death, it's all in His hands, I thank Him that he has chosen me.

Blessed are you, my father, that they will say: I am the ram for the burnt offering of the living God, let your anger be stronger than your compassion, my father, and be the man who does not spare his son.

Like a merciless man, take up your knife, and slaughter me, lest I shall become unclean, do not weep, I shall not stop you, I take myself not because of you.

Why do you cry, said Isaac to Abraham his father, blessed is the Lord of the world who has chosen me, the spirit will rest [by] the blood of the altar, how we went out together in sincerity.

Give me your knife, my father, that I can touch it, I beg you not to defile me, my eyes see the woodpiles put in order, a burning fire on the day of my sacrifice.

Open your mouth and say a blessing, father, I will listen and say amen, my throat is stretched out to you, my father, whatever you please to do please do.[115]

NOTES

1. Hebrew Scriptures, Genesis 22:7.
2. Delaney, *Abraham on Trial*, 33–34.
3. N. R. Goldenberg, "A Theory of Gender as a Central Hermeneutic in the Psychoanalysis of Religion," in *Hermeneutical Approaches in Psychology of Religion,* ed. J. A. Belzen (Atlanta, GA: Rodopi, 1997), 76.
4. Hebrew Scriptures, Genesis 2:22–23.
5. Delaney, *Abraham on Trial*; David Blumenthal, "The Images of Woman in the Hebrew Bible," in *Marriage, Sex, and Family in Judaism,* eds. Michael Broyde and Michael Ausubel (Lanham, MD: Rowman and Littlefield, 2005), 47.
6. Delaney, "The Abrahamic Theory of Procreation," 2–3.
7. Delaney, "The Legacy of Abraham," in *Anti-Covenant*, 32; Stephanie Dalley, ed. and trans., *Myths from Mesopotamia: Creation, the Flood, Gilgamesh, and Others* (Oxford: Oxford University Press, 2000), 278.
8. Delaney, "The Abrahamic Theory of Procreation," 3.
9. Delaney, *Abraham on Trial*, 28; Montagu, *Coming into Being*, 326.
10. Delaney, *Abraham on Trial*, 251.
11. Delaney, *Abraham on Trial*. For the sake of brevity, I abbreviate statements such as 'according to the myth, Abraham did x' to 'Abraham did x.'
12. Carol Delaney, "Cutting the Ties That Bind," in *Relative Values: Reconfiguring Kinship Studies,* eds. Sarah Franklin and Susan McKinnon, 445–467 (Durham, NC: Duke University Press, 2001), 446.
13. Mark Sheridan and Thoman Oden, *Ancient Christian Commentary on Scripture, Old Testament II, Genesis 12–50* (Downers Grove, IL: InterVarsity Press, 2002).
14. Carol Delaney, "The Meaning of Paternity and the Virgin Birth Debate," *Man* New Series 21, 3, (1986): 496.
15. Risa Levitt Kohn and Rebecca Moore, *A Portable God: The Origin of Judaism and Christianity* (Lanham, MD: Rowman and Littlefield, 2007), 189.
16. Delaney, "The Legacy of Abraham," in *Anti-Covenant*, 32; Lerner, *Creation of Patriarchy*; David Hillel, *The Natural History of the Bible* (New York: Columbia University Press, 2006), 234.
17. Kenneth Mathews, *The New American Commentary: An Exegetical and Theological Exposition of Holy Scripture Genesis 11.27–50.26,* vol. 1B. (Nashville, TN: Broadman and Holman, 2005), 91.
18. Hebrew Scriptures, Genesis 17.

19. Carol Delaney, "Sacrificial Heroics: The Story of Abraham and its Use in the Justification of War," (Lecture, New York: Columbia Law School, April 4, 2005), 10; Delaney, "The Legacy of Abraham," in *Anti-Covenant*, 28.

20. Delaney, *Abraham on Trial*, 31.

21. Delaney, "The Abrahamic Theory of Procreation," 3: 'Although Jews maintain that women transmit the religion, it is like the 'mother church,' the nurturing medium in which a child is brought up. It is hardly a matrilineal or matriarchal culture as many proclaim and, indeed, all personal status laws of marriage, divorce, child custody, and inheritance, favor men.'

22. Hebrew Scriptures, Exodus 15:3–15.

23. Hebrew Scriptures, Exodus 34:24.

24. Delaney, *Abraham on Trial.*

25. E.g., Dresen Grietje, "The Better Blood: On Sacrifice and the Churching of New Mothers in the Roman Catholic Tradition," in *Wholly Woman, Holy Blood: A Feminist Critique of Purity and Impurity,* eds. Kristin de Troyer et al. (Harrisburg, PA: Trinity Press International, 2003).

26. Rubin, *La Source Inconsient de Misogyne*.

27. Stone, *Queer Commentary,* 133; Inhorn, *Local Babies, Global Science*, 64.

28. Kenneth Stone, *Practicing Safer Texts: Food, Sex, and Bible in Queer Perspective* (London: T & T Clark International, 2005), 123.

29. Inhorn, *Local Babies, Global Science*, 64; Stone, *Practicing Safer Texts*, 123; Marcia Inhorn, *Quest for Conception: Gender, Infertility and Egyptian Medical Traditions* (Philadelphia: University of Pennsylvania Press, 1994), 54–55; Samuel Brandon, *Creation Legends of the Ancient Near East* (London: Hodder and Stoughton, 1963); Marta Weigle, *Creation and Procreation: Feminist Reflections on Mythologies of Cosmogony and Parturition* (Philadelphia: University of Pennsylvania Press, 1989). The period under discussion is from the Pharaonic period beginning around 3150 BCE to when Egypt fell under Greek rule in 332 BCE.

30. Inhorn, *Local Babies, Global Science*, 64. Preformation theory held that all offspring are preformed by God. Sperm (or, for some, ova) contained miniaturized humans ready for 'planting' in the uterus.

31. David Leeming, *Creation Myths of the World: An Encyclopedia,* 2nd ed. (Santa Barbara, CA: ABC-CLIO, LLC, 2010), 103. Some versions have Atum creating himself.

32. George Hart, *Egyptian Myths* (London: British Museum Press, 1990), 12–13; *New World Encyclopedia* contributors, "Atum," *New World Encyclopedia.*

33. Harold Hays, *The Organization of the Pyramid Texts* (Leiden: Koninklijke Brill NV, 2012). Leeming, *Creation Myths of the World,* 103. Hays estimates the starting date of the Pyramid Texts as about two centuries prior to the end of Egypt's Old Kingdom in the third millennium BCE.

34. Brandon, *Creation Legends of the Ancient Near East,* in Weigle, *Creation and Procreation,* 74; Dalley, *Myths from Mesopotamia*, 233.

35. Lerner, *Creation of Patriarchy,* 145.

36. Bob Becking and Meindert Dijkstra, introduction to *Only One God? Monotheism in Ancient Israel and the Veneration of the Goddess Asherah,* eds. Bob Becking, et al. (London: Continuum, 2001); Meindert Dijkstra, "El, the God of Israel—Israel, the People of YHWH: On the Origins of Yahwism in Ancient Israel," in *Only One God? Monotheism in Ancient Israel and the Veneration of the Goddess Asherah,* eds. Bob Becking, et al. (London: Continuum, 2001); Avner Falk, *A Psychoanalytic History of the Jews* (London: Associated University Presses, 1996), 12.

37. Delaney, *Abraham on Trial*, 31.

38. Delaney, *Abraham on Trial*, 34.

39. Huber, "Comparative Gender Stratification," 73.

40. Blumenthal, "Images of Woman in the Hebrew Bible," 21.

41. Delaney, *Abraham on Trial*, 32.

42. Hebrew Scriptures, Deuteronomy 22:13–21.

43. Delaney, *Abraham on Trial*, 93.

44. Hebrew Scriptures, Genesis 4.

45. Delaney, *Abraham on Trial*, 93.
46. Delaney, *Abraham on Trial*, 94; Ali Shariati and Hamid Algar, trans., "The Philosophy of History: Cain and Abel," in *On the Sociology of Islam*, 97–110 (Berkeley, CA: Mizan Press, 1979), 98.
47. Delaney, *Abraham on Trial*, 95, 94.
48. Mathews, *The New American Commentary*, 91.
49. In this, the Abrahamic story repeats the pattern of the passing over of the firstborn son.
50. Hebrew Scriptures, Genesis 22: 1–9.
51. Hebrew Scriptures, Genesis 22:17–18.
52. Delaney, *Abraham on Trial*, 122.
53. Mathews, *New American Commentary*, 80.
54. Dennis Shulman, *The Genius of Genesis: A Psychoanalyst and Rabbi Examines the First Book of the Bible* (Lincoln, NE: iUniverse Inc., 2003), 61.
55. F. Leemhuis, "Ibrāhīm's Sacrifice of his Son in the Early Post-Koranic Tradition," in *The Sacrifice of Isaac: The Aqedah (Genesis 22) and its Interpretations,* eds. Ed Noort and Eibert Tigchelaar (Boston: Brill, 2002), 127.
56. Louis Jacobs and Ari Sagi, *Encyclopedia Judaica,* vol. 1, 2nd ed., eds. Michael Berenbaum and Fred Skolnik (Detroit: Macmillan References USA, 2007), 555–60.
57. Delaney, *Abraham on Trial*, 163.
58. Delaney, *Seed and Soil*, 7.
59. Søren Kierkegaard, *Fear and Trembling,* trans. C. Evans and S. Walsh (Radford, VA: Wilder Publications, 2008).
60. Jerome Gellman, *Abraham! Abraham! Kierkegaard and the Hasidim on the Binding of Isaac* (Hants, UK: Ashgate Publishing, 2003).
61. Louis Berman, *The Akedah: The Binding of Isaac* (Lanham, MD: Jason Aronson, 1997), 70.
62. Leemhuis, "Ibrāhīm's Sacrifice."
63. Riffat Hassan, "Eidal-Adha (Feast of Sacrifice) in Islam: Abraham, Hagar and Ismael," in *Commitment and Commemoration: Jews, Christians, Muslims in Dialogue,* ed. Andre Lacocque (Chicago: Exploration Press, 1992), 148.
64. Pickthall, Quran, Sura 37:102.
65. Pickthall, Quran, Sura 4: 125.
66. Delaney, *Seed and Soil*, 20.
67. Abdul Hamid el-Zein, "Beyond Ideology and Theology: The Search for the Anthropology of Islam," *Annual Review of Anthropology* 6, (1977): 247, in Delaney, *Abraham on Trial.*
68. Timothy Winter, "Abraham from a Muslim Perspective," in *Abraham's Children: Jews, Christians and Muslims in Conversation,* eds. Norman Solomon, Richard Harries, and Timothy Winter (New York: Continuum, 2005), 32, in Delaney, *Abraham on Trial.*
69. Delaney, *Abraham on Trial*, 72–73.
70. Delaney, *Abraham on Trial*, 72–73.
71. Kierkegaard, *Fear and Trembling*, 38, in Delaney, *Abraham on Trial*, 123.
72. Carol Delaney, "Was Abraham Ethical? Should We Admire his Willingness to Sacrifice His Son?" (Markkula Ethics Center Lecture Series, Santa Clara University, April 18, 2002).
73. Kierkegaard, *Fear and Trembling*, 43.
74. Augustine, *The City of God Books I–VII, The Fathers of the Church,* vol. 8, trans. Demetrius Zema and Gerald Walsh (Washington, DC: Catholic University of America Press, 1950), 54.
75. Shlomo Riskin, "The Akedah: A Lesson in Martyrdom," *Jerusalem Post* (Oct 21, 1994): 11.
76. Delaney, *Abraham on Trial*, 118.
77. Flavius Josephus, "Antiquities of the Jews. Book 1, Section 232," in *The Works of Flavius Josephus,* trans. William Whiston (Buffalo, NY: John Beardsley, 1895), 1.232.
78. Delaney, *Abraham on Trial*, 123.
79. Delaney, *Abraham on Trial*, 148.
80. Immanuel Kant, *The Conflict of the Faculties,* trans. Mary Gregor (Norwalk, CT: Abaris Books, 1979), 115.

81. Jacques Derrida, *The Gift of Death,* trans. David Wills (Chicago: University of Chicago Press, 1995), 67, 72.
82. Derrida, *Gift of Death*, 72.
83. Carol Gilligan, *In a Different Voice: Psychological Theory and Women's Development* (Cambridge, MA: Harvard University Press, 1993), 104–105.
84. Delaney, "Sacrificial Heroics."
85. Delaney, *Abraham on Trial.*
86. David Janzen, *The Social Meanings of Sacrifice in the Hebrew Bible: A Study of Four Writings* (Berlin: Walter De Gruyter Gmbh, 2004).
87. Magda Romanska, "Gender and Jewish Culture: Performing the Covenant: Akedah and the Origins of Masculinity," *Gender Forum an Internet Journal for Gender Studies* 21 (2008): 1.
88. Lori Rowlett, *Joshua and the Rhetoric of Violence: A New Historicist Analysis* (Sheffield, England: Sheffield Academic Press, 1996), 13.
89. Janzen, *Social Meanings of Sacrifice*, 118.
90. Grace Jantzen, *Violence to Eternity*: *Death and the Displacement of Beauty,* vol. 2. (London: Routledge, 2009), 107.
91. Hebrew Scriptures, Joshua 1:18.
92. Jantzen, *Violence to Eternity*, 111, 116.
93. Rowlett, *Joshua and the Rhetoric of Violence*, 5, 182–83.
94. Rowlett, *Joshua and the Rhetoric of Violence*, 115–16.
95. Yael Feldman, *Glory and Agony: Isaac's Sacrifice and National Narrative* (Stanford, CA: Stanford University Press, 2010), xii.
96. Michael Brown, "Biblical Myth and Contemporary Experience: The Akedah in Modern Jewish Literature," *Judaism* 31, (1982): 111.
97. Brown, "Biblical Myth and Contemporary Experience," 100.
98. Yoseph Milman, "The Sacrifice of Isaac and its Subversive Variations in Contemporary Hebrew Protest Poetry," *Religion and Literature* 23, (1991): 61.
99. Jantzen, *Violence to Eternity*, 132.
100. Susan Sered, "Towards a Gendered Typology of Sacrifice: Women and Feasting, Men and Death in an Okinawan Village," in *Sacrifice in Religious Experience,* ed. Albert Baumgarten (Boston: Brill, 2002), 15.
101. Nancy Jay, *Throughout Your Generations Forever: Sacrifice, Religion and Paternity* (Chicago: Chicago University Press, 1992), xxiii.
102. Jantzen, *Violence to Eternity*, 133.
103. Sered, "Towards a Gendered Typology of Sacrifice," 21.
104. Delaney, *Seed and Soil*, 34.
105. Jantzen, *Violence to Eternity*, 132.
106. Jenny Diski, "The Will of the Fathers," *London Review of Books* (December 10, 1998).
107. Theodore Blumoff, "*Genesis*, Gender and Community," *Southern California Review of Law and Women's Studies* 5, (1999); Diski, "Will of the Fathers."
108. Diski was completing a novel on Abraham at the time of the review.
109. Harvey Goldberg, "Review of *Abraham on Trial* by Carol Delaney," *American Anthropologist* 102, 1, (2000): 188.
110. Goldberg, Review of *Abraham on Trial*, 189.
111. Delaney, *Abraham on Trial*, 32.
112. Delaney, *Abraham on Trial*, 33–34.
113. Delaney, "The Abrahamic Theory of Procreation," 2.
114. Delaney, "The Legacy of Abraham," in *Anti-Covenant*, 38–39.
115. Wout van Bekkum, "The Aqedah and its Interpretations in Midrash and Piyyut," in *The Sacrifice of Isaac: The Aqedah (Genesis 22) and Its Interpretations,* eds. Ed Noort and Eibert Tigchelaar (Boston: Brill, 1999), 94.

Chapter Four

Covering the Field

'Judith,' a young Australian woman of Middle Eastern descent, was sent to Australia as a child bride. After 20 years of abuse and several futile attempts to flee her marriage, Judith escaped with her children to a refuge. After Judith's divorce, the Australian Family Court granted her ex-husband equal custody of the children.

Judith has remarried, but her ex tells her that he owns the children and he'll kill her if she doesn't come back. After the school holidays, Judith's ex refuses to return the children and then sets fire to her house. She takes out another intervention order. He tells her it makes no difference, that he 'knows people,' and is sending someone around to slit her throat.

Judith's current husband says her troubles are her fault.

'If I left you,' she says to him, 'would you send someone around to kill me?'

'Of course not,' he says. 'I would do it myself.' 'I picked you up from the gutter,' he tells her, referring to her divorce. She ought to be grateful to him, he says; no one else would have her.

'Can't you arrest my ex for setting fire to my house?' Judith asks the detective. 'Intervention orders don't stop him.'

'Any evidence would've been destroyed in the fire,' says the detective. 'You're right that nothing can stop him if he's hell bent.'

Suburban Australia, 2011.[1]

As an old Afghan saying has it, 'A woman belongs in the house—or in the grave.'[2] What is the role of seed-and-soil in ancient and contemporary honor codes?[3] As in Genesis, seed-and-soil is embedded in culture-specific societal codes for the bestowal and withdrawal of honor. Honor codes grant men control over reproduction, and male honor depends on men's ability to control the sexual behavior of female relatives.[4] In other words, men who cannot protect the boundaries of 'their' women lose honor.[5] Hence, central to honor codes are chastity codes that govern the way in which women must demonstrate sexual unavailability so as to avoid bringing shame on themselves and their families.[6] Women can be beaten, even killed, on the basis of suspicion alone. This chapter draws on anthropological and sociological studies of honor codes, particularly in contemporary Turkish society and its Swedish diaspora.

MEN'S HONOR, WOMEN'S SHAME

Honor systems maintain social order through shaming and the threat of ostracism. According to one definition of shame in this context, it is a reaction to the criticism of others, an acute personal chagrin at failing to live up to our own obligations as well as others' expectations. Under honor codes, self-respect is maintained not by choosing right from wrong but by doing what is expected of one.[7] In the mainstream scholarly literature, honor is understood as the person's 'estimation of his own worth, his *claim* to pride,' but also 'his excellence recognized by his society, his *right* to pride.'[8] Within honor systems, the individual symbolizes the group, whose collective honor is upheld by every single member.[9]

In the 'vast but unsatisfactory' literature on honor societies, says Delaney, observers of Middle Eastern and Mediterranean communities have long remarked on the vigilance with which honor societies carry out the 'protection' of women. Although social scientists agree that sexuality is a prominent aspect of the code, politics, economics, and ecology are the usual focus, none of which explains why male honor has been inextricably tied to women.[10] The literature, says Delaney, reflects commentators' confusion between two interrelated but separable forms of honor. The honor bestowed on a man through the chastity of 'his' women is more fundamental than that bestowed through his reputation as a participant in the community. A man can have the former without the latter, but not vice versa.[11] Another point of contention is whether honor can ever be attributed to women. The Turkish villagers in Delaney's study, for example, perceived women as having inherited Eve's shamefulness and devoid of the 'seeds of honor.'[12] Some groups, however, hold that women may share the honor acquired by their male relatives.[13]

Although many scholars accept the mainstream literature as explanatory, it fails to explain why male honor is inseparably linked to women.[14] Much of the literature is merely descriptive, with scholars tending to take honor and shame for granted as aspects of masculine competition: men compete in the public arena for scarce resources, including the 'resource' of women's chastity.[15] But contemporary feminist analyses point to how honor codes construct women as objects by which men measure their own community standing. In this sense, masculinity cannot be constructed without women's participation as objects.[16] Men set the rules of the game in male-only social arenas and objectify women to reflect men's images of their own ideal selves.[17] Of course, my foregoing summary oversimplifies the matter, and many dialectical tensions complicate the way that honor systems co-construct men and women.

As an 'index' for masculine reputation, says Lama Abu-Odeh, the main function of female virginity is 'the production of the *public effect* of virginity' through an elaborate performance that signifies the physical attachment of the hymen to the body.[18] Such signification displaces the hymen from the vagina onto the body as a whole and displaces it again onto the social space, where it encircles the body as a social 'hymen' and delimits its borders. Transgression of one or more of these borders may result in honor killing. And the man who commits that murder is epitomizing the performance of his gender. His task is to defend 'his' women's virginity against incursions; to fail is to lose his gender: 'he is no longer a man (therefore castrated, a "bottom", a woman).'[19]

Honor systems require the honor killing of women for such transgressions as refusing to enter into an arranged marriage, being the victim of a sexual assault, seeking a divorce (even from an abusive husband), or being accused of committing adultery. The mere perception that a woman has behaved in a way that dishonors her family is sufficient to trigger an attack on her life.[20] In Turkey, says Ayşe Parla, girls are strictly socialized into codes of 'modest demeanor' in speech, dress, and gesture 'that will protect and affirm their chastity' and avoid 'staining the family honor.' In some parts of Turkey, she says, the mere 'possibility of unsupervised interactions with boys' may provide sufficient grounds for an honor killing.[21]

'Severe disruptions of the moral order occur when sexual impropriety is alleged or takes place. To restore the social equilibrium and avoid feuds, the local culture required the shedding of blood to wash away the shame of dishonor.'[22] For a long time, says Carolyn Fluehr-Lobban, this perspective dominated anthropological thinking on the practice of honor killing, and many judges defended the practice 'as less serious than premeditated murder because it stems from long-standing cultural traditions.'[23] However, says Fluehr-Lobban, those very traditions encompass entrenched patterns of cultural discrimination against women.[24] To defend honor killing, particularly on a cultural-relativist basis, is to underestimate intracultural heterogeneity and the prevailing power structures. Moreover, such defense may unwittingly lend support to punitive regimes.

The Christian chastity code advocated in the sixteenth century by Juan Luis Vivres is a particularly lucid example of the genre. However, says Fredrik Ljungqvist, commentators on Vivres typically minimize the patriarchal underpinnings of his espoused rules for women's dress, conduct, and seclusion.[25] Vivres's code lays bare the relationship between the honor code and Christian doctrine on sex differences. For Vivres, any attempt to minimize these differences is tantamount to a crime against nature. In every way, says Vivres, men are superior to women. The proof of women's spiritual weakness is Eve's instigation of the Fall of Man; hence, in the lives of women, men stand in for God. Although honor is reserved for men, women, by their chastity, are the guardians of what men acquire.[26] Women, says Vivres, must accept their subordination just as the blind and the invalided accept their disadvantage. They must cover their heads and remain silent. Women own neither their bodies nor their chastity; these are male possessions. If women defile their chastity, they defile men's honor. As shame alone protects chastity, women must be made to feel ashamed of their bodies and their sexuality. The woman 'who does not fear the suspicion of adultery,' says Vivres, 'does not fear adultery.'[27] Through their sexuality, women can bring dishonor to all their relatives, stigmatizing their families for life.[28]

The solitary goal for a woman's learning, says Vivres, is the protection of her chastity: 'When this has been thoroughly elucidated, she may be considered to have ample instruction.'[29] Given that a woman's natural inclination is to satisfy her insatiable lust, she must be kept busy, but not with learning. Intellectual work is dangerous, says Vivres: 'A woman's thoughts are inconsistent,' and she can readily 'descend from good thoughts to bad.'[30] Families must protect 'their' women's chastity by secluding them from public life. 'On the girl hidden away in her house no one passes judgements,' says Vivres. 'On one who is often seen in public everyone has something to say, whence arises dishonor.'[31] And because men do not want such women as wives, all marriages must be arranged. Essential excursions, to attend Mass, for example, entail the woman taking her private sphere with her in the form of the veil—'a kind of portable domestic space.'[32]

HONOR AND SHAME, SEED AND SOIL

The covering of women, Delaney argues, relates to the protection of seed. In traditional patriarchal and patrilineal societies, the men of the family or clan seek control over fertility, Woman is a factory for making men, and a man's honor depends on ensuring that his children are from his own seed.[33] In Delaney's study, villagers described the woman's body as a 'field,' which, more so than 'soil,' connotes enclosure by ownership. The man must enclose and cover the female soil in order to ensure the seed-child is his own.[34] Moreover, it follows that—just as plant seed can sprout long after being sown—so 'foreign seed' from a woman's previous relationship may 'sprout' into a pregnancy at any time. Such foreign seed contaminates the field permanently; the woman is permanently defiled. Conversely, if a man has intercourse with a defiled woman, he is defiled only until he performs a ritual ablution.[35]

The honor code constructs social intercourse between unrelated men and women as almost equivalent to sexual intercourse, and women's fecundity as indiscriminate. Hence, because women possess no power of self-restraint, men must impose it on them. Enclosing women is a way of controlling male sexuality, to which women have no resistance. 'Women are as easily seduced as Eve was by Satan,' said one villager.[36] Even a desiring look from a man can pollute a woman, who is blameworthy for receiving it. In this schema, bodily attributes take on moral attributes: 'Showing of the hair is the ruination of families,' said another villager.[37] Menstruation is God's punishment for Eve's disobedience, and Eve's vulnerability to satanic temptation justifies male protection of women.[38]

To preserve her family's honor, a girl must participate in her own enclosure: she must avert her eyes around unrelated men, wear the headscarf and other voluminous coverings, and stay indoors. By keeping to the rules, the girl shows that she is closed and covered as opposed to open and uncovered and that she is clean and pure as opposed to dirty and defiled: 'She is preserved in this state until marriage, when her husband has the right to open her and thereafter control the times and places of her opening.'[39]

For the villagers, says Delaney, the relationship between the sexes symbolically represents cosmic order. To change that relationship would be to threaten the nature and order of the universe. Villagers are aware of the immensity of this threat, says Delaney, and all agree on the necessity of covering women. When they discuss East-versus-West or Islam-versus-secularism issues, it is the purity of the Eastern-Islamic women that marks the difference, a purity controlled and enforced by men.[40]

Among findings from contemporary research into honor codes in Turkey, two-thirds of a sample of women surveyed in the eastern and southeastern regions said they would be killed if they committed adultery and, on the mere suspicion of an affair, beaten up or divorced.[41] Two-thirds were unaware of their legal right to divorce their husband for adultery.[42] In some regions, people regarded men's performance of an honor killing 'as a rite of passage into manhood.'[43]

In other findings, more than half of the men surveyed from the western (and three-quarters from the eastern) region said they have absolute legitimate authority over their wives. One-third of those from the west (and half from the east) said they have the right to beat their wives for disobedience. Nearly three-quarters of surveyed rural, married women (and one-third of urban, married women) agreed with the men, exemplifying, perhaps, the psychological and political dialectics of identification with the system. A woman fleeing an abusive husband may be delivered back to him by the authorities even when her life is in danger.[44]

Women in diasporic communities may become doubly burdened. The issue of violence against migrant women in Sweden, says Aylin Akpinar, is caught between two problematic philosophies: on the one hand, the assumption that patriarchal violence is what it is and that notions of 'honor' have nothing to do with it and, on the other hand, the cultural fundamentalist argument that the problem will disappear with assimilation.[45] The latter position, which is prevalent in Swedish academic and public discourse, says Akpinar, is a new construction of exclusion that has replaced the old hierarchy of cultures with spatial difference. The new schema puts each culture in its place and assumes that localized political communities are culturally homogeneous. But violence against women, she says, particularly honor killings, cannot be resisted by ignoring the existence of the honor code: what is required is a cultural analysis without a cultural reductionist trap.[46]

Violence toward women, says Akpinar, is integral to the construction of masculinity in all forms of patriarchal orders, albeit under different structures, and we can read the honor code as a subcategory of patriarchal domination.[47] Traditionally, in Middle Eastern and Mediterranean cultures, as in Latin American peasant societies, southwest Asia, various Indian castes, and Chinese elites, the community holds parents and parents-in-law collectively responsible for infringements of the moral code by daughters. Women are given in marriage at a very young age, says Akpinar, in order to attach them to men who will take over control from their fathers.[48] With the increasing transition to the nuclear-family structure, however, control is shifting to the husband alone and, in the absence of collective support, his personal privilege becomes both more explicit and more vulnerable to being seen as unwarranted—a potentially shameful position.[49]

In the migration context, says Akpinar, where many men feel discriminated against, those holding patriarchal values tend to 'turn inwards' and exert increased pressure on women and girls.[50] Patriarchal beliefs do not disappear, she says, but persist for a long time among immigrant communities. For some immigrant men, controlling their space by controlling their women may compensate for feelings of powerlessness in their vulnerability to unemployment and/or the loss of their authority over children.[51] In addition, the fact that young Turkish boys are socialized to know their place in relation to their father and elder brothers is a component in later violence in which they force others to take the one-down position.[52]

Whereas younger women of Turkish immigrant families typically want to become freer than their mothers are, older women generally want to maintain control over their children. While communities idealize women as mothers, they also perceive them as dangerous because of women's so-called inherent readiness to bring dishonor to the entire community.[53] Hence, the family and community must protect and control women. Women's premarital virginity is an asset, not just for the woman but for the family and the group. Through their sexuality, women represent the privacy of the group, says Akpinar, and the group is vulnerable to defilement and can be polluted by women having sexual relations with outsiders.[54]

Akpinar's case study of immigrant women divorcees of Turkish origin living in Sweden is briefly prefaced with the tragic story of 'Fadime.'[55] After four years of death threats from her father and brother for noncompliance with their wishes for her to marry a cousin, in 2002, Fadime's father shot and killed her.

One of Akpinar's interviewees was fifteen when her natal village in Turkey began pressuring her mother: 'Turkish immigrants who came from the village told my mother to give me in marriage,' said 'Ayla.' 'Turks in the community started spreading gossip about me that I was meeting boys. The gossip soon reached the town in Turkey.' When she visited Turkey, Ayla said, 'my grandmother and my aunts convinced my mother to give me in marriage while in Turkey by mentioning the gossip which had reached them. . . . So they gave me to an elderly man despite all my protests.'

Another interviewee, 'Semra,' left school at her father's bidding when she was fourteen: 'I waited for a year at home,' she said, 'and married in the summer of my fifteenth birthday.' Abused by her husband, Semra begged her father to take her back. 'I cannot welcome you back,' he said. 'I will be ashamed. What will everyone say? I will rather kill myself than call you back.' When visiting neighbors with her in-laws, Semra said, 'I was wearing a skirt and I had to bend my knees. When we came back my father-in-law beat me by saying I had shown my legs. I said a bad word to him. He called my husband and he beat me too. . . . They brought me to [hospital].' Semra's brother warned her to keep quiet about the abuse. Eventually, a teacher and social worker helped Semra to find refuge, but her brother broke in to her flat and beat her. 'He said I had become a prostitute,' said Semra, and that 'everyone was talking about me. Then he took me to his house by force. . . . The rumours about me reached Turkey.'[56]

Gossip and stigma are mechanisms for controlling women, not only by men as a group but also by women to punish nonconforming women.[57] Honor societies enculturate women to internalize the honor code and accept the abuse of women who deviate from the norm. In traditional rural Turkish culture, it is said that 'When a woman leaves her family, only her corpse can come back.'[58] Ayla and Semra rebelled against their surveillance and consequently were beaten. When they don't surrender, says Akpinar, they pay the price by becoming outcasts.[59]

OBJECTIFYING THE SUBJECT

In order to show that the division between male and female is simultaneously arbitrary, contingent, and sociologically necessary, says Bourdieu, we must adopt the perspective of the anthropologist.[60] Only a very particular use of ethnography, he says, can objectify the division between the sexes and treat androcentric society as an archaeology of our unconscious. Because researchers are included in the objects they investigate, their schemas embody those objects' historical structures and modes of thought. Breaking this cycle requires a way of objectifying the subject, of making a laboratory experiment of a society that is 'both strange and familiar.'[61] For Bourdieu, this culture is the Kabyle, who, he says, have preserved a paradigmatic form of the phallo-narcissistic vision and the androcentric cosmology that survives in our contemporary cognitive and social structures. While European culture shares the masculinist chimera, Kabyle society is a still-functioning system relatively untouched by written traditions.[62]

For the Kabyle, as for Delaney's Turkish villagers, sexual differences are immersed in the dualisms that organize the cosmos. Bourdieu and Delaney agree that the ethnographer must approach such notions of sexual differences indirectly so as not to miss their deeper significance.[63] In both societies, the underlying social relation of domination gives the impression of being independent of power relations and the division between the sexes as normal, natural, and seemingly inevitable.[64] The Turkish villagers, says Delaney, embrace both a cosmological aspect to gender and gendered aspects of cosmology. Their theory of conception and their conception of the deity are indissoluble elements of the same system.[65] Hence, rather than asking villagers directly about the differences between the sexes, Delaney enquired into their beliefs about 'coming-into-being' and their values, attitudes, structures, and practices around the body, gender, and procreation. Delaney uses the term 'procreation' to include how life comes into being; what it is composed of; who or what the agents are; what the male or female person is; and how persons are related to one another, the nonhuman world, and the cosmos.[66]

Like Bourdieu, Delaney wanted to study a still-functioning system.[67] Unlike Bourdieu, she was enquiring into worldview as much as into habitus. Bourdieu's criticism of the observer's stance, says Delaney, may have taken him to another extreme. In his emphasis on the construction of knowledge through practical activity, Bourdieu may not have appreciated the impact of Islam, for example, in constructing habits of mind as well as habitus. Bourdieu, says Delaney, does not discuss the influence of Islam on ideas and practices regarding honor, authority, inheritance, kinship and parallel cousin marriage, personal hygiene, or orientations in time and space. Rather, he gives the impression of perceiving the structures, values, and attitudes of everyday life as arising from the ground, the exigencies of the seasons, and the sexual division of labor, which are accepted as given.[68]

Perhaps, says Delaney, we should look less at the observable ethos and practice side of things, and more at worldview. After all, practical activities evolve in relation to a worldview that is far more than a local construction. Universalistic theories about change through practical activity cannot account for the place of work, for example, within a specific worldview. People do change their societies over time; the question is, how does this happen? Where do we locate the forces of change, resistance, and reaction?[69] In Turkey, for example, as men's relationship to the land changes, so does their relationship to women. As development occurs, people begin to view land not as a way of life but as something to exploit. More and more products that women once produced become cheaper to buy elsewhere, and women are increasingly enclosed in the house. Even in the small towns in Turkey, says Delaney, while men are becoming more and more integrated into public life, women are more covered up and shut in small apartments, rarely venturing out.[70]

According to the Sunni-Muslim religion of the Turkish village, as in Abrahamic monotheism, generally, creativity is both masculine and divine. In the procreative division of labor, says Delaney, 'men provide the creative spark of life, the essential identity of a child, while women, like soil, contribute the nurturant material that sustains it.'[71] In the village, religion is inseparable from social life itself. Anthropology as a discipline, says Delaney, is partly responsible for a persistent gap in our understanding of this cultural oneness. Whereas anthropology has presupposed religion as a universal human experience of which Islam is a particular instance,[72] many Muslims—for whom Islam is not one religion among many but purports to be the one true way given in the beginning—reject this. And if Muslims and Christians, for example, do not make similar distinctions between the sacred and the secular, research cannot properly apply categories from one religion to study the other. Religion as Geertz would have it, for example, is only relevant for the faithful or only some of the time, but in fact worldviews flow into the cultural mainstream, affecting the everyday lives of believers and nonbelievers alike and influencing even apparently secular ideas.[73]

Rarely do Western anthropologists investigate procreation beliefs in their own cultures, says Delaney. Equally problematic, they tend to assume that notions of kinship arise from beliefs about procreation and that differences in kinship systems must therefore relate to differences in knowledge about the 'facts' of procreation.[74] Nineteenth-century theorists such as Bachofen, Frazer, Smith, Durkheim, Morgan, and McLennan, Delaney says, misconstrued these relations as evidence of stages of social evolution rather than as different cultural systems with different theories of procreation.[75] Notions of 'paternity,' for example, are entirely absent in theories in which intercourse is understood as merely opening the path for a fetus that comes from elsewhere (e.g., the spirit world), or in which the male ejaculate simply supplies food to the already existing fetus. Paternity, in other words, is not the semantic equivalent of maternity.[76]

As early as 1927, Bronislaw Malinowski warned that the term 'father' must be understood as entirely specific to the society under study.[77] Among the matrilineal society of the Papuo-Melanesian peoples of the Trobriand Islands, says Malinowski, 'descent, kinship, and all social relations are reckoned by the mother only.' As regards pregnancy, 'it is solely and exclusively the mother who builds up the child's body, while the man does not contribute in any way to its production.'[78] The centrality of matrilineality in Trobriand society was highly problematic for Christian missionaries, whose morality turned on the patriarchal and patrilineal family. In order to promote the sacredness of the Father to Son relationship, they had to demonstrate its natural foundation.[79]

Despite Malinowski's admission that the term 'father' had little or no meaning for the Trobrianders, says Delaney, he continued to use it in his conversations with them about their procreation beliefs, hence further confusing the issue he was explaining.[80] Moreover, in spite of his own warning, Malinowski inadvertently reinscribed seed-and-soil by paralleling it with modern embryology. The Trobrianders were very curious about this simile, said Malinowski, 'and asked whether this was "the white man's way of doing it."' In the Trobrianders' view, male ejaculate served 'merely the purposes of pleasure and lubrication,' and it was characteristic 'that the word *momona* denotes both the male and female discharge.'[81] How, then, from a Trobriand perspective, did pregnancy come about? Understood universally among the community, said Malinowski, was that 'the real cause of pregnancy is always a [matrilineal ancestor of the mother], who is inserted into or enters the body of a women, and without whose existence a woman could not become pregnant.'[82]

Recall that only in the twentieth century in the West was scientific evidence of the genetic contribution of the ovum assimilated, and even then only among educated people.[83] Not only that, says Delaney, but scientific revelation plays a relatively minor role within wider systems of meaning and value. Although we may assume that changes in biological knowledge would immediately affect notions of gender, gender cannot be reduced to biology and sex role.[84] In addition, conceptual-change research categorically demonstrates the fundamental neurobiological challenge of accommodating new and complex ideas that conflict with established ones.[85]

In the Turkish village, says Delaney, male and female meanings of the body are neither self-evident nor derived from the facts of reproduction. An equivalent word for 'procreation' does not exist; the ancient terms used translate as 'increase and multiply.' The salient distinction the villagers make is between male and female roles in procreation, roles paralleled in the division of labor and the division of the universe itself.[86]

It follows, says Delaney, that the claim made by many social scientists that women's inferior status is related to their childbearing capacity may be missing the point. Perhaps it is not procreation per se that is devalued but women's perceived role in the process, as in the Quran: 'Your women are a tilth for you [to cultivate] so go to your tilth as ye will.'[87] This directive from God, Delaney says, establishes an 'I-thou' relationship between God and men 'from which women are excluded and in which they are objectified. God talks to men about women.'[88]

Given the thorough naturalization of seed-and-soil in Greco-Roman/Judeo-Christian traditions, European social scientists may be too close to it to 'see' its specific significance in Islamic or Turkish society. The Turkish villagers, for example, view procreation as the human analogue to divine creation.[89] Contrary to the assumption that some familiarity with the biological facts of procreation is universal, the Turkish villagers actually had no alternative theory with which to contrast their seed-field image. Their symbolic understanding of the procreative process amounted to seed-and-soil, a culture-specific, socially constructed notion of procreation from which embedded notions of paternity and maternity cannot be abstracted.[90] Rather than the notion of paternity having a primarily physiological meaning, villagers use its bio-physical elements to express social meanings of gender, authority, kinship, and so forth.[91] One Turkish villager explains the meaning of procreation thus: 'If you plant wheat, you get wheat. If you plant barley, you get barley. It is the seed which determines the kind of plant which will grow, while the field nourishes the plant but does not determine the kind. The man gives the seed, and the woman is like the field.'[92]

According to this theory, blood and milk can be supplied by any woman without affecting the child's essential identity, which comes from the father. The seed that carries the spark of life provided by God is eternal but only fathers have the capacity to transmit it intergenerationally: 'A boy is the flame of the line,' goes a Turkish saying, 'a girl the embers of the house.' If a man has no sons, his hearth is extinguished.[93] Islam views the family as a miniature version of Muslim society; the father's authority symbolizes God's authority in the world.[94] In the Turkish village, because the goal of marriage is to produce children, wives have no right to refuse to have sex with their husbands. Because villagers perceive the father as 'the one who gives life,' they refer to him as the 'child owner.' In the case of divorce, says Delaney, children belong to the father.[95] Village mothers breastfeed their infant boys for two years—six months longer than for girls—in order to strengthen them for later military service.[96] Typically, a young man wants to marry prior to his military service and on returning home be presented with his first child.

Inhorn's study of men and women of the Egyptian urban underclass echoes some of Delaney's findings. As among the Turkish villagers, many of Inhorn's interviewees regarded men's patriarchal power in the family and socially as natural, given that men, not women, are the procreators. In Egypt, says Inhorn, notions of male monogenesis commonly are embedded in preformation theory—the idea that men have preformed fetuses in their spermatic fluid. Even a more 'modern' version—women produce eggs that mix with the male sperm—has it that the male alone gives life.[97] In this view, the 'fact' of women's supposed natural inability to procreate is what makes women inferior to men. And just as men give life it is women who take life away 'by virtue of wombs that fail to facilitate . . . male creation.'[98] In the Turkish village, says Delaney, 'if a man could ejaculate, it was assumed he could produce a child. If none were forthcoming, it was the woman's fault, she was obviously barren.' The man could divorce her or take another wife. If a woman's husband died, his brother could take her as wife in order to continue the seed line.[99]

What of the counternarratives to Delaney's reading of seed-and-soil? Chris Hann, for example, has it that Delaney's work is 'typical of a feminism that seems redolent of Frazerian anthropology,' whose faults include making 'overt appeals to values'; refusing to affect 'disinterested observation'; and describing too much of the trivia of village life (e.g., tea rituals). As to Delaney's remark about the 'leisurely pace' at which villagers prefer to work, Hann says she comes close 'to reproducing old and regrettable stereotypes of the lazy Oriental peasant,' despite her awareness that 'male [sexual] passivity is strongly devalued' in the village.[100] Here, however, Hann is conflating two different meanings of passivity. In addition to such flaws in reasoning, Hann fails to acknowledge the clearly ideological basis of his own critique.

Other critics have it that Delaney overstates the role of Islam in the Turkish village. Inhorn, for example, claims that Delaney, in dismissing the case that Islam promoted duogenesis, emphasizes Islamic philosophy's support for monogenetic theory at the expense of Hadith sources. Delaney counters that she has found no Islamic sources that promulgate the idea that 'both men and women contribute equally and *the same kind of thing*' to conception. Although some Islamic texts mention the term 'female seed,' says Delaney, they typically construct it as inferior to male seed.[101] On this point, Leila Ahmed agrees. Muslim philosophy rejected the jurists' position that men and women contribute equally to conception, she says. Instead they endorsed Aristotle's theory of the superiority of the male seed.[102] While the prominent Persian Aristotelian, ibn Sina (latinized as Avicenna) admitted the existence of ovaries and female semen, he reinstated Aristotle's position, particularly as regards the male contribution to soul.

In further refuting Delaney's position, Inhorn notes that Egyptian theories of male monogenesis predate Islam itself by several millennia: 'In Egypt at least, a long history of monogenetic theorizing is not specifically linked to Islam per se.'[103] While this is true, it does not undermine Delaney's argument that Islamic philosophy upholds male monogenetic theory in some respects.

In slightly different terms, Parla suggests that Delaney overstates the role of Islam's influence in the contemporary link between women's purity and family honor. The notion of honor, says Parla, is contingent on the nationalist histories and social policies of each nation-state.[104] However, the literature (including Parla's own work) demonstrates considerable overlap between Islamic and national-historical elements in Turkish society. It is true that honor systems are culturally specific (and not always Islamic),[105] but all are undergirded by sociocultural assumptions about the 'natural' weaknesses of women.

Parla herself is aware of this overlap. While the state 'granted women legal rights and fashioned the image of the public modern woman,' she says, it simultaneously reaffirmed 'the importance of women's virtue and chastity.'[106] Although the state no longer confined women to the private sphere, their honor came under the surveillance of the state, which until 1999 enforced virginity examinations for girls in state orphanages and high schools, political detainees, and women suspected of prostitution.[107] As such, the then secular state was appropriating religiously based surveillance typically carried out by local Muslim communities. As an aside, Parla's point helps to explain the fact that few girls from Delaney's Turkish village attended the town high schools, which, at the same time as making virginity tests compulsory, forbade girls from wearing headscarves. In village terms, says Delaney, a girl whose head is uncovered is seen as open to men, the equivalent of entering prostitution.[108]

In addition, Delaney's explicit description of seed-and-soil as a cross-cultural construct seems to further erode Parla's criticism that Delaney overstates the influence of Islam. Finally, the argument about whether the Turkish honor code is determined more by socio-historical factors or by Islam is neither resolvable nor central to the main issue, namely, that seed-and-soil remains as a gendered construct incorporating problematic notions of honor and chastity. Recent research suggests that even tertiary educated, American reformist Muslim men may accept Islamic values of 'modesty' for women as given. Findings from a case study of such men's gender-role beliefs highlight their concerns about 'the sexual objectification of women' and beliefs that 'the hijab can prevent this.'[109] The men described women as 'physically weaker than men' and 'better equipped to be nurturers.'[110] And although the researcher—an educated Muslim man—interrogated the latter assumptions, he appeared to accept the modesty code as unproblematic. To reiterate this chapter's central point, the key purpose of the practice of covering women is to protect seed, that is, to protect the purity and honor of the patriline in spite of women's intrinsic and shameful vulnerability to seduction.

Honor and chastity codes are very much alive among diasporic communities.[111] Indeed, some core assumptions about women's 'natural' shame exist tacitly in the contemporary West and perhaps universally. Chapter 10 of this book includes a high-profile example from Australia, one of the first Western industrialized nations to introduce women's suffrage.

NOTES

1. A. Jones, "Rougher than Usual Handling," in *Joining the Dots: Essays and Opinion* (Melbourne: Publish in Ink, 2014).
2. Ducat, *Wimp Factor*, 219.
3. E.g., John Peristiany, ed. *Honour and Shame: The Values of Mediterranean Society* (Chicago: Chicago University Press, 1965); Julian Pitt-Rivers, "Honour and Social Status," in *Honour and Shame: The Values of Mediterranean Society*, ed. John Peristiany (Chicago: Chicago University Press, 1965).
4. Fredrik Ljungqvist, "Female Shame, Male Honor: The Chastity Code in Juan Luis Vivres' *De Institutione Feminae Christianae*," *Journal of Family History* 37, 2, (2012): 139.
5. Suzanne Ruggi, "Commodifying Honour in Female Sexuality: Honour Killings in Palestine," in *Women and Islam: Social Conditions, Obstacles and Prospects*, ed. Haideh Moghissi (New York: Routledge, 2005), 125–30.
6. Ljungqvist, "Female Shame, Male Honor," 139–40.
7. Hiebert, *Anthropological Insights*, 212.
8. Pitt-Rivers, "Honour and Social Status," 21.
9. Victoria Goddard, "Honour and Shame: The Control of Women's Sexuality and Group Identity in Naples," in *The Cultural Construction of Sexuality*, ed. Pat Caplan (London: Tavistock, 1987), 167.
10. Delaney, *Seed and Soil*, 38.
11. Carol Delaney, "Seeds of Honour. Fields of Shame," in *Honour and Shame and the Unity of the Mediterranean*, vol. 22, ed. David Gilmore (Washington, DC: American Anthropological Association Special Publication, 1987), 36.
12. The village selected was Anatolian-Turkish Sunni-Muslim (the predominant ethnic and religious group), economically viable, not depleted by outmigration to the cities, and relatively unexposed to modern Western scientific theories.
13. Ljungqvist, "Female Shame, Male Honor."
14. Delaney, *Seed and Soil;* Goddard, "Honour and Shame," 168.
15. Goddard, "Honour and Shame," 168.
16. Aylin Akpinar, "The Honour/Shame Complex Revisited: Violence against Women in the Migration Context," *Women's Studies International Forum* 26, 5, (2003): 431.
17. Akpinar, "The Honour/Shame Complex Revisited," 431.
18. Lama Abu-Odeh, "Crimes of Honour and the Construction of Gender in Arab Societies," *Comparative Law Review* 2, (2011): 12.
19. Abu-Odeh, "Crimes of Honour," 12–13.
20. Human Rights Watch, "Item 12—Integration of the Human Rights of Women and the Gender Perspective: Violence against Women and 'Honour' Crimes," (UN Commission on Human Rights. April 6, 2001).
21. Ayşe Parla, "Honor: Turkey and the Caucasus," in *Encyclopedia of Women and Islamic Cultures: Family, Law and Politics*, vol. 2, eds. Suad Joseph and Afsaneh Najmabadi (Leiden: Koninklijke Brill NV, 2005), 218.

22. Carolyn Fluehr-Lobban, "Cultural Relativism and Universal Rights," *The Chronicle of Higher Education* (June, 1995): 34.
23. Fluehr-Lobban, "Cultural Relativism and Universal Rights," 34.
24. Fluehr-Lobban, "Cultural Relativism and Universal Rights," 34.
25. Ljungqvist, "Female Shame, Male Honor."
26. Juan Luis Vivres, *De Institutione Feminae Christianae*, 2 vols. eds. and trans. C. Fantazzi and C. Matheeussen (Leiden: E. J. Brill, 1996), vol. 1, 41.
27. Vivres, *De Institutione Feminae Christianae*, vol. 2, 129.
28. Ljungqvist, "Female Shame, Male Honor," 144.
29. Vivres *De Institutione Feminae Christianae*, vol. 1, 5.
30. Vivres, *De Institutione Feminae Christianae*, vol. 1, 107.
31. Vivres, *De Institutione Feminae Christianae*, vol. 1, 201.
32. Llewellyn-Jones, *Aphrodite's Tortoise*, 200.
33. Ruggi, "Commodifying Honour in Female Sexuality"; Delaney, *Seed and Soil*, 39.
34. Delaney, "The Meaning of Paternity," 498.
35. Ulla-Britt Engelbrektsson, *The Force of Tradition: Turkish Migrants at Home and Abroad* (Göteborg: Acta Universitatis Gothoburgensis, 1978), 137; Delaney, *Seed and Soil*, 41.
36. Delaney, *Seed and Soil*, 41.
37. Delaney, *Seed and Soil*, 41.
38. Delaney, "The Meaning of Paternity," 499.
39. Delaney, *Seed and Soil*, 42.
40. Delaney, *Seed and Soil*, 278.
41. Pinar İlkkaracan, *Deconstructing Sexuality in the Middle East,* (Burlington, VT: Ashgate Publishing, 2008).
42. Pinar İlkkaracan, ed., *Women and Sexuality in Muslim Societies* (Istanbul: Women for Women's Human Rights, Kadinin Insam Haklari Projesi, 2000).
43. Akpinar, "The Honour/Shame Complex," 428.
44. Akpinar, "The Honour/Shame Complex," 428.
45. Akpinar, "The Honour/Shame Complex," 426.
46. Akpinar, "The Honour/Shame Complex," 426.
47. Akpinar, "The Honour/Shame Complex," 427.
48. Akpinar, "The Honour/Shame Complex," 425.
49. Akpinar, "The Honour/Shame Complex," 427.
50. Akpinar, "The Honour/Shame Complex."
51. Akpinar, "The Honour/Shame Complex," 428–29, 435.
52. Deniz Kandiyoti, "The Paradoxes of Masculinity: Some Thoughts on Segregated Societies," in *Dislocating Masculinity: Comparative Ethnographies,* eds. Andrea Cornwall and Nancy Lindesfarne (London: Routledge, 1994), 207.
53. Akpinar, "The Honour/Shame Complex," 429–30.
54. Akpinar, "The Honour/Shame Complex," 431.
55. Akpinar, "The Honour/Shame Complex," 433.
56. Akpinar, "The Honour/Shame Complex," 435–36.
57. Ulf Hannerz, *Exploring the City: Inquiries toward an Urban Anthropology* (New York: Columbia University Press, 1980), 187.
58. Akpinar, "The Honour/Shame Complex," 436.
59. Akpinar, "The Honour/Shame Complex," 438.
60. Bourdieu, *Masculine Domination*, 2.
61. Bourdieu, *Masculine Domination*, 5.
62. Bourdieu, *Masculine Domination*, 6.
63. Bourdieu, *Masculine Domination*, 7.
64. Bourdieu, *Masculine Domination*, 8.
65. Delaney, *Seed and Soil*, 8–11.
66. Delaney, *Seed and Soil*, 3.
67. Bourdieu, *Masculine Domination*, 6.
68. Delaney, *Seed and Soil*, 19–20.
69. Delaney, *Seed and Soil*, 20.

70. Delaney, *Seed and Soil*, 267–68.
71. Delaney, *Seed and Soil*, 3.
72. E.g., Clifford Geertz, "Religion as a Cultural System," in *The Interpretation of Cultures: Selected Essays* (New York: Basic Books, 1973).
73. Delaney, *Seed and Soil*, 18–19.
74. Delaney, *Seed and Soil*, 9.
75. Delaney, *Seed and Soil*, 9–10; Stephen Beckerman and Paul Valentine, eds., "The Concept of Partible Paternity among Native South Americans," in *Cultures of Multiple Fathers: The Theory and Practice of Partible Paternity in Lowland South America* (Gainesville: University Press of Florida, 2002).
76. Delaney, "The Meaning of Paternity," 495.
77. Bronislaw Malinowski, *The Father in Primitive Psychology* (London: Routledge, 2002), 12.
78. Malinowski, *The Father in Primitive Psychology*, 8.
79. Malinowski, *The Father in Primitive Psychology*, 57.
80. Delaney, "The Abrahamic Theory of Procreation," 9.
81. Bronislaw Malinowski and Robert Redfield, *Magic, Science and Religion and Other Essays 1948* (San Francisco: The Free Press, 1948), 197.
82. Malinowski and Redfield, *Magic, Science and Religion*, 190.
83. Delaney, *Seed and Soil*; Jonathon Harwood, *Styles of Scientific Thought: The German Genetics Community, 1900–1930* (Chicago: University of Chicago Press, 1993).
84. Delaney, *Seed and Soil*, 13.
85. George Posner et al., "Accommodation of a Scientific Perception: Toward a Theory of Conceptual Change," *Science Education* 66, 2, (1982); Norman Doidge, *The Brain That Changes Itself* (New York: Viking, 2007). Such substantive psychological accommodation requires actual neural restructuring.
86. Delaney, *Seed and Soil*, 27. A term has yet to be invented, says Delaney, 'to refer to culturally constructed notions of the process by which life comes into being.'
87. Pickthall, Quran, Sura 2: 223.
88. Delaney, *Seed and Soil*, 27, 30.
89. Delaney, *Seed and Soil*, 3.
90. Delaney, *Seed and Soil*, 31–32.
91. David Schneider, "What is Kinship all About?" in *Kinship Studies in the Morgan Centennial Year*, ed. P. Reining (Washington, DC: The Anthropological Society, 1972), 45, in Delaney, "Meaning of Paternity," 495.
92. Meeker, "The Black Sea Turks," 157, in Delaney, "Virgin Birth," 497.
93. Delaney, "Meaning of Paternity," 498.
94. Sayyed Hossein Nasr, *Ideals and Realities of Islam* (London: Allen and Unwin, 1985), 110.
95. Delaney, *Seed and Soil*, 45, 36. Until the early twentieth century in Europe and the United States, after divorce, children legally belonged to and remained with their father.
96. Delaney, *Seed and Soil*, 72.
97. Inhorn, *Infertility and Patriarchy*, 22.
98. Inhorn, *Infertility and Patriarchy*, 24.
99. Delaney, "The Abrahamic Theory of Procreation," 6.
100. Chris Hann, "Culture and Anti-Culture: The Spectre of Orientalism in New Anthropological Writing on Turkey," *Journal of the Anthropological Society of Oxford* 24, 3, (1993): 227–28.
101. Delaney, *Seed and Soil*, 48.
102. Leila Ahmed, "Arab Culture and Writing Women's Bodies," in *Women and Islam: Social Conditions, Obstacles and Prospects*, ed. Haideh Moghissi (New York: Routledge, 2005), 209.
103. Inhorn, *Infertility and Patriarchy*, 25.
104. Parla, "Honor: Turkey and the Caucasus," 219.
105. Ljungqvist, "Female Shame, Male Honor."

106. Parla, "Honor: Turkey and the Caucasus," 219; Akpinar, "The Honour/Shame Complex Revisited." In 2002, Turkish men and women were accorded full legal equality in the family.

107. Parla, "Honor: Turkey and the Caucasus," 219.

108. Delaney, *Seed and Soil*, 91.

109. Elham Bagheri, "A Qualitative Investigation of Religion, Gender Role Beliefs and Culture in the Lives of a Select Group of Muslim Men," (PhD diss., University Of Iowa, 2012), 83.

110. Bagheri, "Investigation of Religion ," 95.

111. E.g., Mary Gearin et al., "UK Arranged Marriages: Kidnapping, Rape and Murder in the Name of Family Honour," ABC *Foreign Correspondent* Program (Australian Broadcasting Corporation, November 12, 2013).

Chapter Five

The Science of Sex: Aristotle's Seed and Soil

This chapter examines the Aristotelian reconstruction of seed-and-soil as science. With the reintroduction to Europe of the Aristotelian corpus in the thirteenth century, Aristotle's theory of procreation entered the natural-philosophical canon and remained there for five hundred years. Ancient Greek and Hebrew creation myths are pertinent here; both cultures took story elements from the ancient Middle East, including Babylonian, Phoenician, Egyptian, and Canaanite cosmogonies, but the stories function differently in their particular contexts.[1] Whereas, for example, the coming into being of men is of little or no interest to Homer and Hesiod, it is the raison d'être of Genesis. More than three centuries before Aristotle, Hesiod's purpose in the theogonic and cosmogonic narrative, *Theogony,* was to map Zeus's rise to power and the inception of patriarchal rule.[2] Roughly a century before Aristotle, Aeschylus (via the role of Athena) recalled and celebrated patriarchy's final and decisive overthrow of Mother right: 'For there was no mother who gave me birth, and in all things, except for marriage, whole-heartedly I am for the male and entirely on the father's side.'[3] Hence, by the time of Aristotle, received knowledge included the doctrine of male monogenesis.

The finer points of Aristotle's theory of procreation are specialized as well as complex and here I set down the relevant arguments without attempting to adjudicate. This chapter abstracts some continuities that run throughout Aristotle's theory, including the notion of gender hierarchy.

ACTIVE AND PASSIVE, SEED AND SOIL

The current literature on gender in antiquity is alive with questions. Did the ancient Greeks view the body as inherently 'feminine' or 'masculine'? Did they see women and men as having different 'natures'? If so, where is the difference located and how was it explained? Does the ancient material have anything new to offer contemporary debates about gender, matter, and the body? How well do contemporary writers understand the nuances of Aristotle's so-called 'single-seed' theory? How did women fare in the sociopolitical stratification? Was the Greek attitude toward women similar to our own, namely, that women must be respected for their 'womanly' virtues, but kept subordinate? After all, says classicist Brooke Holmes, we are still dealing with the 'seamier legacies' of antiquity, among them patriarchal bias and systemic misogyny.[4]

I begin my enquiry at the intersection of gender, sexuality, and power in ancient Greece. The Greeks expressed sexual relations, as well as power relations more generally, in gendered terms regardless of whether men, women, or boys were involved.[5] For Michel Foucault, following Kenneth Dover, the key theme of sex involving at least one adult male was domination and submission constituted by phallic penetration; this theme has been dubbed the 'penetration model' of gender and power relations.[6] Just as imperial status has to do with the penetration and appropriation of the land of foreigners, so gender status is about the ability to sexually penetrate the other, bodily and psychologically.[7] A similar model is implicit in Genesis as a sliding scale from male exemplar to 'unmen.'[8] Penetrating men top the hierarchy above women, eunuchs, hermaphrodites, and girls. Boys occupy a transitional space between the world of women and that of adult men. The monstrous—that is, penetrated men and penetrating women—are at the lowest level. According to this logic of conquest, men can penetrate without shame everyone on the levels below them, including foreign males and slaves. The purpose of raping male prisoners of war is to deny them their male status.[9]

And what of intellectual penetration? Just as the Greek patriarchs guarded the entrance to the literary-philosophical canon, granting special privileges, for example, to Aeschylus's tragedies alone,[10] so have successive generations of gatekeepers penetrated the body politic, reinscribed their position at the top of the hierarchy, and assimilated or marginalized dissent.

Apropos ancient Greek male same-sex relations, the penetration model contains an inherent paradox, namely, the juxtaposition of an ethos of male superiority with the notion of intercourse as involving only one dominant partner.[11] That ethos was of no consequence when a woman or slave was involved; the problem arose only when the 'passive' partner was a free male. While, according to the patriarchal ethos, men 'naturally' found pleasure in a boy, it was 'unnatural' for a boy to enjoy being an object of pleasure, as that desire feminized him. The boy must not enjoy and identify with the subordinate role; rather he must appear to adopt it for reasons of gratitude and kindness toward his older lover.[12] Plato even questioned whether courage could be nourished in the soul of the seduced, concluding that no one would expect so, 'blaming the softness of the one who gives in to the pleasures and is incapable of mastering them, and reproving the resemblance in image of the one who undertakes the imitation of the female.'[13] Plutarch was more direct: 'Those who enjoy playing the passive role, we treat as the lowest of the low, and we have not the slightest degree of respect or affection for them.'[14]

The profound significance of the distinction between the active and the passive role in male same-sex relationships hinged on what the Greeks understood by the idea of 'nature.'[15] Nature allocated the active, superior role to men and the passive role to women: 'the role that could be imposed by force.'[16] Women's natural passivity, said the Greeks, leaves them lacking the moral insight and firmness of purpose that empower men to resist the temptations of comfort and pleasure.[17] The term 'unnatural,' says John Winkler, did not refer to same-sex relationships or to the contravention of some kind of necessary order of the world but rather to behavior that was 'self-indulgent, luxurious, and exceedingly appealing.'[18] Such behavior precluded the self-restraint characteristic of manliness and thus superiority. The Greeks, says Foucault, viewed the submissive role as antithetical to a man's civic duty: 'When one played the subordinate partner in the game of pleasure relations, one could not be truly dominant in the game of civic and political activity.'[19] The penetration model requires men to be on permanent standby to carry out the civic, political, and territorial needs of the state. Thus, only by way of a strict regime of self-discipline will the superior man be ready to kill and be killed on demand. Women's duty, on the other hand, is to bear and raise such men.

Given that, according to Greek gender norms, men do not naturally choose submission, the Greeks viewed the male prostitute as subordinating himself unnaturally to other men.[20] Greek moralists perceived men's use of prostitutes as a sign of such self-indulgence and the beginning of a slippery slope ending with the seduction of boys from elite families: 'The ultimate crime against nature' was 'to treat the city's future leaders as if they were slaves available in a common brothel.'[21] This 'ultimate' crime, then, was an offence against class, an upsetting of the social hierarchy. Sexual appetite, said the Greeks, arose from an instinctive force that by nature was liable to the kind of excess that could extend to civic rebelliousness.[22] This interpenetration of gender and class makes sense of why states typically regard any threatened destabilization of patriarchal gender norms as a threat to the state itself.[23]

Some scholars have questioned the usefulness of Foucault's reading of the penetration model; more seriously, some rebuke Foucault himself for undertaking a major work in an area (classics) in which he self-admittedly is not accredited.[24] Criticism ranges from chastisement for beginning a history of sexuality with the Platonists, thus omitting the likes of Hesiod and Aeschylus,[25] to scathing remarks on the work being 'so ill-informed' as to be 'not worth reading.'[26] In addition, much feminist commentary points to startling gaps in Foucault's volumes. The main problems, says Amy Richlin, are that his so-called history of sexuality is a history 'of the desiring subject, who is male'; women figure only as objects, and then rarely at all; he reinscribes the marginality of men who prefer the passive role in same-sex relations; and he fails 'to ask where women's subjectivity was.'[27] Similarly, Silvia Federici criticizes Foucault for collapsing female and male histories into an undifferentiated whole and, in regard to the disciplining of women, for ignoring 'one of the most monstrous attacks on the body perpetrated in the modern era: the witch-hunt.'[28] In my view, the potency of the penetration model as metaphor lies in its evocative imagery of male-dominant social relations and power-based behaviors such as physical and psychological rape. While Holmes criticizes simplistic uses of the model, she argues for the 'language of penetration' as 'vivid shorthand' for talking about power relationships in general.[29] And although there can be no definitive answer to the perennial question of whether Greek male same-sex relationships were based more on prosocial attributes than on power disparities,[30] a plausible alternative hypothesis lies in a dialectical view.

As to the man demonized by Plutarch as 'the lowest of the low,' he is the cultural figure defined by his desire to be penetrated, the inversion of historically specific ideals of masculinity that embody deep cultural anxieties about gender.[31] Such anxieties, says Holmes, are the projections of men whose greatest phobia is of 'becoming a woman,' that is, men afflicted by femiphobia. Moreover, femiphobia relates not only to the man's fear of being treated as women are treated, but also to his fear that the Other will penetrate him psychologically, that is, know him intimately for who he is in the same way that society permits women to know children and other women.

The penetration model still operates in the twenty-first century, perhaps at its most complex in the context of the sexual abuse of boys and men. Many male abusers, for example, gender themselves as heterosexual and their boy victims as feminine.[32] Men who fear women may seek out the feminine aspects in a boy, perceiving boys as 'safe' substitutes. Many perpetrators 'reason' that, as the boy is not a man, the perpetrator himself is not 'homosexual.' Similarly, many priest-abusers convince themselves that having sex with boys does not violate their priestly vow of celibacy. Boy victims often conclude that they themselves cannot be 'real boys' because they attracted men who were not normally drawn to men.[33] The boy who believes he has actively participated in the abuse may feel that he has revealed himself to be feminine or homosexual.[34] All these ideas are embedded in the penetration model. In the context of male-on-male rape in prison settings, for example, inmates and staff typically label the victim, but not the perpetrator, as feminine.[35]

To return to ancient Greece and the Woman question, did Woman function as Other to the male norm? Hesiod, for example, constructed a form of sexual difference that denied women a proper place in the natural order and the human community.[36] Counter-readings, however, argue that women played active, and sometimes prestigious, roles in ancient Athens, and that they were integrated into the community particularly via public religious cults in a range of roles, for example, wife, prostitute, and priestess.[37] duBois, however, suggests that perceptions of Woman changed from alien Other (in archaic Greece) to inferior Other within the species of men (in classical Greece).[38] Also complicating matters is that the roles of wife, prostitute, and priestess have typically been cast ambiguously, often as threatening to men as well as fascinating. Other scholars characterize the Greek religious arena as one in which Greek women assumed roles equal to those of men. But the evidence—namely that the Greek pantheon included goddesses—is no more persuasive than that of the Catholic Church's inclusion of Mary makes it egalitarian.

Furthermore, to what extent did religious positions held by women originate in earlier, more egalitarian traditions? Was it precisely such traditions that nascent patriarchy had to marginalize in order to establish and maintain itself? The evidence of women's participation in religious cults certainly adds depth to the historical picture, but to what degree does it destabilize the significance of their exclusion from most aspects of public life or undermine how the canon operated to justify that exclusion? And what if we view these aspects of women's lives—agency, resistance, and exclusion—as in dialectical tension rather than as mutually exclusive? Because ritual activity entrusts historical women with their own production, says Barbara Goff, they are subjects as well as objects of the process. Moreover, she says, ritual activity must rehearse women in the contradictions necessary for the operation of patriarchy.[39]

SEED AND SOIL IN ANCIENT GREECE

The primeval notion of Woman as abundance began with perceptions of Earth-Gaia as limitless, spontaneously all-giving, and parthenogenic. Hesiod describes how mortal men originally lived as gods, 'free from toil and grief' and having 'all the good things; for the fruitful earth unforced bore them fruit abundantly and without stint.'[40] But, over millennia, the Greeks constructed

Woman as enclosed field requiring insemination by the life-giving male in order to produce his crop. duBois argues that with the gradual onset of agrarianism the once-communal lands were appropriated, fenced, plowed, and seeded by men, and that Hesiod marks the difference between an economy of gathering and one of cultivation. While the gathering culture receives gifts from the earth, says duBois, the cultivated earth sometimes appears to hide the sown seeds, begrudging farmers the crops she produces.[41]

Delaney, on the other hand, has it that seed-and-soil—in Genesis, for example—preexisted agrarianism. It is not agrarianism that is foundational, says Delaney, but rather men's appropriation of generation as a male-only activity, which entailed a shift from understanding the male role in procreation to assuming its primacy. In analyzing cultural transformations, Delaney's major concern is with worldview, whereas duBois's focus on the transition to agrarianism has more in common with Bourdieu's emphasis on habitus.[42]

Turning to Hesiod's worldview, *Theogony* is concerned with the origins of the cosmos, the gods, and the world, whereas mortal men play a relatively insignificant role.[43] Part way into the story, the sky god Ouranos is holding his children in Gaia's womb in an effort to prevent usurpation by his sons. But Gaia and her son Kronos castrate and overthrow Ouranos. Kronos, fearing usurpation in his turn, swallows his sons—all except Zeus, who is hidden by his mother, Rhea, in a cave protected by Gaia. Zeus subsequently overthrows his father. In order to prevent his own overthrow by a future son, Zeus swallows his wife, the powerful goddess Metis, who is pregnant with Athena. When Athena springs fully formed from his head, Zeus claims that he alone parented her.[44] In a notable paradox, while Hesiod proclaims Zeus's feat of monogenesis, he also leaves no doubt that Metis is Athena's mother: 'But when she was about to bear gray-eyed Athena . . . [Zeus] deceived the mind of Metis with guile and coaxing words and lodged her in his belly . . . then from his head he himself bore gray-eyed Athena.'[45] In addition, not only does the oracular element of the story emphasize the patriarch's dread of usurpation, it also renders problematic the role of the Oracle in catalyzing that dread.

Hesiod's narrative deals with a process of increasing male power—from a supreme matriarch, Gaia (who, without a partner, gives birth to Sky, Sea, and Mountains) to a supreme patriarch, Zeus who, allegedly without a partner, creates offspring. *Theogony* tells of a gradual transformation from Chaos to Law, from the physical to the nonphysical, and from feminine to male dominance (which Hesiod equates with law and order).[46] Hence, the narrative charts Zeus's rise to the top of the Greek pantheon, which, says Lerner, reflects changes in religious and gender concepts that had already taken place in Greece.[47]

In the story, Zeus, in retaliation for the theft of fire, 'makes an evil thing' and sends it to wreak havoc among men. According to Hesiod, Zeus's 'dear heart' was so angered by the theft that he created Pandora, 'formed of earth [in] the likeness of a shy maiden,' an object of 'sheer guile.' From her 'is the deadly race and tribe of women,' a race that causes endless trouble for men and is interested only in taking their wealth. Pandora is a sign of sexual difference: Woman is an alien, crafted species, Other to men. In addition, she is primarily destructive, bestial, and voracious, not generative like Gaia.[48]

Wives, says Hesiod, 'do mischief' all day while their husbands work. Just as drone bees 'stay at home in the covered hives and reap the toil of others into their own bellies,' Zeus 'made women to be an evil to mortal men, with a nature to do evil.'[49] As an aside, it is unlikely that Hesiod knew that his worker-husband/drone-wife analogy was inverted in reality. Even 350 years later, Aristotle assumed that the bee triad comprised king, male soldiers, and female drones. Why? Because Nature, said Aristotle, does not arm females.[50]

Rehearsing the patriarchal narrative, *Theogony* has it that the arrival of Zeus's instrument of revenge brings evil into the hitherto idyllic lives of mortal men and forces them into grueling labor in order to obtain food.[51] Worse, Pandora is intrinsically programmed with the desire to manipulate, deceive, and otherwise injure men, and this malicious desire is indivisible from her sexuality. In other words, women's malevolence is both cosmologically and biologically determined and, hence, beyond their control.

Debate continues about the degree to which Hesiod's construction of Woman permeated Greek culture generally.[52] In duBois's deconstruction, pre-Platonic Woman comprises a different species from men, whereas, under post-Platonic monotheistic metaphysics, she becomes a deficient male.[53] Radical difference gives way to sameness-deformed-by-lack, which Aristotle elevates into scientific theory and Freud later inscribes into his theory of female as castrated male. For duBois, Hesiod's Woman-as-separate-species is at least about difference than about mutilated sameness.[54] This is important to duBois because her purpose in seeking differences among the Greeks is to arm her criticism of Freud's biased interpretation of the Greeks in crafting his own variation on the Aristotelian theme.[55]

But whether or not Hesiod's misogynous construction characterized the wider Greek culture, the canon remains. Aristotle's corpus came to be among the most influential in the Western natural-philosophical canon and thereby pivotal to the construction of powerful male elites throughout millennia. Prior to examining Aristotle's reformulation of seed-and-soil as science, I turn briefly to Aeschylus's *Eumenides*, a sequel to *Theogony* that some interpret as depicting the final, decisive defeat of the female principle by patriarchy.[56]

Aeschylus takes up Hesiod's narrative: Agamemnon has sacrificed his daughter Iphigenia in the service of war against Troy. On his return, Iphigenia's mother, Clytemnestra, kills him; her son Orestes then kills her. When the Furies hound Orestes, he says the killing of his mother was justified and that they should have punished her for her crime. The Furies respond that mother-right has primacy:

> Furies: 'She was not related by blood to the man she killed.'
> Orestes: 'Then am I my mother's kin by blood?'
> Furies: 'How else could she have nurtured you, murderer, beneath her belt? Do you reject the nearest kinship, that of a mother?'
> Apollo: 'The mother of what is called her child is not the parent, but the nurse of the newly-sown embryo. The one who mounts is the parent, whereas she, as a stranger for a stranger, preserves the young plant, if the god does not harm it. And I will show you proof of what I say: a father might exist without a mother. A witness is here at hand, the child of Olympian Zeus who was not nursed in the darkness of a womb, and she is such a child as no goddess could give birth to.'
> Athena: 'It is my duty to give the final judgment and I shall cast my vote for Orestes. For there was no mother who gave me birth; and in all things, except for marriage, whole-heartedly I am for the male and entirely on the father's side. Therefore I will not award greater honor to the death of a woman who killed her husband, the master of the house.'
> Furies: 'Younger gods, you have ridden down the ancient laws and have taken them from my hands!'[57]

Athena's deciding vote frees Orestes and banishes the Furies and with them the claims of the mother-goddess.[58] Arguably, the primary purpose of Hesiod and Aeschylus's narratives was to inscribe Father right, annihilate Mother right, and behave as if this new order had always existed as 'god-given' and 'natural.' The poets portray the Father principle that stops at nothing to avoid usurpation, whether by sons or by the female principle represented by the goddess.

Following Aeschylus, Plato describes womankind as comprising reincarnated badly behaved men: 'All those creatures generated as men who proved themselves cowardly and spent their lives in wrong-doing were transformed at their second incarnation into women.'[59] Next comes Aristotle, whose thesis, says duBois, goes beyond Plato and changes everything. While Plato appropriates the maternal power to the male philosopher—constructed as resembling the good, the sun, the father, and the son—Aristotle pits the male body, whole, complete, and perfect, against the female, a mere part of the whole, 'a thing lacking completion, lacking heat, lacking soul.'[60] This establishment of the male subject as superior is not a universal fact of culture, says duBois; it occurs in the philosophical tradition at the point when 'women's humanity is conceded, when they are named as defective, partial men.'[61] Hence, by the classical period the Greek conceptualization of Woman had shifted from Woman-as-different-species (as in Hesiod) to Woman-as-misbegotten-male (as in Aristotle).[62]

ARISTOTLE ON THE GENERATION OF ANIMALS

Hesiod (c.750–650 BCE); Aeschylus (525–456 BCE); Hippocrates (460–377 BCE); Plato (c.428–c.348 BCE); Aristotle (384–322 BCE)

Aristotle was born seventy-four years after the inaugural performance of *Eumenides*. His parents claimed to be descended from Asclepius, god of medicine.[63] Scholars still debate the precise nature of Aristotle's theological ideas; some suggest they are monotheistic. The Platonists, Peripatetics, and Stoics used the term 'God,' as did Aristotle.[64] What the term meant, however, is contested. 'God,' says Aristotle, 'is a living being, eternal, most good,' who brings into being things that are divine and eternal and things that have existence and nonexistence.[65] The first unmoved mover, he says, is 'the God.'[66] Is this evidence of monotheism? Or might Aristotle be referring to Fate, the supreme power to which everything, including the gods, is subject? Perhaps the facts of the matter have been lost in translation.

Aristotle equates maleness with activity, heat, dryness, purity, and divinity as opposed to the passivity, coldness, dampness, pollution, and profanity of femaleness. While Aristotle, the pre-Socratics, and the Hippocratics agreed that, in comparison to Man, Woman is wetter and spongier, they reached no consensus about her coldness. The case for Woman's dampness was that her excess moisture sets in with menstruation when her body begins to accumulate excess blood (a byproduct of food).[67] If the blood vessels expand properly, the excess blood moves to the uterus and is evacuated. However, excess blood trapped in the body can cause night terrors and suicidal delusions. The Greeks held that Woman's 'natural inactivity' compounds the problem. The recommended cure was sexual intercourse, thought to dilate the internal passageways.[68] More than five hundred years later, Galen followed the Hippocratics on the nature of the body, and the tradition persisted into the modern era.

In its Aristotelian form, the doctrine of male monogenesis shaped Western science and philosophy and prevailed in Europe and the United States into the twentieth century.[69] Male monogenetic theory, says David Blumenthal, held sway for millennia and 'remains as a central underlying motif of all three biblical religions.'[70] Jesus of Nazareth was known by the name 'Monogenes,' says Delaney, which meant 'begotten of the Father'; 'Mary provided the physical body, but she was not co-Creator'.[71]

Contemporary Western duogenetic theory holds that men and women contribute equally to the hereditary substance of the fetus.[72] However, we cannot read the term 'two-seed' (particularly in its polysemous contemporary use) as implying equally generative male and female contributions to the offspring.[73] Aristotle's view is that women's contribution to the formation of offspring consists of matter, specifically, menses, which is neither generative nor of equal value to the male seed. Aristotle grounded this counterfactual account in a broad-ranging philosophical system, says Lerner, and thereby elevated a myth to the level of science.[74] Whether scholars can properly describe Aristotle's theory as monogenetic, however, is contested,[75] as is whether the two-seed theory of Hippocates and, later, Galen is duogenetic. Given the specialized nature of these questions and the contested meaning of the key terms, the following overview inevitably oversimplifies matters.

Take, first, the question of whether Aristotle's theory is truly monogenetic. The answer is not immediately apparent, particularly as Aristotle discusses three different types of soul and variously describes (seemingly interchangeably) the female contribution to conception as 'secretion,' 'seed,' 'semen,' 'catamenia,' and 'menses.'[76] Aristotle sometimes refers to female 'seed,' says Holmes, which is insufficiently cooked or concocted to become semen.[77] In Aristotle's view, both male and female seed are 'residues' that begin as blood concocted from food. The male alone has sufficient innate heat to concoct the blood further into a seed capable of transmitting the higher forms and movements that define human life. The female's blood is too cold to transform catamenia into semen.

Catamenia, says Aristotle, 'are analogous in females to the semen in males'; they 'are semen not in a pure state but in need of working up.' Moreover, as 'it is not possible that two such discharges should be found together, it is plain that the female does not contribute semen to the generation of the offspring.'[78] 'It is the male that has the power of making the sensitive soul.'[79] In other words, by Aristotle's reckoning, female semen appears not to possess the full generative potency of male seed. On the other hand, the female contribution is not inert; it contributes to the embryo's nutritive soul and some physical characteristics.[80]

Might Aristotle's theory of causation speak to the question? Four causes underlie everything, says Aristotle: 'the final cause, that for the sake of which a thing exists'; 'the formal cause, the definition of its essence'; the material cause; and the moving principle or efficient cause.[81] Axiomatic to Aristotle is that all living things have soul and that soul is superior to body. The soul is the source of a set of powers or faculties to engage in nutrition, reproduction, appetite, perception, locomotion, and cognition.[82] Humans alone have what Aristotle calls the highest form or 'rational' soul.[83] Of the offspring, says Aristotle, 'the physical part, the body, comes from the female and the soul from the male.'[84] As to the four causes, the male provides the definition, the movement, and the final cause of the embryo, whereas the female provides the material cause. 'While the body is from the female, it is the soul that is from the male, for the soul is the reality of a particular body.'[85]

Furthermore, says Aristotle: 'the secretion of the female contains material alone';[86] the semen is the mover and the menses is the moved;[87] 'the first principle of movement, or efficient cause, whereby that which comes into being is male is better and more divine than the material whereby it is female';[88] the male represents the 'effective and active' and the female 'the passive'; the male is 'the one who is able' and the female 'the one who is unable.'[89]

All this activity in the male, says Aristotle, is a result of his greater heat:[90] the male 'is much superior to the female in natural heat.'[91] For generation to occur, 'semen must be made first, and that is the function of the generating parent.'[92] In the semen is 'the spiritus conveying the principle of soul . . . in and with which comes away from the male.'[93] This principle 'is not connected with matter, and belongs to those animals in which is included something divine (to wit, what is called the reason).' In the absence of this principle of soul, the fetation 'does not develop into a living creature, for this is introduced by the semen of the male. When such a principle has been imparted to the secretion of the female it becomes an embryo.'[94] In summary, Aristotle has it that when the male semen comes together with the female menses, the semen, by way of its generative vital heat, concocts the matter of the menses into a heart, which develops into an embryo.[95]

The potential for a new embryo is in the menses, but the father alone contains the vital heat required to concoct the female menses,[96] and only the male semen provides the moving cause.[97] In addition to vital heat, the fifth celestial element is required, the *aither/thermon*, which informs matter and generates new beings. This element is the connate *pneuma*, which is the result of vital heat on the blood of the living creature.[98]

The more perfect animals are those that are hotter, says Aristotle, and only the male is the product of normal fetal development. The non-male fetus is a result of deformity due to one or more inauspicious variables, including, for example, a cold south wind blowing at the time of conception. The cold conditions cause too much secretion, which 'is harder to concoct.'[99] Semen with the greatest amount of vital heat yields an offspring most like its father—male and resembling him. Less vital heat results in less resemblance; less again creates a female embryo, and less still a degenerative monster.[100]

How does Aristotle's theory compare with that of Hippocrates, whose perspective, incidentally, prefigured that of Charles Darwin two millennia later?[101] The question of parent-child resemblance drove Hippocrates's notion of duogenetic inheritance. Both male and female, says Hippocrates, contribute semen, which occur during sexual excitation, intermix to form offspring, and pass on inheritance. Countering this position, Aristotle reiterates that 'the female does not contribute semen to the generation of the offspring,' and that too much female principle gives birth to monstrosities: 'For the female is, as it were, a mutilated male, and the catamenia are semen, only not pure; for there is only one thing they have not in them, the principle of soul.' When Nature departs from type, the result is a monstrosity; 'the first departure' is that 'the offspring should become female instead of male.'[102] 'We must look upon the female character as being a sort of natural deficiency.'[103]

Although Holmes is aware of Aristotle's belief that male seed alone can 'generate' and 'ensoul' the embryo, and that female 'seed' is incapable of generating life, form, or movement, she argues that Aristotle's theory cannot properly be described as 'single-seed' because it refers to 'female seed.'[104] But how often, for example, does Aristotle use the term 'seed' for menses? Dean-Jones, for one, claims that he 'did away with the notion of female seed altogether' and constantly refers to the female contribution as no more than menses.[105]

How, then, to properly describe Aristotle's theory? While 'duogenetic' hardly applies, some scholars claim that the term 'monogenetic' underrates Aristotle's attitude toward the female input. Unlike Plato, at least Aristotle acknowledges the female contribution as more than incubation. However, base matter and nutritive soul hardly compare to the male's gift of movement, definition, raison d'être, and the life spark. The fact that the formal and informal terms (e.g., 'duogenesis' and 'two-seed') are not semantically equivalent further complicates an already complex debate.

Holmes also suggests that Aristotle cannot properly be described as an essentialist.[106] Aristotle can be read in different ways, she says, and his corpus contains a mixture of theory and observation, some conflicting. On the question of essentialism, says Holmes, Aristotle acknowledges similarities as well as differences, for example, that male and female are members of the same species. In addition, the male seed does not predetermine the sex of the embryo but interacts with the female material. If the female matter overcomes the 'motions' of the male seed (e.g., if the seed is too cold), the male seed becomes female.[107] Thus, according to Holmes, in Aristotle's view, anatomical and other differences between men and women are to do with matter not essence (i.e., sameness but lacking).[108] Among the intricacies of these debates, the continuities remain: themes of male superiority and female subordination.

What are the political ramifications of Aristotle's theory of procreation? The biology of women, says Aristotle, makes them inferior to men, particularly in reasoning and decision-making. Hence, 'the superior principle should be separated from the inferior. . . . So far as it is possible, the male is separated from the female.'[109] Men must exclude women from the political realm; just as men must rule over their families, men must rule over the state.

Of the courage and justice of men and women, says Aristotle, 'the one is the courage of command, and the other that of subordination.'[110]

The revival of Aristotelian thought in thirteenth-century Europe further legitimated the notions of Woman as unfit for public life and, apropos procreation, as providing no more than base matter to be ensouled by men. Embedded in the European natural-philosophical canon, the doctrine survived for more than five hundred years. And as Corliss Lamont remarked of Aristotle, the 'first great naturalist' marred the purity of his Naturalism by redefining supernaturalist concepts such that the Catholic Church could readily assimilate Aristotelian thought with apparent logic into its theology.[111]

Something else that changed after Aristotle was the Greek attitude to menses. The archaic Greeks did not think of menses as particularly polluting.[112] However, Aristotle's view is that because the fetus is made of menstrual blood, menses signifies a failed conception, that is, an undeveloped 'corpse.'[113] Aristotle warns against conceiving during menstruation because the prime matter for the embryo is polluted and produces abominations. Aristotelian notions of menstruation, says O'Grady, construe it not as cleansing or regenerating but as 'commensurate with death.'[114]

How accurate is much of what comprises Aristotle's corpus today? In the centuries after Aristotle's death, says Edward Grant, material attributed to him clearly shows encyclopedic authors repeating and frequently distorting what earlier compilers had already repeated and distorted from their predecessors.[115] The Aristotelian corpus underwent multiple redactions, some clearly inaccurate, over a period of 1,500 years before entering Latin Christendom. During that period almost all original Greek texts were lost to the Latin West, not in the sense of being unavailable but because few could read them. As the Western Roman Empire disintegrated, knowledge of Greek all but disappeared.[116] During the period from the ninth until the twelfth century, Europe knew little of Aristotle's corpus. Then everything changed—to the extent that, in Murat Aydemir's words, the 'manifest endurance' of Aristotle's On the Generation of Animals can still be read in contemporary struggles to come to terms with how conception, gender, and reproductive substances are constructed. Aristotle's seed-and-soil lingers still.[117]

NOTES

1. Jan Bremmer, *Greek Religion and Culture, the Bible, and the Ancient Near East* (Leiden: Koninklijke Brill NV, 2008), 1-2.
2. Bremmer, *Greek Religion and Culture*, 17. Scholars have also noted similarities between Genesis and Plato's *Timaeus*.
3. Aeschylus, *Eumenides*, 605-778.
4. Holmes, *Gender Antiquity*, 67, 16-17, 43, 5, 7.
5. Holmes, *Gender Antiquity*, 25.

6. Kenneth Dover, *Greek Homosexuality* (Cambridge, MA: Harvard University Press, 1978), 100-109; Michel Foucault, *The History of Sexuality*, vol. 2 *The Use of Pleasure*, trans. Robert Hurley (New York: Random House, 1986); John Winkler, *The Constraints of Desire: The Anthropology of Sex and Gender in Ancient Greece* (New York: Routledge, 1990), Introduction, 7; David Halperin, *One Hundred Years of Homosexuality and Other Essays on Greek Love* (New York: Routledge, 1990).

7. Michael Carden, "Endangered Ancestress Revisited: Sarah's Miraculous Motherhood and the Restoration of Eden," *The Bible and Critical Theory* 1, 3, (2005); Dover, *Greek Homosexuality,* 100–109; Foucault, *The Use of Pleasure*; Amy Richlin, *Garden of Priapus* (New York: Oxford University Press, 1992); Davidson, *Greeks and Greek Love*; Halperin, *One Hundred Years of Homosexuality.*

8. Carden, "Endangered Ancestress Revisited."

9. Carden, "Endangered Ancestress Revisited."

10. Alan Sommerstein, *Aeschylean Tragedy,* 2nd ed. (London: Bloomsbury, 2013), 8.

11. Foucault, *Use of Pleasure*, 220.

12. Foucault, *Use of Pleasure*, 220-21.

13. Plato, *Laws*, 8, 228, 836d-e.

14. Plutarch in David Gilmore, *Manhood in the Making* (New Haven, CT: Yale University Press, 1990), 155; Plutarch, *Dialogue on Love*, 768a.

15. Dover, *Greek Homosexuality.*

16. Foucault, *Use of Pleasure*, 46.

17. Dover, *Greek Homosexuality.*

18. Winkler, *Constraints of Desire*, ch. 1.

19. Foucault, *Use of Pleasure*, 220.

20. Dover, *Greek Homosexuality*, 67.

21. Winkler, *Constraints of Desire*, 22.

22. Foucault, *Use of Pleasure*, 49-50.

23. E.g., Lerner, *Creation of Patriarchy.*

24. E.g., Richlin, *Garden of Priapus.*

25. duBois, *Sowing the Body*; Bourdieu, *Masculine Domination.*

26. Richlin, *Garden of Priapus*, xxix.

27. Richlin, *Garden of Priapus*, xiv.

28. Silvia Federici, *Caliban and the Witch* (Brooklyn, NY: Autonomedia, 2004), 8.

29. Holmes, *Gender Antiquity*, 98.

30. James Davidson, *The Greeks and Greek Love* (London: Weidenfeld, 2007); Marilyn Skinner, *Sexuality in Greek and Roman Culture,* 2nd ed. (Hoboken, NJ: Wiley-Blackwell, 2013).

31. Holmes, *Gender Antiquity*, 97.

32. Michel Dorais, *Don't Tell: The Sexual Abuse of Boys* (Montreal: McGill-Queens University Press, 2002), 115.

33. Dorais, *Don't Tell,* 115.

34. Dorais, *Don't Tell,* 119.

35. Dorais, *Don't Tell,* 115.

36. Holmes, *Gender Antiquity*, 21.

37. Holmes, *Gender Antiquity*, 22-23.

38. duBois, *Sowing the Body*, 42.

39. Barbara Goff, *Citizen Bacchae: Women's Ritual Practice in Ancient Greece* (Berkeley: University of California Press, 2004), 159.

40. Hesiod, *Works and Days*, 117-18.

41. duBois, *Sowing the Body*, 29, 52.

42. Delaney, *Seed and Soil*, 20.

43. Hesiod and Apostolos Athanassakis, trans., *Hesiod: Theogony, Works and Days, Shield,* 2nd ed. (Baltimore, MD: Johns Hopkins University Press, 2004), 1, 7.

44. Hesiod, *Theogony*, 205.

45. Hesiod and Athanassakis, *Hesiod: Theogony,* 3-4.

46. Hesiod and Athanassakis, *Hesiod: Theogony,* 5.

47. Lerner, *Creation of Patriarchy*, 204.
48. Holmes, *Gender Antiquity*, 19, 21.
49. Hesiod, *Theogony,* 590-600.
50. Danielle Allen, "Burning *The Fable of the Bees*: The Incendiary Authority of Nature," in *The Moral Authority of Nature,* eds. Lorraine Daston and Fernando Vidal (Chicago: University of Chicago Press, 2004), 95.
51. duBois, *Sowing the Body*, 79.
52. Holmes, *Gender Antiquity*, 6.
53. duBois, *Sowing the Body*, 24.
54. duBois, *Sowing the Body*, 24.
55. duBois, *Sowing the Body*, 29.
56. Erich Fromm, *The Crisis of Psychoanalysis* (New York: Holt, Rinehart and Winston, 1970), 115; Millett, *Sexual Politics,* 111–15; Marilyn Arthur, "Greece and Rome: The Origins of the Western Attitude toward Women," in Lerner, *Creation of Patriarchy*, 280.
57. Aeschylus, *Eumenides,* 605-778.
58. Lerner, *Creation of Patriarchy*, 205.
59. Plato, *Timaeus*, 90e-91a.
60. duBois, *Sowing the Body*, 184.
61. duBois, *Sowing the Body,* 187.
62. duBois, *Sowing the Body*, 42.
63. Pierre Pellegrin, "Aristotle," in *Greek Thought: A Guide to Classical Knowledge,* eds. Jacques Brunschwig and Geoffrey Lloyd, trans. Rita Guerlac and Anne Slack (Cambridge, MA: President and Fellows, Harvard College, 2000), 554.
64. Michael Frede, "Monotheism and Pagan Philosophy in Later Antiquity," in *Pagan Monotheism in Late Antiquity,* eds. P. Athanassiadi and M. Frede (New York: Oxford University Press, 2002), 44.
65. Aristotle, *Generation of animals*, bk. I, 1.
66. Aristotle, *Metaphysics*, bk. 8, 1074a26-7.
67. Holmes, *Gender Antiquity*, 29.
68. Holmes, *Gender Antiquity*, 30.
69. Tine Tjørnhøj-Thomsen, "'It's a Bit Unmanly in a Way': Men and Infertility in Denmark," in *Reconceiving the Second Sex: Men, Masculinity, and Reproduction,* ed. Marcia Inhorn, et al., (New York: Berghahn Books, 2009), 248.
70. Blumenthal, "Images of Woman in the Hebrew Bible," 47.
71. Delaney, *Abraham on Trial*, 32.
72. Inhorn, *Quest for Conception*, 67.
73. P. Van der Horst, "Sarah's Seminal Emission: Hebrews 11:11 in the Light of Ancient Embryology," in *Greeks, Romans, and Christians: Essays in Honor of Abraham J. Malherbe* (Minneapolis, MN: Fortress Press, 1990) in Delaney, *Abraham on Trial*, 157.
74. Lerner, *Creation of Patriarchy*, 205–206.
75. Holmes, *Gender Antiquity*, 41; Robert Mayhew, *The Female in Aristotle's Biology: Reason or Rationalisation* (Chicago: University of Chicago Press, 2004), 35; Lesley Ann Dean-Jones, *Women's Bodies in Classical Greek Science* (Oxford: Clarendon Press, 1996).
76. Aristotle, *De Anima*, 2, 1, 413a23, in Robert Bolton, "Definitions of The Soul: *De Anima* II, 1–3," *Phronesis* 23, 3, (1978); Mayhew, *The Female in Aristotle's Biology*, 35; Dean-Jones, *Women's Bodies in Classical Greek Science*. The three types of soul are the nutritive, the sensitive, and the rational.
77. Holmes, *Gender Antiquity*, 43.
78. Aristotle, *Generation of Animals*, bk. 1, 19.
79. Aristotle, *Generation of Animals*, bk. 2, 5.
80. Catherine McKeen, "The Female in Aristotle's Biology," *Philosophical Books* 48, 1, (2007): 64–65.
81. Aristotle, *Generation of Animals*, bk. 1, 1.
82. Mark Schiefsky, "Galen's Teleology and Functional Explanation," in *Oxford Studies in Ancient Philosophy 33*, ed. D. Sedley, 369-400 (Oxford: Oxford University Press, 2007).

83. Aristotle, *De Anima*, 2, 1, 413a23. The nutritive soul is shared by humans, other animals, and plants. The sensitive soul is shared by humans and other animals.

84. Aristotle, *Generation of Animals*, (2004 edition) 49.

85. Aristotle, *Generation of Animals*, (2004 edition) 49.

86. Aristotle, *Generation of Animals*, 4, s. 1.

87. Aristotle, in Peter Eichman, "Sex, Blood and Soul: The Transmission of Form in Aristotle's Biology," 10.

88. Aristotle, *Generation of Animals*, (2004 edition) 36.

89. Aristotlc, *Generation of Animals*, 765b36.

90. This may explain why Aristotle's theory requires the female to be cold as well as damp.

91. Aristotle *Generation of Animals*, 4, s. 6.

92. Aristotle, *Generation of Animals*, (2004 edition) 40.

93. Aristotle, *Generation of Animals*, (2004 edition) 46.

94. Aristotle, *Generation of Animals*, (2004 edition) 46. In the same passage, Aristotle differentiates the rational soul, which comes from the male, from the other principle of soul that 'is inseparable from matter.' The 'fetation' is the result of semen acting on menses.

95. Eichman, "Sex, Blood and Soul," 3.

96. Eichman, "Sex, Blood and Soul," 4. On all accounts, Aristotle's term 'secretion' refers to the menses, e.g., Eichman, "Sex, Blood and Soul."

97. Aristotle, *Generation of Animals*, 2, 3, 737a28-30, in Eichman, "Sex, Blood and Soul," 3.

98. Eichman, "Sex, Blood and Soul," 6-7. Pneumaticization occurs when 'the hot' is applied to 'the moist' as in the concocting of food into blood.

99. Aristotle, *Generation of Animals*, 4, s. 2.

100. Eichman, "Sex, Blood and Soul," 4.

101. Inhorn, *Quest for* Conception, 56-57.

102. Aristotle, *Generation of Animals*, (2004 edition) 27, 50, 46, 100.

103. Aristotle, *Generation of Animals*, (2004 edition) 4, 6.

104. Holmes, *Gender Antiquity*, 43.

105. Dean-Jones, *Women's Bodies in Classical Greek Science*, 14-15, 177; Mayhew, *The Female in Aristotle's Biology*, 31-35. Mayhew argues against this and much other feminist interpretation of Aristotle, but his thesis omits the more specialized criticism. In addition, Mayhew's notion of the 'ideological' is highly problematic (see McKeen, "The Female in Aristotle's Biology," 64-65).

106. Holmes, *Gender Antiquity*, 42.

107. Holmes, *Gender Antiquity*, 43.

108. Holmes, *Gender Antiquity*, 42-43.

109. Aristotle, *Generation of Animals*, (2004 edition) 36.

110. Aristotle, *Politics*, in *Aristotle in 23 Volumes,* vol. 21, trans. H. Rackham (Cambridge, MA: Harvard University Press, 1944), 1260a, 24-25.

111. Corliss Lamont, *The Philosophy of Humanism,* 8th ed. (New York: Basic Books, 1997), 37.

112. Sue Blundell, *Women in Ancient Greece* (Cambridge, MA: Harvard University Press, 1995), 100.

113. Kathleen O'Grady, "The Semantics of Taboo: Menstrual Prohibitions in the Hebrew Bible," in *Wholly Woman, Holy Blood: A Feminist Critique of Purity and Impurity,* ed. Kristin de Troyer et al. (Harrisburg, PA: Trinity Press International, 2003), 8.

114. O'Grady, "The Semantics of Taboo," 8.

115. Edward Grant, *The Foundations of Modern Science in the Middle Ages: Their Religious, Institutional and Intellectual Contexts* (Cambridge: Cambridge University Press, 1996), 13.

116. Paul Vincent Spade, "Medieval Philosophy," in *The Stanford Encyclopedia of Philosophy,* ed. Edward Zalta (Stanford, CA: Stanford University, 2010).

117. Murat Aydemir, *Images of Bliss: Ejaculation, Masculinity, and Meaning* (Minneapolis: University of Minnesota Press, 2007), 4.

Chapter Six

The Seeds of Christianity

Twelfth-century author Walter Map, on hearing that St. Bernard had thrown himself on the body of a dead boy, praying for his revival, quipped: 'He was the most unlucky of monks. I have never heard of any monk lying on a boy without the boy immediately getting up afterwards.'[1]

Hesiod (c.750–650 BCE); Aeschylus (525–456 BCE); Hippocrates (460–377 BCE); Plato (c.428–c.348 BCE); Aristotle (384–322 BCE); Galen (129–c.200 CE); Tertullian (c.160–c.220 CE); Lactantius (250–325); ibn Sina (980–1037); ibn Rushd (1126–1198); Albert the Great (c.1200–1280); Aquinas (1225–1274)

How did early Christianity, outlawed until the year 313, express its Aristotelian legacy of seed-and-soil? The second-century founder of Western theology, Tertullian, one of few remaining Greek-literate theologians, elaborated on both Aristotle and Genesis in constructing Woman as cold-blooded, evil temptress. The Christian fathers used this notion to forge bonds among the Church brotherhood and to warn it off women. This chapter compares and contrasts Galeno-Hippocratic and Aristotelian theories of procreation and examines the Church's connection to *Women's Secrets*, a particularly misogynous instruction manual for clerics on how to advise women about their 'infirmities' and to devise appropriate penances for them.

SEED AND SOIL IN EARLY CHRISTIANITY

Early medieval knowledge about Greek philosophy came from patristic authors such as Tertullian, who drew his ideas about procreation largely from Aristotle: 'The sperm of the male consists of a "corporeal" and a "psychic" element,' says Tertullian, 'a hot, aerial essence, just like the soul itself.' And by its generative heat, the male seed is 'convected into the *coagulum* of the woman's blood' and thus forms the embryo.[2]

Lambasting his predecessor Marcion's portrayal of an incorporeal Christ, Tertullian teases him about the corporeal gestation and birth of Jesus: the 'uncleanness of the generative elements within the womb, the filthy concretion of fluid and blood, of the growth of the flesh for nine months long out of that very mire.'[3] In stark contrast, an essay by third-century Christian apologist Lactantius speaks to the 'marvellous works of God,' including the process of conception. How much he could explain, he says, about 'the marvelous workings of the genital parts of the body' were he not held back by modesty.[4]

Is modesty, politics, or both responsible for the editorial note—'It has been judged advisable not to translate this and the first part of the next chapter'—in a 1905 edition of Lactantius's essay? Several subsequent editors heeded the advice, figuratively placing the chapter in plain wrap for everyone bar the classicist. The chapter remains untranslated in the current online English edition of the essay.[5] Is Lactantius's appreciation of the 'marvellous workings' of procreation immodest? There was indeed much patristic contempt for these workings, which, said the fathers, were not in God's original plan but were foisted on humankind because of Eve's evildoing.

Jerome treated Lactantius ambivalently at best: 'Would that he had been as good at affirming our doctrines,' said Jerome, 'as he was at demolishing those of others.'[6] Some contemporary scholars, says Edgar Foster, also doubt Lactantius's proficiency as a Christian and theologian, describing him as 'shallow, naïve, or inept in the matter of articulating sound Christian doctrine' and his theology as essentially derivative, 'an isolated phenomenon' of ancient theology that 'has always been regarded as heterodox,' if not 'much nearer to the suspicion of Arian heresy.'[7] Yet Lactantius's thought, Foster says, is neither an isolated phenomenon nor deemed heretical by any conciliar body.[8]

Lactantius's essay targets the 'senselessness of the philosophers who follow Epicurus, who blame the works of nature that they may show that the world is prepared and governed by no providence.' Whether speaking of God's construction of the human being ('How beautiful the other parts are can scarcely be expressed') or non-human creatures ('How is it with birds . . . that they sometimes appear to have something of human intelligence?'), Lactantius's tone is manifestly appreciative.[9]

At least one translator, Sister Mary Francis McDonald, has taken the chapter out of its plain wrap. Following is a substantial extract from her translation of Lactantius's account of conception, which he ascribes to Aristotle as well as to Varro, most of whose putatively prodigious oeuvre has been lost.

> The vessel in males which contains the semen is two-fold. . . . Just as there are two kidneys, there are, likewise, two testes; so there are two seminal vesicles, but cohering in one structure. ... The right one contains the male seed, the left one the female (and in the entire body on the whole the right side is the masculine, the left feminine). . . . Likewise in women, the womb is divided into two parts. . . . The part which twists to the right is the masculine part, that which is on the left, feminine. Varro and Aristotle think that conception takes place thus. They say that the seed is not in males only, but also in the females, and that from this fact those like to the mothers are procreated, but that their blood semen has been purged. Then, if this is rightly mixed with the male's, the two, congealed and coagulated at the same time are formed. First, then, the heart of a man is fashioned, because in it resides life and all wisdom, and then the whole is consummated by the fortieth day. . . . They think that the likenesses (to parents) are brought about in the bodies of children in this way. When the seeds which have been mixed among themselves coalesce, if those of the male are in the ascendency, then the offspring, whether male or female, will be like the father; if the woman's prevail, the offspring of either sex will resemble the mother. Now that of the two parts prevails which is the richer; for in a certain way it embraces and includes the other. . . . Different natures also are thought to come about in this way. When it chances that a male seed from a male parent falls into the left part of the uterus, the opinion is that a male is begotten, but since it is conceived in the female part, it suffers some female characteristics to hold sway in it more than its masculine splendour: either a beautiful figure, or exceeding whiteness or lightness of the body, or delicate limbs, or short stature, or a soft voice, or a weak mind, or several of those characteristics. Likewise, if seed of a feminine stock flows into the right part, a female is, of course, begotten, but, since it is conceived in the masculine part, then some characteristics of maleness hold sway more than the usual sex classification would permit: either strong limbs, or excessive height, or a ruddy complexion, or a hairy face, or an unlovely countenance, or a heavy voice, or a daring spirit, or several of these. If, however, a masculine seed comes into the right part and a feminine into the left, the two fetuses come forth rightly, so that for the feminine the beauty of its nature holds throughout all things, and for the masculine manly strength is preserved both as to the mind and the body. . . . By the union of the two sexes by the excitement of pleasure, a young child was to be produced. . . . But greater strength was assigned to the males so that women might be more easily forced to suffer the marital yoke. [10]

Did Lactantius's account—in tone more Galeno-Hippocratic than Aristotelian—constitute a counter-narrative to what became orthodox Christian doctrine? Perhaps, in Lactantius's own terms, he was begotten of a male seed that fell in the wrong side of the uterus, hence undermining his ability to meet the demands of patristic orthodoxy. Some scholars suggest that Lactantius was misinformed about Aristotle's actual views,[11] perhaps from his reading of Varro, who may have mistakenly attributed the Hippocratic theory of male and female seed to Aristotle as well.[12] Lactantius's theory of variations in the features of offspring, while probably not drawn directly from Galen,[13] may have come from Hippocrates, particularly the notion of conjoint seminal emission. Since the discovery of ovaries, scholars had speculated that males and females may contribute more similarly to procreation than Aristotle had allowed.[14] As an aside, Lactantius shared a dubious honor with one of the enemy—the long dead Epicurean poet Lucretius—namely that several English translators had thrown down their pens at the prospect of translating the more erotic passages of their works.[15] In addition, Lactantius's essay raises questions about the issue of celibacy in Christian antiquity. What were the competing ideas about celibacy as doctrine? Was the main point of priestly celibacy to distinguish Christianity from its predecessors, namely Judaism and paganism? What drove the Church's misogyny far beyond Aristotle's?

Such questions are beyond the scope of this book, but it is interesting to note that Christianity, as Susan Ashbrook-Harvey explains, displayed neither unity nor a dominant orthodoxy until late in the fourth century. Thus, from the mid-second century in the Greco-Roman sphere, among others, the gnostic Marcionist movement was influential. Following the Pauline doctrine, 'in Christ there is neither male nor female,' the Marcionists glorified asceticism and celibacy and permitted women to teach, exorcize, and baptize, a stance that horrified the orthodox Tertullian.[16] The Syrian Christians, too, allowed women a handful of options apart from motherhood. Unlike Western Christianity, the Syriac Church consecrated celibacy 'as vital for the lay as well as vocational believer, both male and female.'[17] In contrast to the Greco-Latin cultures, says Ashbrook-Harvey, the Syriac Church encouraged 'feminine symbols in response to Christian revelation.'[18] Under Ishtar's various names, her cult spread across the Greco-Roman world and endured into the early Christian era. In Syrian-Oriental paganism she was part of the 'holy trinity' of Mother, Father, and Son.[19]

By the late third century, however, the Western Church was pressuring Syrian Christians to conform to Greco-Latin dogma, which viewed asceticism as a priestly, not lay, vocation. And indeed how would Christianity multiply if its flock did not? Arguments about modes of celibacy, contempt for the physical realm, and glorification of the perfect life of the ascetic turned to hatred for the female as the source of all pollution.[20] This turn to misogyny in the Syrian Orient constituted a backlash against Marcionist ideas that would have permitted more equality for women.[21] Increasingly, the Church introduced rigid restrictions on women and on female religious symbols and, by the fifth century, constructed even ascetic women as perilous for men.[22]

As to celibacy, the Western (but not Eastern) Church instituted clerical celibacy in the fourth century and ratified it in 1139.[23] Yet, as neither scripture nor doctrine mandates celibacy, why did the Church make it law? The primary reason appears to be sexual asceticism, followed by the Christians' desire to set themselves apart from rival groups, whether Jewish, heretical, or pagan.[24] As Judaism constructed marriage as divinely instituted, Christianity's emphasis on controlling sexuality pointedly set it apart.[25]

THE BROTHERHOOD

In *On the Apparel of Women*,Tertullian's letter to his 'best beloved sisters,' he bids each woman to dress to 'affect meanness of appearance, walking about as Eve mourning and repentant' to atone for the 'odium' and 'ignominy' she has inherited from Eve:

> The sentence of God on this sex of yours lives in this age: the guilt must of necessity live too. Do you not know that you are (each) an Eve? You are the devil's gateway: you are the unsealer of that (forbidden) tree: you are the first deserter of the divine law: you are she who persuaded him whom the devil was not valiant enough to attack. You destroyed so easily God's image, man. On account of your desert—that is, death—even the Son of God had to die.[26]

More than a thousand years later, the Church would still hold that the cause of human misery and malevolence was the sinfulness of Eve and, by extension, all women. Notwithstanding that according to Christian principles Eve's slate had been wiped clean by the death of Jesus, she was still the Church's favored scapegoat. When, later, it came to constructing the late-medieval witch as female, nothing surpassed Eve as prototype. Like the women she spawned, said the Church, Eve was childlike, narcissistic, deceitful, impulsive, and, most tellingly of all, given to consorting with devils.

What was the early Church saying about women and, more importantly, to what end? I situate the question in the second century, a time when the Romans were accusing one of myriad out-groups—the Christians—of every known crime from sexual licentiousness to ritual murder and cannibalism. The Christian fathers faithfully recorded these fabricated charges, preserved the manuscripts in monastic libraries, and endlessly recopied them. When came the Christians turn to accuse some new religious out-group of heresy, it was as simple as drawing from the 'stock of defamatory clichés,' which always included participation in orgies and child-eating.[27]

The Church resurrected such charges repeatedly across centuries, territories, and out-groups. During several hundred years before the early modern witch hunts, various popes oversaw the burning of thousands of so-called heretics at the stake. What were their heresies? Most of the victims were from groups of peasant and working-class Christians (Waldensians, for example) who had publicly criticized the Church for its un-Christ-like interest in wealth accumulation.[28] Their preoccupation with orgies and child-eating was not apparent even to their prosecutors. Then, as now, the insider turned public critic was the most dangerous of all heretics. 'The great scandal of

Christians,' says Garry Wills, 'is the way they have persecuted fellow Christians . . . driving out heretics, shunning them, burning their books, burning them.'[29]

Prefiguring the early modern Church's construction of the witch as female (witchcraft enjoyed gender equality up until that time), from the beginning, the church fathers used the problem of lust as an opportunity to denounce women. The fathers turned Jesus's warning—'lusting after a woman in the heart is an adulterous sin'—into grounds for blaming women for their sexual attractiveness.[30] Despairing perhaps of his capacity to avoid this particular kind of lusting, third-century Alexandrian theologian Origen allegedly castrated himself. And Jerome, in the fourth century, 'discovered that only by working on his Bible translation could he sublimate his passion and be rid of the tormenting visions of dancing girls.'[31] Patristic blame for all of this fell on women: 'What is the difference,' writes Augustine, 'whether it is in a wife or a mother; it is still Eve the temptress that we must be aware of in any woman.'[32]

Also in the fourth century the so-called golden-mouthed Archbishop of Constantinople, John Chrysostom, sublimated his passion by cultivating both his rhetoric and his young priests. In 'Two Exhortations to Theodore after his Fall' (into the arms of one comely Hermione), Chrysostom begs Theodore to extricate himself and return to his 'heavenly bridegroom' against whom he is currently committing adultery, not to mention desertion.

'I know that you are now admiring the grace of Hermione,' writes Chrysostom, 'and you judge that there is nothing in the world to be compared to her comeliness.' But if Theodore chooses correctly, *he*, Theodore, says Chrysostom, shall exceed Hermione in comeliness and grace, 'as much as golden statues surpass those which are made of clay.' 'Your beauty,' Chrysostom continues, excels hers 'as heaven is superior to earth; or rather it was much better and more brilliant than this.' The beauty of men, who God designed in His image, is incomparable to that of women, who were fashioned from a piece of rib—thus forever bent—and a bit of clay. Hermione's 'corporeal beauty,' says Chrysostom, 'is nothing else but phlegm, and blood, and humor, and bile, and the fluid of masticated food. So that if you consider what is stored up inside those beautiful eyes, and that straight nose, and the mouth and the cheeks, you will affirm the well-shaped body to be nothing else than a whited sepulchre; the parts within are full of so much uncleanness.'[33]

As Theodore remains both fallen and unmoved to right himself, Chrysostom pleads again. It is not grievous for a wrestler to fall or a warrior to suffer a wound, he writes, but it is to despair or to neglect the wound! A soldier who deserts his post may still become a champion and prevail over the enemy. Just because 'the enemy has shaken you a little from your position,' says Chrysostom, 'do not thou give yourself an additional thrust into the pit, but stand up bravely, and return speedily to the place from which you have departed.'[34]

'But marriage is right,' Theodore protested. 'God has not forbidden to marry.'

God has not forbidden to marry; I know this as well as you, retorts Chrysostom:

> For marriage, we read, is honorable and the bed undefiled, but fornicators and adulterers God will judge. He has forbidden to commit adultery. . . . It is no longer possible for you to observe the right conditions of marriage. For if he who has been attached to a heavenly bridegroom deserts him, and joins himself to a wife the act is adultery. . . . May you be preserved from ever engaging yourself in marriage![35]

The idea of a male divinity, says Nancy Frankenberry, both legitimates male authority and deifies masculinity. In addition, the idea of homoerotic desire becomes problematic in a religious system that both privileges heterosexual complementarity and imagines the male deity as equipped with a phallus: it is men, after all, not women, who enter into the covenantal marriage with God.[36] Attempted solutions to the dilemma, says Frankenberry, included prohibiting the depiction of God and exaggerating women's Otherness. When the same dilemma arose for Christian men in relation to the body of a male Christ, they avoided it by speaking of the Christian community collectively as a woman.[37]

During the period from the ninth until the twelfth century, Europe knew little of Aristotle's corpus, and the thirteenth-century reintroduction of Aristotle's thought to Europe occurred largely through the commentaries of ibn Rushd (latinized as Averroës). From Arabic translations of Aristotle's work, ibn Rushd wrote extensive commentaries that, when translated into Latin, spearheaded the recovery of Aristotle's thought in the Latin West.[38] Ibn Rushd reversed the Galeno-Hippocratic trend toward acknowledging the woman's contribution to procreation and essentially repeated Aristotle's view of the female as inferior, feebleminded, and having the particular virtues of silence and obedience.[39] In *Colliget*, ibn Rushd reinscribes Aristotle's view that the female contributes nothing but matter to the generation of offspring.[40]

Whereas philosophers and theologians tended to favor Aristotle's position, physicians leaned toward the Galeno-Hippocratic.[41] Knowing about the existence of ovaries, Galen constructed them as homologues of male testes and, like the Hippocratics, assumed they were storage containers for sperm collected from all over the body.[42] The philosopher and theologian Giles of Rome, true to Aristotelian and Averroën form, announced his intention 'to refute the [Galeno-Hippocratic] idea that the woman's sperm contributes to generation.' 'It needs in addition,' Giles says, 'the generative spirit included in the male semen.'[43]

Although many European theologians welcomed the teachings of Aristotle, papally sanctioned bans beginning in 1210 kept them from being read at the University of Paris for around forty years.[44] The Church later condemned Aristotle anew for positing—among other ideas contrary to Genesis—an eternal world. However, medieval scholars wishing to promulgate the outlawed ideas (known as the core of the new pagan and Arabic learning) employed creative doublespeak during their lectures and avoided excommunication. Medieval theology and medicine gradually integrated Aristotle's natural philosophy in more or less original form or as embedded in Galeno-Hippocratic theory, and Aristotelian ideas dominated Western European thought for the next five hundred years.

Albertus Magnus (St. Albert the Great), influential promulgator of Aristotle's theories, relied on Aristotle's physiology to draw conclusions about women's psychology and intelligence, for example, that the woman's menstrual flow weakens her cerebrum.[45] Albert tried and failed to weld Galenic terminology onto Aristotelian ideas, and concluded that the female does not in any sense actively inform the conceived being: 'A generative power emanates from the soul into the testicles,' whereas 'the womb does not have a formative power,' and 'It is the masculine [semen] that works and makes, while the feminine is worked on and made.'[46]

Thomas Aquinas, much revered former pupil of Albert's, follows Aristotle to the letter: 'The active power in the seed of the male,' says Aquinas, 'tends to produce something like itself, perfect in masculinity.' The female, however, results from 'the debility of the active power,' 'unsuitability of the material,' or a change effected by 'external influences, like the south wind . . . which is damp, as we are told by Aristotle.'[47] On the polluting nature of menses, Aquinas says that menstrual blood has a 'certain natural impurity of corruption,'[48] and that the Levitical menstrual regulations are 'moral' as well as ceremonial precepts. By what reasoning? Aquinas defers to Jerome's claim that offspring conceived during menstruation are 'deformed, blind, lame, [and] leprous.'[49]

Integral to God's plan for creation, says Aquinas, is that Woman be subjected to Man—not in the sense of slavery but as a ruler manages his subjects for their own advantage and benefit. In *Summa Theologia,* which remains a classic text of Scholasticism, Aquinas combines Aristotle's (Woman as 'misbegotten male') with Tertullian's (Woman as Eve) theories of generation.[50] Today's *Catholic Encyclopedia* still has it that the subordination of women is God-given:

> The female sex is in some respects inferior to the male sex, both as regards body and soul. . . . Man is called by the Creator to this position of leader, as is

> shown by his entire bodily and intellectual make-up. . . . [A] certain social subordination . . . which in no way injures her personal independence is assigned to woman . . . in every form of society that springs . . . from the family.'[51]

WOMEN'S SECRETS

Chrysostom's Woman as 'whited sepulcher' might have been the subtitle of the treatise *De Secretis Mulierum* (*Women's Secrets*), putatively written by one of Albert's students around 1300 and which generated enormous interest among clerics, physicians, and the educated laity.[52] *Women's Secrets*, says translator Helen Lemay, contained an exaggerated and popularized form of Albert's ideas on women.[53] *Secrets* was a more misogynous continuation of Aristotle's discussions of generation; it also reflected a profound change in the way men read gynecological literature. The speculative character of *Secrets* differed from that of formal gynecological texts but as the speculative works gained ground the genres became virtually interchangeable. Country gentlemen, urban merchants, noblemen, physicians, and clergy were reading gynecological texts in the vernacular to satisfy their curiosity about women's bodies and procreation.[54]

Chaucer's Wife of Bath decries a 'book of wikked wyves' that her husband likes to read from 'every nyght and day . . . / Whan he hadde leyser and vacacioun / From oother worldly occupacioun.'[55] As variations on tales of wikked wyves, the vernacular gynecological texts fed into the clerical tradition of 'compiling litanies of the alleged evils and deceits of women' in order to dissuade celibate clergy from consorting with them.[56] The appeal of *Women's Secrets* was such that, two centuries after its first publication, the printed copy underwent fifty editions in its first twenty-five years. And a novelistic version of *Secrets* claimed to tell 'of all the infirmities of women which cause displeasure to men, be they infirmities of the vagina, of the hands, mouth, teeth, or skin.'[57] Pope Innocent III, too, decried the lack of opportunity for men to 'look over' a potential bride:

> When you buy a horse, an ass, an ox, a dog, clothes and a bed, even a cup and a pitcher, you first get a chance to look them over. But no one displays a bride, lest she displease before the marriage. Whatever her condition one takes her as she is. No matter if she be ugly or foul-smelling, if she be sick or foolish, if she be proud or easily angered, or if she have any vicious trait; only for adultery can a man put away his wife.[58]

Later, almost two centuries after its publication, *Women's Secrets* would directly influence the authors of the witch hunting tract, the *Malleus Maleficarum*. *Secrets*, says Lemay, evidenced 'a widely circulating epitome of the sixteenth-century view of women that resulted in witch hunting.'[59] And it was officials and learned men, rather than commoners, who promulgated that view. For Europe's intellectual elite, the link between women and witchcraft came from Aristotle and Christianity.[60] The early patristic writings of clerics such as Tertullian had reinforced church suspicions about women's sexuality.

The ostensible purpose of *Women's Secrets* was to instruct clergy about women's 'infirmities' together with appropriate advice and penances. *Secrets* claims to 'bring to light certain hidden, secret things about the nature of women'—things that women hide from men. These secrets, says Karma Lochrie, are 'women's bodies, their sexualities, and their devious natures, which it is the task of religious men (the narrator and reader) to expose and defuse.'[61] Addressed 'to my dear companion and friend in Christ,' *Secrets* includes chapters on 'The signs of corruption of virginity,' 'The signs of chastity,' 'A defect in the womb,' and 'The impediments to conception.' *Secrets*, says a commentator of the time, will enable clergy: 'to provide a remedy for [women's] infirmities and so that in confessing them we might know how to give suitable penances for their sins. . . . Women are so full of venom in the time of their menstruation that . . . whenever men have sexual intercourse with them they are made leprous and sometimes cancerous.' Those who wish to avoid these leprous, cancerous evils, says the commentator, 'must abstain from this unclean coitus and from many other things which are taught in this book.'[62]

And, as *Secrets* describes them, one might think twice about consorting with the women who conceal iron in their vaginas to harm the penis during intercourse or who choose to have intercourse at the most dangerous times for men. But how is one to tell the deadly from the less deadly woman? Even the women themselves do not know, says the author, although many 'are familiar with the effect. For when men have sexual intercourse with these women it sometimes happens that they suffer a large wound and a serious infection of the penis because of iron that has been placed in the vagina, for some women or harlots are instructed in this and other ill deeds.' At this point the author becomes coy: 'And if it were right to talk about this,' he says, 'I would say something about them, but because I fear my creator I shall say nothing more about these secrets at present.'[63]

Who holds these secrets about women's sexuality? Certainly not women, who, according to the narrative, neither know them nor have the power to withhold them or possess them discursively. Rather, says Lochrie, it is the author who '*has* the feminine secret, while Woman *is* the secret that men know. Through their exchanges women become the "open secret", a product and object of masculine ways of knowing.'[64]

'To what end?' is the question. If the author's sole intention is to dissuade clergy from courting women, why does he first whet the clerical appetite by using explicitly sexual material in an explicitly titillating way? Furthermore, why does he arouse this clerical desire for knowledge and more and then break off abruptly, leaving his reader hanging? How might all this benefit the Church?

The author hints about the existence of dangerous (i.e., exciting) secrets about female sexuality, which, although not explicated, the priestly reader knows relate to woman's insatiable desire for the phallus and her animalistic capacity at best to diminish it, at worst, rip it to shreds. Does the author hope, as did Chrysostom, to coax the young priest back into the arms of his heavenly bridegroom and his earthly father? What is the nature of that relationship anyway? And, more importantly, how does the Church (mis)use the notion of Woman to foster it?

In Jehan Bonnet's *Placides et Timèo,* Timèo instructs Placides on what happens during the act of heterosexual intercourse. The woman (who, as we know from Aristotle, is naturally cold) naturally begins to steal the man's heat: 'The more she feels [his heat], the more she desires,' says Timèo. He then breaks off abruptly and addresses Placides directly: 'When you have been more obedient, when I have received from you more marks of affection, then I will tell you the deep secrets that must not be revealed to anyone, except to one's dearest friend.'[65]

At this point, the reader's urgent desire to know encounters, in Lochrie's words, *discursus interruptus*.[66] The result is that the seductive and awful secrets remain intact as backdrop for the next intimate encounter between men—the pupil who is desperate to know and the master who pulls his strings. In this context, the condition of homoerotic desire is the gift of knowledge, while the hidden secret of female desire becomes its site. In order to increase the pupil's desire, the master withholds the secret; he may even destroy and revive it in the interest of masculine intimacy—to stimulate the pupil's desire to know and his reverence for the master.[67] And in the process, the danger of man's refrigeration by she-who-steals-heat is deferred.[68]

Plato's *Symposium* contains another explanation about why the manliest men are not naturally disposed toward women. In a putatively fictitious and satirical speech, Aristophanes tells how man originally had eight limbs and two faces, but Zeus punished him for his insolence by splitting him in half. Now each half yearns for his lost other. 'They who are a section of the male follow the male, and these are the best of boys and youths because they have the most manly nature.' And, says Aristophanes in an apparent *coup de grace*:

> These when they grow up become our statesmen, and these only, which is a great proof of the truth of what I am saying. When they reach manhood they are lovers of youth, and are not naturally inclined to marry or beget children. . . . And when one of them meets with his other half, the actual half of

> himself, whether he be a lover of youth or a lover of another sort, the pair are lost in an amazement of love and friendship and intimacy, and one will not be out of the other's sight.[69]

Satire or not, that was the view from on top. When it came to the relationship between master and pupil in ancient Athens, as in Christendom, there was no social equality;[70] the pupil accepted the 'naturally' inferior position in both didactic and erotic terms. What, then, of that homoerotically charged association? Who or what protects the pupil—and his natural desire for knowledge, acceptance, and connection—from intentional or unintentional exploitation? How easy is it for the master, wittingly or not, to coax his pupil into the adoringly passive, dangerously vulnerable, 'feminine' position? And how might this tradition speak to the issue of pederasty in the Church, as well as to the modern scourge of clerical sexual abuse in which, on average, 80 percent or more of victims are prepubertal boys? (In the general population, 25 percent of victims of sexual abuse are boys.)[71]

To stay in the modern era momentarily, one particular historical event among others inadvertently increased the number of children exposed to potential clerical abuse. Throughout the history of the Catholic Church, says John Cornwell, the Latin Christian tradition required Christians from the age of puberty onward to confess their sins at least once a year. That changed in 1910 when Pius X, putatively to strengthen the church from within, decreed that confession would begin at the age of seven and take place weekly where possible. This meant not only that priests instructed children as young as five or six on the nature of sin—particularly sexual sin—hence inculcating intense shame about their bodies, but also that clerical sexual predators had weekly private access to very young children.[72] And priest-abusers encouraged their victims to believe (wrongly) that the seal (secrecy) of confession applied also to the penitent, meaning that to report the abuse was to commit a mortal (grave) sin, which, the Church taught, led to eternal damnation.

Is clerical training a recipe for moral immaturity, as Cornwell has it? What accounts for the pattern of arrested intellectual and moral development not only in the perpetrators but also in the members of the hierarchy who did little or nothing to stop them? Pivotal factors, says Cornwell, include 'the priest's own upbringing as a Catholic child . . . aspects of his seminary formation . . . [and] the abuser's own use of the sacrament of confession for himself.'[73] The point of Cornwell's second factor is that years of seminary training in cloistered residence leave clerics ill-equipped to take responsibility for their own lives, even for day-to-day decision-making. From 1910 onward, seminaries became 'spiritual bootcamps': students rarely read original texts or asked questions and had no access to newspapers, radio, visitors, or home visits. Too often, vocational training became a matter of habit rather than choice, in such cases amounting to training in irresponsibility.[74] As to the third factor, the confession protocol enabled priest-abusers to repeatedly receive absolution for their crimes: one such priest admitted to having confessed roughly 1,500 times to thirty or so different priests over a period of twenty-five years and receiving no more severe penance than prayer.[75]

Priest-abusers, says Garry Wills, employ religion as part of the seduction process: prayer serves as foreplay and the abuse often occurs in the sacristy, the confessional, the rectory, Catholic schools, and Catholic clubs with sacred pictures on the walls. The predator disarms his victim by sophistication, including 'a special arsenal of stun devices,' and often claims that sex is part of his priestly ministry, representing his sexual predation as an act of 'healing.'[76] Recall that femiphobia and fear of women can drive some men to perceive boys as a safe substitute. This is particularly advantageous to priest-abusers who construct sex with boys as not breaching their vow of celibacy.[77]

Historical enquiry into the antecedents of sexual abuse in the modern Catholic Church, as with that into witch hunting, often meets the criticism that it is anachronistic. For example, despite Nicholas Cafardi's expressed keenness to avoid falling into the trap of anachronizing, his research on clerical sexual abuse has thus been charged.[78] In defense of such enquiry, Cafardi emphasizes: 'There has not been an era in the Church's long history when it was not aware of the evil of the sexual abuse of children, especially by clergy. . . . This condemnation was repeated throughout the ages, which indicates that the sexual abuse of children by priests has been a problem throughout the Church's existence.'[79] 'There was no question,' says Diarmaid MacCulloch, that 'society was perfectly aware of what child abuse, as we would understand it, was. That had been a concern in monastic Orders since the days of the Desert Fathers.'[80]

To return to the Woman question, the sixteenth-century Catholic essayist, Michel de Montaigne, wrote that the friendship between intimate equals that he enjoyed with his late friend La Böetie is not to be found within the homosexual tradition of the Greeks, 'since as they practised it, it required a great disparity of age and divergence of favors between the lovers.' Nor is egalitarian friendship to be found with women, says Montaigne, as women are 'not normally capable of responding to such familiarity and mutual confidence as sustain that holy bond of friendship, nor do their souls seem firm enough.' There is, he says, 'no example yet of woman attaining to it and by the common agreement of the Ancient schools of philosophy she is excluded from it.'[81]

Coming back to my initial question, what did the Church gain by constructing Woman as concealing dreadful secrets about the extent of her capacity and desire to harm men? Did the Church, wittingly or not, exploit male fear of women and of being feminized (perhaps via loss of heat) in order to reinforce in-group cohesion? After all, men must stick together in order to resist the pull of the 'whited sepulchre.' What is the real and open secret about Woman? In the tradition of concocting women's secrets to foster bonding among men (and simultaneously rattle them about their own 'feminine' propensities), the secret is that Woman is the instrument by which men both bond with and punish other men in order to keep them in the fold. How men are broken—in military training and war, for example—is by placing them in the 'passive' position of Woman. Femiphobia manifests as the entirely legitimate fear of being treated as women are treated in patriarchal cultures, worse, of being excluded from the world of men. Without the use of Woman as measure of abasement, might such practices lose some of their traumatogenic potency?

ARISTOTLE'S LEGACY

Most medical textbooks of the sixteenth and seventeenth centuries espoused Aristotelian or Galeno-Hippocratic theory, the latter holding that women's semen is scantier, colder, and wetter and their testicles smaller and less perfect than men's.[82] Indeed, one reason that Galen gives for women's inferiority is that they do not produce enough heat to push their sex organs out the right way: 'The male's greater supply of innate heat causes his organs . . . to protrude from the body. . . . The female, on the other hand, less richly supplied with heat, is unable to give them the final eversion which makes them protrude, and so they remain inside.'[83]

Galen's theory of procreation integrated and expanded Aristotelian and Hippocratic ideas. Like Hippocrates, for example, Galen believed that, given that a woman without ovaries cannot conceive, both sexes must contribute to conception.[84] However, Galen harbored no doubt as to the superiority and unique spiritual qualities of the male seed. He described the inferior female seed as an infertile and non-generative glandular secretion.[85]

In the same vein, sixteenth-century physiologist Jean Fernel devoted an entire chapter in his *Physiologia* to 'the nature, excellence, and powers of semen,' declaiming 'how great the powers of the semen are from what Aristotle laid down.' Male semen, says Fernel, 'possesses a movement and origin of such a sort' and 'is equipped with the powers of the soul' and 'a divine and heavenly heat . . . in which the whole of life has its dwelling.'[86] It 'acts on its own and is the origin of all motion'; testicular heat stirs the vital spirit, which then 'readily swells into foam' And this, says Fernel, is why it is said that Venus was born of foam.[87]

Fernel wrote at a time when Aristotle, still venerated as 'the master of those who know,' was enjoying a revival. Scholars of the 1500s looked to newly translated works of Aristotle to resolve noted theoretical inconsistencies. Such resolution was essential in medicine: not only did Galen draw on some of Aristotle's ideas, but also some Aristotelians had adapted Galen's work. Fernel's *Physiologia* exemplifies a strict Aristotelian concern for logic and the search for causes.[88]

One competing theory of procreation gained credence during the seventeenth century. Preformation theory held that all offspring are preformed by God; sperm or ova contained 'the entire miniaturized human ready to be "planted" into the womb.'[89] With the invention of the microscope in 1677, the 'spermists' conquered the 'ovists,' named spermatozoa 'homunculi' and, hence, furthered the idea that women's role in conception was purely nutritive.[90] Preformation theory remained popular into the second half of the eighteenth century, with the dominant medical texts omitting opposing views by the learned women of the early 1700s. Some physicians claimed that women were entirely superfluous to the procreative process. Physicians could simply deposit human male seed into the belly of a horse and nurture it to term with human blood.[91]

The historical importance of Aristotle is that his counterfactual assumptions became, in the medieval and early modern periods, a source of standard Western arguments for the subordination of women at home and in society.[92] The church fathers merely repeated Aristotle's theory that the female's passive, cold, damp, polluted, and profane nature was responsible for her infirmities, failings, and inferiority. Aristotelianism had an extraordinary impact on European thought over a vast period, says Horowitz; the greatest number of Aristotelian works was produced during five hundred years from 1150 to 1650.[93] Legitimized by the Academy, Aristotle's theory of generation persisted into the modern period.[94]

To summarize two other critical points, the publication of *Women's Secrets* took Christian Aristotelian notions of gender and procreation to new heights of misogyny. It provided inspiration and source material for the witch hunters' guide, the *Malleus Maleficarum,* almost two centuries later. Second, in the early modern period, misogynous vernacular gynecological material written for a male audience became embedded in the Western natural-philosophical canon as medical 'truths' about women. The next chapter examines the ramifications of these developments.

NOTES

1. Nicholas Orme, *Medieval Children* (New Haven, CT: Yale University Press, 2003), 104.
2. Tertullian, *De Anima,* trans. Jan Hendrik Waszink (Leiden: Koninklijke Brill, 2010), 344.
3. Tertullian, *On the Flesh of Christ,* trans. Peter Holmes (Christian Classics Ethereal Library online, 1970), 19.3, 4.21; Delaney, "The Abrahamic Theory of Procreation," 4: 'In Christian tradition, a newborn child is baptized in the spirit to wash away the dross of earthly birth.'

4. Lactantius (250–325), "On the Workmanship of God," in *Lactantius, the Minor Works, The Fathers of the Church: A New Translation,* vol. 54, trans. Mary Francis McDonald (Washington, DC: Catholic University of America Press, 1965), 41.

5. Lactantius, "On the Workmanship of God," In *Ante-Nicene Fathers,* vol. 7, *Fathers of The 3rd and 4th Century,* ed. Philip Schaff, trans. William Fletcher (Grand Rapids, MI: Christian Classics Ethereal Library, 2009).

6. Jerome, in Edgar Foster, "Metaphor and Divine Paternity: The Concept of God's Fatherhood in the Divinae Institutions of Lactantius (250–325 CE)," (PhD diss., University of Glasgow, 2008), 6. The works to which Foster refers are in Lactantius, *Divine Institutes,* trans. Anthony Bowen and Peter Garnsey (Liverpool, UK: Liverpool University Press, 2003), 3.

7. Foster, "Metaphor and Divine Paternity," 6. 'Arian heresy' refers to the so-called Arian theology of Donatus.

8. Foster, "Metaphor and Divine Paternity," 4–5, 7, xii.

9. Lactantius, "On the Workmanship of God" (Catholic University of America Press version), 39–41.

10. Lactantius, "On the Workmanship of God" (Catholic University of America Press version), 39–41.

11. Nancy Siraisi, *Medicine and the Italian Universities 1250–1600* (Leiden: Brill, 2001), 189.

12. Robert Sharples, *Peripatetic Philosophy 200 BC to 200 AD* (New York: Cambridge University Press, 2010).

13. Peter Roots, "The De Opificio Dei: The Workmanship of God and Lactantius," *Classical Quarterly* 37, (1987).

14. Sharples, *Peripatetic Philosophy*, 256.

15. Titus Lucretius Carus, *The Nature of Things: A Didactic Poem in Two Volumes,* vol. 2, trans. John Mason Good (London: Longman, Hurst, Rees, and Orme, 1805), 202, 161.

16. Susan Ashbrook-Harvey, "Women in Early Syrian Christianity," in *Images of Women in Antiquity,* ed. Averil Cameron and Amélie Kuhrt (New York: Routledge, 1993), 294.

17. Ashbrook-Harvey, "Women in Early Syrian Christianity," 295.

18. Ashbrook-Harvey, "Women in Early Syrian Christianity," 289.

19. Ashbrook-Harvey, "Women in Early Syrian Christianity," 289.

20. Arthur Vööbus, *History of Asceticism in the Syrian Orient* (Louvain: CorpusSCO, 1958, in German), I, 83, in Ashbrook-Harvey, "Women in Early Syrian Christianity," 295.

21. Ashbrook-Harvey, "Women in Early Syrian Christianity," 295.

22. Ashbrook-Harvey, "Women in Early Syrian Christianity," 295.

23. Carl Triebs, *About Celibacy I Have No Instructions from the Lord: An Inquiry into the Origins of Clerical Celibacy* (Victoria, BC: Trafford Publishing, 2004), 122; Diarmaid MacCulloch, *Silence: A Christian History* (London: Allen Lane, 2013).

24. Triebs, *About Celibacy*, 123.

25. Triebs, *About Celibacy*, 124.

26. Tertullian, "On the Apparel of Women," in *Ante-Nicene Fathers,* vol. 4, eds. Alexander Roberts, James Donaldson, and A. Cleveland Coxe, trans. S. Thelwall (Buffalo, NY: Christian Literature Publishing, 1885), bk. 1, ch. 1.

27. Stephen Benko, *Pagan Rome and the Early Christians* (Bloomington: Indiana University Press, 1984) (regarding Romans); Norman Cohn, "The Demonisation of Medieval Heretics," in *The Witchcraft Reader* ed. Darren Oldridge (London: Routledge, 2002), 49 (regarding Christian fathers).

28. Cohn, "The Demonisation of Medieval Heretics."

29. Garry Wills, *Why Priests? A Failed Tradition* (New York: Penguin, 2013).

30. Joseph Klaits, *Servants of Satan: The Age of the Witch Hunts* (Bloomington: Indiana University Press, 1985) .

31. Klaits, *Servants of Satan*, 67.

32. Augustine, *Letters 211–270, 1–29 of St Augustine* (New York: New City Press, 2005), 243:10.

33. Chrysostom, John, "Two Exhortations to Theodore after his Fall," in *Nicene and Post-Nicene Fathers, First Series,* vol. 9, ed. Philip Schaff, trans. W. R. W. Stephens (Buffalo, NY: Christian Literature Publishing, 2009), Letter 1, 14.

34. Chrysostom, "Two Exhortations to Theodore," Letter 2, 1.

35. Chrysostom, "Two Exhortations to Theodore," Letter 2, 3.

36. Nancy Frankenberry, "Philosophy of Religion in Different Voices," in *Feminist Philosophy of Religion: Critical Readings,* ed. Pamela Sue Anderson and Beverly Clack (New York: Routledge, 2004), 15; Howard Eilberg-Schwartz, "The Father, the Phallus, and the Seminal Word: Dilemmas of Patrilineality in Ancient Judaism," in *Gender, Kinship, and Power: A Comparative and Interdisciplinary History,* eds. Mary Jo Maynes, et al. (New York: Routledge, 1996). As Eilberg-Schwartz has shown, says Nancy Frankenberry, the ancient Israelites did image God in human form but veiled the divine phallus. The Hebrew Scriptures discursively constitute Israelite men as 'she' and describe them as 'whoring' when they stray into idolatry (adultery).

37. Frankenberry, "Philosophy of Religion in Different Voices," 15.

38. Prudence Allen, *The Concept of Woman* (Cambridge, UK: Wm B Eerdmans Publishing, 2006), 110–11.

39. Allen, *Concept of Woman*, 110–11.

40. Allen, *Concept of Woman*, 111.

41. Faith Wallis, trans. *Medieval Medicine: A Reader* (Toronto: University of Toronto Press, 2010), 222.

42. V. C. Medvei, *The History of Clinical Endocrinology: A Comprehensive Account of Endocrinology from the Earliest Times to the Present Day* (New York: Parthenon Publishing, 1993). For his part, Aristotle thought that testicles were intended as weights to keep the seminal passages tight.

43. Wallis, *Medieval Medicine*, 223, 226, 231.

44. Grant, *Foundations of Modern Science*, 71.

45. Yvonne Merrill, "'In Silence My Tongue is Broken': The Social Construction of Women's Rhetoric Before 1750" (PhD diss., University of Arizona, 1994), 97.

46. Albertus Magnus, *De homine*, 17.3 and *De mort et vita,* 2.6 in Philip Reynolds, trans. *Food and the Body: Some Peculiar Questions in High Medieval Theology* (Leiden: Koninklijke Brill, 1999), 250, 246, 248.

47. Thomas Aquinas, *St Thomas Aquinas: Summa Theologica,* trans. E. Hill (London: Cambridge University Press, 1964), Ia. 92, 1.

48. Aquinas, *Summa Theologica*, Question 31, Article 5.

49. Aquinas, *Summa Theologica*, Question 64, Article 3.

50. Beverly Clack, *Misogyny in the Western Philosophical Tradition: A Reader* (New York: Routledge, 1999), 74.

51. Augustin Rössler and William Fanning, "Woman," in *Catholic Encyclopedia* (New Advent, 2013).

52. Monica Green, *Making Women's Medicine Masculine: The Rise of Male Authority in Pre-Modern Gynecology* (New York: Oxford University Press, 2008).

53. Helen Lemay and St Albertus Magnus, *Women's Secrets: A Translation of Pseudo-Albertus Magnus's* De Secretis Mulierum *with Commentaries* (Albany: State University of New York Press, 1992), 50.

54. Green, *Making Women's Medicine Masculine*, 205.

55. Chaucer "The Wife of Bath" (c.1400), in Green, *Making Women's Medicine Masculine*, 204.

56. Green, *Making Women's Medicine Masculine*, 206.

57. Green, *Making Women's Medicine Masculine*, 213.

58. Pope Innocent III, *On the Misery of the Human Condition,* trans. Margaret Mary Dietz (New York: Bobbs-Merrill, 1969), 20–21.

59. Lemay, *Women's Secrets*, 2.

60. Daly, *Gyn/Ecology*, ch. 6, pt. III. Wiesner-Hanks, *Women and Gender*, 272.

61. Karma Lochrie, *Covert Operations: The Medieval Uses of Secrecy* (Philadelphia: Pennsylvania University Press, 1999), 8.

62. Lemay, *Women's Secrets*, 59, 33, 60, 60.
63. Lemay, *Women's Secrets*, 88.
64. Lochrie, *Covert Operations*, 123.
65. William Kibler, *Medieval France: An Encyclopedia* (London: Routledge, 1995).
66. Lochrie, *Covert Operations*.
67. Lochrie, *Covert Operations*.
68. Lochrie, *Covert Operations*, 123.
69. Plato, "Symposium," in *Collected Works of Plato*, 4th ed. trans. Benjamin Jowett (Oxford: Oxford University Press, 1953), 189c–193e.
70. According to Erich Fromm, parties in a teacher-student relationship involving rational authority are perceived as *social* equals, notwithstanding that the teacher holds a certain authority of knowledge and experience.
71. Karen Terry, et al., eds., "The Causes and Context of Sexual Abuse of Minors by Catholic Priests in the United States" (A report presented to the United States Conference of Catholic Bishops by the John Jay College Research Team, 2011), 9–10; Sophie Williams, "Facts and Figures Relating to the Childhood Sexual Abuse and Sexual Assault of Males" (Nungar, Queensland: Living Well, 2014).
72. John Cornwell, *The Dark Box: A Secret History of Confession* (London: Profile Books, 2014), Prologue, 2.
73. Cornwell, *The Dark Box,* Prologue, note 26.
74. Cornwell, *The Dark Box,* Author's note.
75. Cornwell, *The Dark Box,* Prologue, note 26.
76. Garry Wills, "Scandal," *New York Review of Books*, May 23, 2002.
77. Dorais, *Don't Tell,* 115.
78. E.g., John Beal, "*Before Dallas: The U.S. Bishops' Response to Clergy Sexual Abuse of Children* (review)," *The Catholic Historical Review* 97, 1, (2011), 188.
79. Nicholas Cafardi, *Before Dallas: The U.S. Bishops' Response to Clergy Sexual Abuse of Children* (Mahwah, NJ: Paulist Press, 2008), 9.
80. MacCulloch, *Silence: A Christian History*, n. 32.
81. Michel de Montaigne (1533–1592), "On Affectionate Relationships," in *The Complete Essays,* ed. and trans. M. A. Screech (New York: Penguin Books, 1989), 210.
82. Galen, *Galen on the Usefulness of the Parts of the Body,* trans. Margaret Tallmadge (Ithaca, NY: Cornell University Press, 1968).
83. Galen, *Galen on the Usefulness of the Parts*, 56.
84. Jeanette Hacker and Kandeel Fouad, "Male Reproduction: Evolving Concepts of Procreation and Infertility through the Ages," in *Male Reproductive Dysfunction: Pathophysiology and Treatment,* ed. Kandeel Fouad (New York: Informa Healthcare, 2007), 4. Galen believed that conception required orgasm on the part of both partners.
85. Allen, *Concept of Woman*, 109.
86. Jean Fernel (1497–1558), *The Physiologia of Jean Fernel (1567)*, trans. John Forrester (Philadelphia: American Philosophical Society, 2003), 543, 547.
87. Fernel, *Physiologia of Jean Fernel,* 6.
88. Fernel, *Physiologia of Jean Fernel.*
89. Cynthia Daniels, *Exposing Men: The Science and Politics of Male Reproduction* (New York: Oxford University Press, 2006), 11; Jean Mancel, "Vie de Nostre Seigneur Jesus Christ," illumination, Bibliothèque Nationale, Paris. Mancel's fifteenth-century depiction of the Annunciation shows baby Jesus as a homunculus descending from heaven to the Virgin.
90. Hacker and Fouad, "Male Reproduction," 4–5.
91. Daniels, *Exposing Men,* 16.
92. Maryanne Cline Horowitz, "Aristotle and Women," *Journal of History of Biology* 9, 2, (1976): 183.
93. Horowitz, "Aristotle and Women."
94. Scientists were still praising Aristotle's *Generation of Animals* in the early 1930s: 'The depth of Aristotle's insight into the generation of animals has not been surpassed by any subsequent embryologist,' wrote embryologist Joseph Needham, *Chemical Embryology,* vol. 1 (Cambridge: Cambridge University Press, 1931), 64.

Chapter Seven

Burning Desire: The Witch as Female

The accuser had been ill for ten weeks. After six weeks of his disease, he asked a seer for advice. The seer said he knew who had bewitched him but would not tell the sorceress's name. The accuser sent his wife to the suspect, whose name was Maria, and had the suspect come to him. Arriving at his house, Maria complained that she, too, was ill. He begged her to forgive him if he had ever aggrieved her, and when she said the same he gave her his forgiveness. They had had an extended history of mutual inhospitality. After Maria had forgiven him, his sickness was cured—and as a result of the healing he knew for sure she was the one who had bewitched him, as he had suspected all along.[1]

'Thou shalt not suffer a witch to live,' says Exodus.[2] This chapter examines the early modern central-European construction of the witch as female. How did the legacy of seed-and-soil underpin the early modern witch hunts? Whereas the medieval witch hunts had been sporadic rather than methodical, the accusations more likely to be of heresy than witchcraft, and the prosecuted just as likely to be men as women, all that changed in the early modern period. After several hundred years of sporadic activity, why did witch hunting become widespread, systematized, and gendered? Why did the Church, from the thirteenth century onward, begin to take seriously a phenomenon it previously had decreed nonexistent? Why did the case types (heresy vs. witchcraft) reverse in prevalence and the witch become, in the main, female? Did the authorities target a certain *type* of woman and, if so, what role did the Christian canon play in that targeting? I suggest that, in general, although not always, the inquisitors did hunt down a certain type of so-called 'deviant' woman and that the canon was crucial to the cultural construction of that deviance. In addition, the early modern construction of the witch as female opened up a new era that criminalized gender nonconformity in both women and men as 'perversion.'

All women, said the medieval Church, carry Eve's inherent sinfulness and propensity to fall for Satan. Unlike the male and female witches of the medieval period, however, consorting with devils became the all-consuming passion of the newly-minted female witch.[3] Everything she thought, said, and did was potentially diabolical. No longer could one safely look at, touch, speak or listen to her, let alone accept her hospitality, buy her wares, or enlist her services as cook, counsellor, midwife, or healer. Thus was the scene set for the early modern witch hunts.

I turn first to the question of why witch hunting became systematic and gendered. Multi-factorial and contested as the issue is, it seems that by the early modern period counter-narratives of reason and fairness had been discursively—if not actually—eliminated. Stereotyping women as devils or angels was as old as history, but the early modern clerics had become obsessed with devils.[4]

First a little context: the early modern European witch hunts started around the 1480s, peaked between 1560 and 1630, and declined in the eighteenth century. The largest hunts took place in Germany, with other major hunts in France, Austria, Bohemia, Poland, Switzerland, Denmark, England, and Scotland.[5] Ronald Hutton estimates the number of executions during the 350-year period from 1428 to 1782 at between forty and fifty thousand; other scholarly estimates vary from thirty-five thousand to one million.[6] The witch hunters interrogated, tortured, or terrorized many more.[7]

What were the triggering events? A major development occurred in 1484 when Pope Innocent VIII issued a papal bull proclaiming that not only did witches actually exist but they were also in league with the devil to destroy Christendom from the inside. This Witch Bull reversed the 906 CE ruling that witchcraft was a fantasy.[8] It gave permission for the inquisition to do whatever was necessary to eradicate the threat.[9] Was this a sudden about-turn? No, says Rosemary Guiley; various papal bulls since the thirteenth century had raised concerns about witchcraft.[10] The Witch Bull, however, aimed to stop at nothing to destroy it. 'Desiring with supreme ardour,' it begins, that the Catholic faith 'grow and flourish' and put 'all heretical pravity' far from its territories, 'it has come to our ears' that many persons 'give themselves over to devils male and female,' 'ruin and cause to perish the offspring of women,' 'hinder men from begetting and women from conceiving, and prevent all consummation of marriage.'[11] Clearly, Christian procreation—or the lack thereof—was a central theme.

The bull also notes that some clergy are not pulling their weight against the witch scourge, some even daring to disagree with the whole idea of taking what were in the main old and destitute women and burning them alive. Certain clerics, says the bull, claim that such persons ought not to be punished, imprisoned, and corrected. But upper Germany 'may not be deprived of the office of the inquisition.' 'Let no man, therefore, dare to infringe this page of our declaration.'

Today's *Catholic Encyclopedia* is still holding the line about witchcraft: 'In the face of Holy Scripture and the teaching of the Fathers and theologians, the abstract possibility of a pact with the Devil and of a diabolical interference in human affairs can hardly be denied.' [12] In other words, today's Catholic hierarchy is claiming that Holy Scripture cannot be wrong about the possibility of persons enacting pacts with the devil. The entry goes on to express bewilderment that many 'confessions of the victims, often involving all kinds of satanistic horrors, have been made spontaneously and apparently without threat or fear of torture.' 'The full admission of guilt seems constantly to have been confirmed on the scaffold when the poor suffered [*sic*] had nothing to gain or lose by the confession.' [13]

With this, the Church is claiming to be bemused by inquisitorial reports of people spontaneously confessing, in the absence of threat, to crimes such as riding broomsticks and causing hailstorms. Does this mean that the Church accepts as given the inquisitorial records? Even at the time, the authorities were skeptical about the accuracy of confessions made under duress. In sixteenth-century France, for example, for a confession extracted under torture to stand as proof, the accused had to repeat it 'freely' a day after the torture finished. [14] How 'free' the accused felt to rescind their story is unclear. Second, from the Catholic Church's twenty-first-century perspective, why would it imagine that the accused were less prone than their persecutors to being swept up in the mass delusion? Might not the victims have harbored any number of legitimate fears, desire notoriety, use hallucinogens, suffer from mental deficiency or psychological illness, and so forth? What today's hierarchy does *not* acknowledge is the medieval Church's production and wide dissemination over more than two hundred years of works that demonized women to an extent capable of creating moral panic. Many historians view Christianity as having creating a framework for the witch hunts. And while approval for witch hunting rested with secular governments, those governments often relied on church advice. [15] In contrast, the Catholic Church is still pointing the finger mainly at secular and Protestant forces. [16]

THE *MALLEUS MALEFICARUM* (HAMMER OF WITCHES)

The late medieval obsession with sexual relations between witches and devils, says Merry Wiesner-Hanks, was new. [17] Moreover, it confined itself loosely to central Europe, the area under the influence of the 1487 *Malleus Maleficarum*, the witch hunters' guide to trying, judging, and sentencing accused witches. The *Malleus*, say its translators, entered the Christian canon 'as a complete, persuasive, massively documented, and seemingly authorized description of the witches in its midst.'[18] Catholics and Protestants down to the eighteenth century cited the *Malleus* to support their beliefs about women and witches.

Much more misogynous than comparable earlier works, the *Malleus* defined male heresy as intellectual and theological and female heresy as demonic and sexual. This emphasis on the demonic came from the educated, Aristotelian political and religious elites, not from ordinary people.[19] To the latter, witchcraft was everyday magic and sorcery, not devilry. Where ordinary people cited instances of magic, the intelligentsia saw signs of Satan. Hence, the courts convicted suspects not of sorcery but of diabolism, a much more serious, sensational charge more likely to carry the death penalty. Trial judges took their lead from the theological literature.[20]

Before the *Malleus,* prosecutors charged at least as many men as women for witchcraft. After the *Malleus,* roughly eight out of every ten prosecuted were women. Moreover, the accused men were often related to a female suspect or charged with an additional crime.[21] The prosecutors never pursued the contemporaneous masculine (and real) activity of necromancy with the same vigilance as they did the feminine (and illusory) activity of witchcraft. They may also have targeted gender nonconforming men, as they did later for so-called male 'perversion.'[22]

Unlike the medieval hunts, where allegations of women consorting with devils were few, such charges were pivotal to the early modern hunts. What brought about this new, gendered, and sexualized version of the witch, and how was it converted into law? It is not as if all of Europe adopted the idea. Non-central-European countries such as Finland, Russia, and Iceland did not feminize witchcraft.[23] What happened in central Europe? The politico-religious elites took the European idea of witchcraft, hitherto understood only in terms of disparate elements (sorcery, heresy, etc.) and reconstructed it into a single concept of satanic activity carried out mainly by women.[24] Then, in the late Middle Ages, the masculine conception of magic evolved into a more gender-neutral one and feminization followed. Finally, highly influential misogynous writings confirmed the association between women and witchcraft.[25] Recall that *Women's Secrets*, entrenched in the canon since the early 1300s, was among the most influential of the earlier misogynist writings. *Secrets* epitomized a view of women that led directly to their persecution on the grounds of their evil and dangerous sexuality.[26]

Prior to *Secrets* and in perfect pitch with Aristotle and Pliny, Aquinas described how witches, with the Devil's assistance, physically commit their crimes. Old women, says Aquinas, are particularly prone to being 'moved to wickedness,' and the air is infected by their eyes in the same way as a mirror is tarnished by 'the look of a menstruating woman, as Aristotle says.' 'Hence then when a soul is vehemently moved to wickedness, as occurs mostly in little old women . . . the countenance becomes venomous and hurtful, especially to children. . . . It is also possible that by God's permission, or from some hidden deed, the spiteful demons co-operate in this, as the witches may have some compact with them.'[27]

Some have asserted, continues Aquinas, that witchcraft is nothing in the world but an imagining of men. 'But this is contrary to the authority of holy men who state that the demons have power over men's bodies and imaginations, when God allows them. . . . We believe that angels fell from heaven and that the demons exist. . . . Wherefore others have maintained that witchcraft can set up an impediment to carnal copulation.'[28]

Aquinas explains the conundrum of how the Devil, reputedly sterile, is able to beget children by copulating with witches: the Devil can discharge as *incubus* only what he has previously absorbed as *succubus.* If some persons are 'occasionally begotten from demons, it is not from the seed of such demons . . . but from the seed of men taken for the purpose; as when the demon assumes first the form of a woman, and afterwards of a man.'[29] The Devil procures human sperm either by having intercourse (as a *succubus*) with a man or by squeezing the organs of the dead.[30]

Nearly two centuries after *Secrets,* entire chapters of the *Malleus* describe 'How Witches Impede and Prevent the Power of Procreation'; 'How . . . They Deprive Man of His Virile Member' (which they supposedly collected by the dozens, kept in nests, and fed oats and corn). The *Malleus*, following Aquinas and *Secrets*, explains 'How in Modern Times Witches Perform the Carnal Act with Incubus Devils and How They Are Multiplied by This Means': The devils assume a body 'that they may by the means of this body preserve human semen, and pass the semen on to another body. . . . Devils cannot give life . . . it is true; but materially life springs from the semen, and an Incubus devil can, with God's permission, accomplish this by coition. And the semen does not so much spring from him, as it is another man's semen received by him for this purpose.'[31]

Why is it women who are chiefly addicted to evil superstitions, asks the *Malleus.* The answer is that women 'know no moderation in goodness or vice . . . when they are governed by an evil spirit, they indulge the worst possible vices.'[32] The *Malleus,* echoing Valerius, likens Woman to the Chimera—with the face of 'a radiant and noble lion,' 'the filthy belly of a goat,' and 'the virulent tail of a viper.' Woman, in other words, 'is beautiful to look upon, contaminating to the touch, and deadly to keep.' She is a liar by nature: 'her voice is like the song of the Sirens, who with their sweet melody entice the passers-by and kill them.' And how do women kill their victims? 'By emptying their purses, consuming their strength, and causing them to forsake God.' Woman, says the *Malleus*, 'is more bitter than death.'[33] Compared to men, women 'are intellectually like children' and 'feebler in mind and body.' They 'have slippery tongues and are unable to conceal from their fellow-women those things which by evil arts they know.' But the natural reason that Woman is prone to evil superstitions is that 'she is more carnal than a man, as is clear from her many carnal abominations' such as menstruation.[34]

The inquisitors alleged that a witch's sexual activity with devils and other witches left an incriminating mark on her body, which they made every effort to locate. Using methods that can only be described as sexual sadism, says Wiesner-Hanks, groups of male officials stripped and shaved the suspect and, if necessary, pricked her body with needles or applied other tortures to locate the insensitive patch of skin allegedly caused by the devil's claw.[35]

'All witchcraft comes from carnal lust,' the *Malleus* warns, 'which is in women insatiable.' 'There are three things that are never satisfied . . . yea, a fourth thing which says not, It is enough; that is, the mouth of the womb.' It is the greedy cervix, in other words, that drives women's voracious lust. And for the sake of fulfilling this lust 'they consort even with devils.' The *Malleus* gives thanks to God for preserving, so far, 'the male sex from so great a crime'; 'He was willing to be born and to suffer for us, therefore He has granted men this privilege.'[36]

According to the *Malleus*, 'certain witches are in the habit of devouring and eating infant children,' and midwives 'surpass all others in wickedness.' Penitent witches have admitted that 'no one does more harm to the Catholic faith than midwives. For when they do not kill children . . . [they] offer them to devils.'[37] The witch hunters, say the translators, most frequently accused, tried, and convicted older women. The *Malleus* held that post-menopausal women, like midwives, were among those most vulnerable to demonic suitors (as the female sex drive allegedly increased with age). In addition, the authorities assumed that witchcraft ran in families, particularly from mother to daughter, and among lower-class women.[38]

Nicolas Rémy, purportedly one of the more gullible prosecutors, outlined the evidence that most witches are women: 'Certainly I remember to have heard of far more cases of women than men, and it is not unreasonable that this scum of humanity should be drawn chiefly from the feminine sex. Fabius says that women are more prone to believe in witchcraft and Pliny that women excel in their knowledge of witchcraft.'[39]

Five years earlier, Henri Boguet, author of the 1590 *Examen of Witches*, was more definitive: 'The Devil uses them so, because he knows that women love carnal pleasures, and he means to bind them to his allegiance by such agreeable provocations; moreover, there is nothing which makes a woman more subject and loyal to a man than that he should abuse her body.'[40]

In the 1597 treatise *Daemonologie*, King James I of England explained how the prevalence of female witches derives from Eve's original encounter with the devil: 'The reason is easie, for as that sexe is frailer than man is, so it is easier to be intrapped in these grosse snares of the Devill, as was over well proved to be true, by the Serpents deceiving of Eve at the Beginning.'[41] And in a 1595 document prepared for Justices of the Peace in Yorkshire, the first item on the 'List of Presumptions Against Witches' is 'that they are most commonly weeke women.'[42]

As to 'weeke women,' François Rabelais knew exactly what it was that sapped women's strength. In his 1546 *The Life of Gargantua and of Pantagruel*, Rabelais has a physician-character, Rondibilis, advise Pantagruel's friend Panurge about Woman and her wandering womb. The section is entitled 'How Rondibilis declares that cuckoldry is naturally one of the attributes of marriage':

> When I say woman, I mean a sex so fragile, so variable, so mutable, so inconstant and imperfect, that Nature (speaking in all honor and reverence) seems to me to have strayed from that good sense by which she had created and formed all things, when she built woman. . . . Certainly Plato does not know in what category he should place them, that of reasonable animals or that of brute beasts. For Nature has placed in their body, in a secret place inside, an animal, a member, which is not in men, in which are sometimes engendered certain salty humors, nitrous, boracic, acrid, biting, tearing, bitterly tickling, by whose pricking and painful titillation (for this member is all nerves and acutely sensitive) their entire body is shaken, all their senses transported, all desires internalized, all thoughts confused, so that if Nature had not sprinkled their foreheads with a little shame, you would see them, as if beside themselves, chasing the codpiece, more frightfully than ever did the Proetids, or the Mimallonids, or the Bacchic Thyades on the day of their Bachanals, because this terrible animal has connections with all the main parts of the body, as is evident in anatomy.[43]
>
> I call it animal, following the doctrine of both the Academics and the Peripatetics. For if automotion is a certain indicator of an animate being, as Aristotle writes, and if all that by itself moves itself is called animal, then Plato rightly calls it animal, recognizing in it independent motions of suffocations, precipitation, corrugation, and indignation, indeed so violent that by them very often is ravished from woman every other sense and movement, as if it were a lipothymy, a swoon, epilepsy, apoplexy, and a real semblance of death.
>
> No small praise is due to upright women, who have lived chastely and blamelessly, and have had the virtue to bring this frenzied animal to obedience, to reason. . . . When this animal is satiated (if satiated it can be) by the food that Nature has prepared for it in man, all its individual motions have reached their goal, all its appetites are put to sleep, all its furies pacified. Therefore don't be astonished if we are in perpetual danger of being cuckolds, we who do not always have in abundance the wherewithal to pay it off and satisfy it to contentment.

'By the powers of others than little fishes!' says Panurge, 'don't you know any remedy for it in your art?' 'Indeed I do, my friend,' replies Rondibilis, 'and a very good one, which I use; and it is written down by a famous author, eighteen hundred years ago. Listen.'[44]

But Panurge is too overwhelmed with excitement and gratitude to listen. As with *Secrets* and similar erotically charged texts, Panurge's (and the reader's) urgent desire to know (in order to satiate the insatiable beast and postpone cuckoldry) encounters *discursus interruptus*.[45] The promised remedy (leave well enough alone) is philosophically wise but, most importantly, keeps intact (for next time) the alarming secrets about women, their bodies, and the trouble they cause men.

WHY WITCHES?

How does history make sense of the early modern European witch hunts? Mooted contributing factors include the machinations of the sixteenth- and seventeenth-centuries' Counter-Reformation and the Thirty Years' War; the desire to restore decimated populations and rid Europe of knowledge about contraceptives; the macroeconomics of early capitalism; and the unusual weather pattern dubbed the 'Little Ice Age.'[46]

The main purpose of the hunts, say some historians, was to divert community blame for the appalling social conditions away from the authorities and onto the witch scapegoat.[47] Others say it was an attempt to wipe out contraception and abortion by criminalizing birth control methods and the women who taught or practiced them.[48] German midwives, for example, had to inform their municipal physicians about clients who used birth control, procured abortions, or gave birth to illegitimate children. Midwives suspected of non-compliance were sometimes charged with witchcraft.[49]

By the fifteenth century, women had joined the ranks of viable scapegoats by proving themselves capable of organized group existence independent of the Church and of men. To salvage male control, the Church framed female collective organization as a conspiracy of witches under Satan's leadership.[50] Contemporaneously with the hunts, the authorities were excluding women (midwives included) from workmen's guilds, hence forcing them out of work and back to the hearth.[51]

Debate continues about whether those accused of witchcraft came mainly from the margins of their communities or whether the marginalized could even qualify as credible scapegoats.[52] Perhaps the otherwise defenseless enjoyed that certain 'status' historically imposed on those perceived as possessing magical powers.[53] Perhaps, given the shifting and tangled web of variables during the hunts, anything could and did happen. All these arguments are plausible and mutually compatible. My main concern here, however, is the role of the Christian canon in the feminization of witchcraft and whether the witch hunters singled out a particular type of woman for persecution.

The role of the canon, say some, was relatively insignificant, and misogyny was not pivotal to witch hunting.[54] Historian Stuart Clark argues thus: 'If witch prosecuting was misogynous we ought to find woman-hating in abundance in those who most actively supported it. The problem is that we do not.' Rather, 'what the witchcraft writers said about women' was to do with female intellectual frailty, inordinate affections, carnal appetites, and vindictiveness compared to men's. Women as such are the Devil's preferred target; some demonologists 'even called Eve the first witch'; 'The association of witchcraft with women incorporates traditional Aristotelian notions . . . of women as "deformed" males, and . . . Christian hostility to women as originators of sin.' For the prosecutors, 'the inferiority of women was self-evident.' Thus, the witch hunters 'were entirely representative of their age and culture . . . they showed little interest . . . in using [witchcraft] to denigrate women.' This means that 'some recent comments on the anti-feminism in demonology seem very exaggerated.' The connection between women and witchcraft was 'so obvious to them . . . they felt no need to elaborate on it or indulge in additional women-hating to back it up'; 'Belief in witchcraft depended on nothing more than conventional misogyny.'[55] In a nutshell: 'Of course the witch hunters were misogynous, but no more so than their peers; hence it warrants no particular attention.' Are there echoes here with other persecutions in which the persecutors' belief in 'the inferiority' of the 'enemy' 'was self-evident' and 'depended on nothing more than conventional' intolerance and hatred?

Also problematic is Bossy's claim that it is nonsense to say 'a whole culture or population was mistaken for what it customarily did.' What may properly be said about people in the past, says Bossy, is only what 'might possibly have made sense to the parties concerned. . . . If you interpret one language game by another language game, you risk standing on the absurd position that "the game has always been played wrong."'[56]

Cultures, however, are always heterogeneous; always, some elements reject the dominant game while the main players pursue it regardless. Is there any such thing, then, as a 'whole' culture behaving customarily? Wherever oppressive customary practices reign (denying women access to education; female genital mutilation; stoning women to death for 'fornication'; capital punishment; child labor, etc.), dissenting voices have declared the custom wrong. To ignore the existence of such opposition is to treat cultures as if they are no more than their ethical lowest common denominators. While Bossy's argument putatively honors difference, it is flawed inasmuch as it delegitimates criticism—even of a kind that acknowledges intracultural heterogeneity—of abusive regimes.

Learned public protest against the witch hunts existed even before the Witch Bull, and dozens of lawyers, theologians, priests, and physicians—including the king's surgeon Pierre Pigray—later joined in.[57] It follows that scholars may reasonably interrogate entire cultures and claim that the dominant game has 'been played wrong' in at least some respects for a very long time. Patriarchy, for example, has been the main game for millennia, and witch hunting is its perennial favorite, whether led by popes or politicians. That it is still the main game is evidence of patriarchy's capacity to marginalize dissent rather than that there is no dissent. To imagine one can see through the eyes of an entire culture is to ignore the fact that history in the main is written and disseminated by entrenched interests.

A CERTAIN TYPE OF WOMAN

Did the witch hunters target and victimize a particular type of woman? Given the lack of records, the question is unresolvable. It is clear, however, that previous claims of several million women being burned, or of pagan women being particularly targeted, are not supported by the available data.[58]

So how *do* women figure in the diverse social and political contexts of the hunts? Witch hunting, says historian Christina Larner, 'is the hunting of women who do not fulfil the male view of how women ought to conduct themselves.'[59] In the central European hunts, says Wiesner-Hanks, witch hunting was the tracking down of a certain type of woman—a type of woman who among other things used words as a weapon.[60] As the inversion of the 'good woman,' the witch was 'argumentative, wilful, independent, aggressive, and sexual rather than chaste, pious, silent, obedient, and married.' The indictment of one Margaret Lister, for example, describes her as 'a witch, a charmer, and a libber,' meaning 'liberated,' a term used then, and sometimes still, pejoratively.[61] The witch's laughter in particular indicted her: 'All laughter is allied with the monstrous,' says Catherine Clément. By her laughter, a woman crosses a dangerous line beyond which she finds herself excluded.[62]

The stereotypical images of 'evil' women, says Clark, were the 'shrews' and 'Amazons' 'who challenged patriarchal rule or were wilful and domineering'; the 'scolds,' 'gossips,' and women preachers 'who usurped male control of language'; and the 'whores,' seductresses, and viragos 'who sought sexual superiority or behaved like men.' These were old motifs that achieved greater prominence and dispersion in the age of the witch trials.[63]

In the Scottish hunts, says Larner, the hostility was toward women who showed independence or aggression or who failed to fulfil 'womanly' functions such as the nurture of men and children.[64] The purpose of punishing deviance, she says, is to demonstrate the dominance not only of rulers but also of conformist values. Witchcraft has the advantage of summing up all forms of nonconformity: 'Witches are evil.' The prosecution of witches, says Larner, 'is a peculiarly economical way of attacking deviance.'[65]

Under the regime of terror, women gained security through conforming to the 'good woman' stereotype and by being seen to criticize those who deviated from it.[66] Yet on this basis—that many accusers and witnesses were women—some historians conclude that the hunts were not aimed particularly at women.[67] However, to reiterate a critical point, the fact is that systemic punishment typically enculturates victims to accept it as valid.[68] It is common for traumatized people to turn against nonconforming neighbors in the hope of averting further terror. In addition, direct or indirect threats from authorities can catalyze community-based attacks on an out-group.[69]

Increasingly, along with the hunts, the authorities criminalized women for formerly tolerated gendered behaviors such as prostitution.[70] The numbers tried for religious and sexual offences in the late-fifteenth and early-sixteenth centuries rose and fell with those of witch prosecutions. In this period, the state criminalized women for the first time on any scale.[71] Previously, women did not possess adult status—their fathers and husbands were legally responsible for them—but now they began appearing in courts, the young charged with infanticide and the old with witchcraft (including advising women about birth control).

In Germany and Switzerland, Church and state began to regulate marriage and sexual conduct.[72] The crackdown included a massive campaign against prostitution that closed municipal brothels and defined all non-marital sexuality as lewdness or fornication.[73] Luther had called for the closure of brothels and, for the Reformers, moral politics was a central means of attaining social power and control. When criminalization of prostitution began in the second half of the sixteenth century, the authorities punished equal numbers of men and women but, by the eighteenth century, they were sentencing many more women than men. Authorities in Zurich and elsewhere penalized women for premarital intercourse and pregnancy even if they had subsequently married. Historians, says Susanna Burghartz, adopted the purity yardstick laid down by the Reformers, which became the political and philosophical basis for the analysis of conflicts between the state and various marginalized groups.[74]

What effect did three hundred years of demonization, witch trials, torture, and executions have on women survivors and those who came after them? The omnipresent threat of false and fantastical accusation, arrest, torture, and being burned alive had terrorized millions. Did women withdraw into the frail protection of their homes, foregoing their work in the community as healers, physicians, midwives, and teachers? Did they stop advising others about birth control? Did older women alter their appearance and behave more submissively, maintaining silence, as set down in 1 Timothy, and looking to their husbands for guidance? ('I do not permit a woman to teach or have authority over a man; she must be silent,' 1 Timothy 2).[75]

The stereotypical older woman of the late Middle Ages—'bawdy, aggressive, and domineering' (attributes also ascribed to witches)—says Edward Bever, became the stereotypical passive, submissive, older woman of the later period.[76] Perhaps because of the witch trials, older women began taking particular care not to appear witchlike.[77] That message remains embedded in contemporary gender norms for older women in particular. No matter how well qualified the older woman, if she appears in public without makeup, manicure, and micromanaged hair, she risks being publicly ridiculed for her appearance and, by association, her scholarship.

As for Maria's accuser, he presented his evidence against her in a municipal court in 1499 during a wave of persecution in Lucerne. Alongside thirty others (virtually all women), Maria was tried for sorcery (weather magic, poisoning, illness to men) and riding on a wolf. Of a population of a few thousand souls, thirty-one—almost exclusively women—went on trial for sorcery and nine were burnt at the stake. I don't know if Maria was among those executed.[78]

THE AFTERMATH

Early modern central Europe not only reinvented the witch as female but also reinforced the idea of women's innate susceptibility to satanic seduction. No longer could people place their trust in women or their work. As a direct consequence, male enterprises began to appropriate women's work as healers and midwives.[79] Barber-surgeons—a trade banned to women—were eligible to attend classes in anatomy and dissection and learn to use the newly invented obstetrical forceps. In mid-seventeenth-century France, barber-surgeons began advertising their services for childbirth.[80] These 'man-wives' presented themselves as more 'modern' and 'scientific' than their women colleagues, and they became increasingly popular.[81]

On the subject of complicated births and the prescription of medicines, sixteenth-century physician Michele Savonarola advised: 'I leave such treatment to male physicians, for it is not the work of women.'[82] Savonarola's texts, written for men only, found an extraordinarily wide audience,[83] and the first obstetrical text written for surgeons appeared by the 1550s. A further development—the rediscovery of the Hippocratic gynecological corpus—secured the field of gynecology and obstetrics as a masculine endeavor free of its medieval (i.e., feminine) past.[84]

Given the evidence, the fomenting role of the Christian canon, particularly *Secrets* and the *Malleus*, in the early modern witch hunts is incontestable. These and other extraordinarily misogynous texts constructed the witch as female and the female as potential witch, innately evil by dint of her sexuality. This construction led directly to the brutal torture and killings of tens of thousands of people, roughly 80 percent of them women. Did the witch hunters disproportionately target women who deviated too far from their prescribed 'feminine' role? Although the answer can never be definitive, evidence from the hunts as well as from the preceding and following periods suggests so.

The early modern central European witch hunts not only set the scene for the male takeover of women's medicine but also opened the way for the criminalization of nonconforming women and men for so-called perversion. Perverse, against-the-grain behavior is as old as humankind, says Robert Nye, but the last decades of the nineteenth century saw a conceptual revolution that reinvented perverse behavior as a 'perversion' enacted by a 'pervert.'[85] The authorities created the notion of perversion, defined as an excess or deficiency in normal organic function, in order to differentiate 'abnormal' from 'normal' behavior and to police the boundaries of respectability. The medical establishment constructed perversion as a 'kind of natural identity, whether inherited or acquired.' 'Inversion' (later known as homosexuality), the Academy agreed, was the most troubling of all the perversions.[86] The next chapter examines the Academy's worst fears, together with its construction of Woman as ineducable, hysterical, and deviant.

NOTES

1. Adapted from Richard Kieckhefer, *European Witch Trials: Their Foundations in Popular and Learned Culture 1300–1500* (Berkeley: University of California Press, 1976), 101.
2. Hebrew Scriptures, Exodus 22:18.
3. Wiesner-Hanks, *Women and Gender*.
4. Wiesner-Hanks, *Women and Gender*.
5. Ronald Hutton, *Triumph of the Moon* (New York: Oxford University Press, 1999).
6. Hutton, *Triumph of the Moon*, ch. 8, 1
7. Brian Pavlac, *Witch Hunts in the Western World: Persecution and Punishment from the Inquisition through the Salem Trials* (Westport, CT: Greenwood Press, 2009), 89.
8. Rosemary Guiley, *The Encyclopedia of Witches, Witchcraft and Wicca* (New York: Infobase Publishing, 2008).
9. Guiley, *Encyclopedia of Witches*.
10. Guiley, *Encyclopedia of Witches*.
11. Guiley, *Encyclopedia of Witches*.
12. Herbert Thurston, "Witchcraft," *Catholic Encyclopedia*, vol. 15 (New York: Robert Appleton Co, 1912), para. 18.
13. Thurston, "Witchcraft," para. 18.
14. Jean Bodin, *On the Demon-Mania of Witches* trans. Randy Scott (Toronto: CRRS University of Toronto, 1995), 25.
15. Brian Pavlac, "Ten common errors and myths about the witch hunts," Myth #2, accessed April 27, 2014, http://departments.kings.edu/womens_history/witch/werror.html.
16. Thurston, "Witchcraft," para. 13–14.
17. Wiesner-Hanks, *Women and Gender*.
18. Alan Kors and Edward Peters, eds. *Witchcraft in Europe 400–1700: A Documentary History*, 2nd ed. (Philadelphia: University of Pennsylvania Press, 2001), 180. Inquisitor Heinrich Kramer, main author of the *Malleus*, prefaced it with the 1483 Witch-Bull, 273.
19. Wiesner-Hanks, *Women and Gender*; Kieckhefer, *European Witch Trials*; Gary Jensen, *The Path of the Devil: Early Modern Witch Hunts* (Lanham, MD: Rowman and Littlefield Publishers, 2007), 149.
20. Kieckhefer, *European Witch Trials*, 104.

21. Hans Peter Broedel, trans. *The Malleus Maleficarum and the Construction of Witchcraft: Theology and Popular Belief* (Manchester, UK: Manchester University Press, 2003).
22. Matthew Moebius, "Clerical Conceptions of Magic and the Stereotype of the Female Witch," *Oshkosh Scholar* 6, (2011).
23. Wiesner-Hanks, *Women and Gender*.
24. Michael Bailey, "The Feminisation of Magic and the Emerging Idea of the Female Witch in the Late Middle Ages," *Essays in Medieval Studies* 19, (2002).
25. Moebius, "Clerical Conceptions of Magic," 9.
26. Lemay, *Women's Secrets*.
27. Aquinas, *Summa Theologica*, Part I, Q 117, article 3.
28. Aquinas, *Summa Theologica*, Part I, Q 58, article 2.
29. Aquinas, *Summa Theologica*, Part 1, Q 51, article 3, reply to Objection 6.
30. King James I, *Demonology*, trans. Paul Tice (San Diego, CA: The Book Tree, 2002), 66 ff.
31. Kors and Peters, *Witchcraft in Europe*, Part I, Question III, 111.
32. Kors and Peters, *Witchcraft in Europe*, 181.
33. Kors and Peters, *Witchcraft in Europe*, 187.
34. Kors and Peters, *Witchcraft in Europe*, 183–84.
35. Wiesner-Hanks, *Women and Gender*.
36. Kors and Peters, *Witchcraft in Europe*, 188.
37. Kors and Peters, *Witchcraft in Europe*, 189.
38. Kors and Peters, *Witchcraft in Europe*, 19.
39. Nicolas Rémy, *Demonolatry: An Account of the Historical Practice of Witchcraft*, ed. Montague Summers, trans. E. A. Ashwin (New York: Dover Publications, 2008).
40. Henri Boguet, *Examen of Witches*, ed. Montague Summers, trans. E. A. Ashwin (London: Kessinger Publishing, 2003).
41. King James 1, *Demonology*.
42. Darren Oldridge, ed., General Introduction to *The Witchcraft Reader* (London: Routledge, 2002).
43. Proetids (daughters of Proetus) were afflicted by a madness that made them believe they were cows. Mimallonids or Thyades were female followers of Bacchus.
44. François Rabelais, *The Complete Works of François Rabelais*, trans. Donald Frame (Berkeley: University of California Press, 1991), bk. 3, ch. XXXI, 356–57.
45. Lochrie, *Covert Operations*.
46. H. C. Erik Midelfort, *Witch hunting in Southwestern Germany 1562–1684: The Social and Intellectual Foundations* (Stanford, CA: Stanford University Press, 1972), 31 (regarding Counter-Reformation [1545–1648] and Thirty Years' War [1618–1648]); Gunnar Heinsohn and Otto Steiger, "The Elimination of Medieval Birth Control and the Witch Trials of Modern Times," *International Journal of Women's Studies* 3, 1, (1982) (regarding contraceptives); Wiesner-Hanks, *Women and Gender*, 268 (regarding capitalism); Wolfgang Behringer, "Weather, Hunger and Fear: Origins of the European Witch Hunts in Climate, Society and Mentality," in *The Witchcraft Reader*, ed. Darren Oldridge (London: Routledge, 2002) (regarding weather).
47. Leonard Berkowitz and James Green, "The Stimulus Qualities of the Scapegoat," *Journal of Abnormal and Social Psychology* 7, (1962): 205.
48. John Riddle, "The Great Witch Hunt and the Suppression of Birth Control: Heinsohn and Steiger's Theory from the Perspective of an Historian," in *Witchcraft, Population Catastrophe and Economic Crisis in Renaissance Europe: An Alternative Macroeconomic Explanation* by Gunnar Heinsohn and Otto Steiger (Discussion Paper, University of Bremen, 2004).
49. Sigrid Brauner, *Fearless Wives and Frightened Shrews: The Construction of the Witch in Early Modern Germany* (Amherst: University of Massachusetts Press, 2001), 17.
50. Jensen, *Path of the Devil*.
51. Brauner, *Fearless Wives and Frightened Shrews*; Merry Wiesner-Hanks, *Working Women in Renaissance Germany* (New Brunswick, NJ: Rutgers University Press, 1986).
52. E.g., Berkowitz and Green, "The Stimulus Qualities of the Scapegoat."
53. Brauner, *Fearless Wives and Frightened Shrews*, 18.

54. Stuart Clark, *Thinking with Demons* (New York: Oxford University Press, 1999); Brauner, *Fearless Wives and Frightened Shrews*, 11; Rainer Decker and H. C. Eric Midelfort, *Witchcraft and the Papacy: An Account Drawing on the Formerly Secret Records of the Roman Inquisition* (Charlottesville: University of Virginia Press, 2010).

55. Clark, *Thinking with Demons*, 112–17.

56. Bossy, "Thinking with Clark," 242.

57. Midelfort, *Witch-hunting in Southwestern Germany*; Kors and Peters, *Witchcraft in Europe*, 173–75, 393.

58. E.g., Ronald Hutton, "Counting the Witch Hunt," (online paper).

59. Christina Larner, *Enemies of God: The Witch Hunt in Scotland* (Oxford: Basil Blackwell, 1983), 100.

60. Wiesner-Hanks, *Women and Gender*, 277.

61. Wiesner-Hanks, *Women and Gender*.

62. Catherine Clément, "The Guilty One," in *The Newly Born Woman*, Hélène Cixous and Catherine Clément, trans. Betsy Wing (London: I. B. Tauris and Co, 1996), 33.

63. Clark, *Thinking with Demons*, 130–32.

64. Christina Larner, "The Crime of Witchcraft in Early Modern Europe," in *The Witchcraft Reader*, ed. Darren Oldridge (London: Routledge, 2002), 209.

65. Larner, "The Crime of Witchcraft," 210.

66. Larner, "The Crime of Witchcraft," 209; Wiesner-Hanks, *Women and Gender*.

67. E.g., Diane Purkiss, *The Witch in History: Early Modern and Twentieth-Century Representations* (London: Routledge, 2012), 94; Brian Levack, *The Witch Hunt in Early Modern Europe* (London: Pearson Longman, 2006), 140.

68. Connell and Messerschmidt, "Hegemonic Masculinity."

69. Whitmer, *Violence Mythos*.

70. Larner, "The Crime of Witchcraft," 91.

71. Larner, "The Crime of Witchcraft," 208–209.

72. Susanna Burghartz, "Ordering Discourse and Society: Moral Politics, Marriage and Fornication during the Reformation and the Confessionalization Process in Germany and Switzerland" in *Social Control in Europe, 1500–1800*, vol. 1, eds. Herman Roodenburg and Pieter Spierenburg (Columbus: Ohio State University Press, 2004).

73. Burghartz, "Ordering Discourse and Society," 80–81.

74. Burghartz, "Ordering Discourse and Society," 80–81.

75. Christian Scriptures, Timothy 2; Daly, *Gyn/Ecology*, ch. 6, pt. III and IV

76. Edward Bever, "Old Age and Witchcraft in Early Modern Europe," in *Old Age in Preindustrial Society*, ed. Peter Stearns (Teaneck, NJ: Holmes and Meier, 1982), 180.

77. Bever, "Old Age and Witchcraft," 180.

78. Adapted from Stefan Jaeggi, "Lucerne Method of Sorcery and Witchcraft until the Mid-Sixteenth Century," *Swiss Journal of History* 52, (2002) .

79. Jensen, *Path of the Devil*, 6.

80. Wiesner-Hanks, *Women and Gender*.

81. Wiesner-Hanks, *Women and Gender*.

82. Savonarola in Green, *Making Women's Medicine Masculine*, 246.

83. Green, *Making Women's Medicine Masculine*.

84. Green, *Making Women's Medicine Masculine*, 248.

85. Nye, "Sexuality," 18.

86. Nye, "Sexuality," 18.

Chapter Eight

The Nineteenth Century: Less Sex, More Lies

How did the legacy of seed-and-soil inform nineteenth- and twentieth-century constructions of Woman as ineducable, hysterical, and deviant? 'The male is by nature superior to the female,' said Aristotle. 'The one governs, the other is governed.'[1] And more than two millennia later? If Woman 'has been a slave,' said psychiatrist Henry Maudsley, 'she has been a slave content with her bondage.'[2] And neurologist Paul Möbius: 'All progress is due to man. Therefore the woman is like a dead weight on him.'[3] And philosopher Otto Weininger: 'The lowliest man ranks infinitely higher than the most distinguished woman.'[4]

Misogynous sexism was the norm in the nineteenth- and twentieth-century European Academy. This chapter argues that the Academy's campaigns against women's suffrage, together with the medicalization (as Freud's hysteria) and criminalization (as perversion) of gender nonconformity, were grounded in Aristotelian constructions of gender. As the fathers of nineteenth-century medicine and social science saw it, the most dangerous threat of their time was the feminization of men through the absence of war and the masculinization of women through education. Hence, the focus of the Academy was less on advancing scientific knowledge than on preventing women's emancipation. These fathers of modern science represented themselves as the protectors of civilization and of the species itself from decay through the encroaching masculinization of women. This chapter examines their arguments as well as their apparent ambivalence about their own masculinity, feminine/homoerotic aspects, and fantasies about women's capacity to undermine or usurp male power. The first section of this chapter sets the scene with a brief historical overview of Western culture's denial of education to women.

WOMAN AS INEDUCABLE

The poet Menander was little taken by the notion of educating women. In the fourth century BCE, he wrote: 'Teach a woman letters? A terrible mistake: Like feeding extra venom to a horrifying snake.'[5] Classical and post-classical cultures, at least up to the fall of Byzantium, provided few opportunities for the vast majority of women and men to learn to read or write.[6] Only the elites, predominantly men, became literate. Whereas institutions denied education to many boys and men because of their class, they denied education to all girls and women on the grounds of gender alone.[7] And yet, says Lerner, this does not mean that women are primarily victims. Women have always been agents in history. They have made history but, without education, they could neither know their own history nor interpret history.[8]

Aristotelian philosophy in the West, like Eastern Confucianism, regarded women not only as inferior to men but also as dangerous to men's spiritual and intellectual development. Hence, educational institutions in ancient Greece excluded women. No women attended Plato's academy or featured in his dialogues.[9] Of women's poetry, few poems of Sappho's survive and a mere handful of scraps from other women poets. Church-based attacks on Sappho's reputation virtually erased her voice from Western tradition; it reappeared with regularity only in the twentieth century.[10]

From ancient Rome no complete works written by women survive, only fragments from a handful. In the few cases when women's discourse did enter the public arena, the dominant discourse either assimilated it or erased it from cultural history.[11] Historically, women's voices emerged at times when the dominant discourse was socioeconomically and ideologically complacent enough to allow women some social power.[12]

In the medieval West, a small number of upper-class women became highly literate, the German philosopher Hildegard of Bingen, for example, who had entered a monastery at the age of eight. Christian and Muslim women who became literate did so mainly for religious reasons. A few Muslim women were known for their interpretations of *hadith*.[13]

Did women artists fare any better? In Byzantium, for example, most women artists were from convents or wealthy families. Generally, the culture dismissed women's creative work as 'craft,' as distinct from the 'fine art' of the male creator. After the Norman conquest of England, convents previously headed by abbesses became subsidiary to male abbots, which further marginalized women.[14]

The Renaissance in Europe saw the founding of universities, humanist academies, and art academies for men only. As in the Middle Ages, women artists were nuns, aristocrats, or daughters of artists. Artemisia Gentileschi, for example, was the daughter and pupil of Orazio Gentileschi, a follower of Caravaggio.[15] The state oversaw the academic institutions and set their standards. Art training required students to depict the male nude—an exercise generally barred to women. This kept the prestigious commissions for the mandatory grand mythic/historical scenes with their obligatory male nudes from falling into the hands of women. The French Académie, refounded after the French Revolution, barred women altogether, say Rozsika Parker and Griselda Pollock, and the English Royal Academy 'systematically excluded women from its schools and privileges' for a century after its 1768 founding.[16]

According to the Academy, men made the only worthy orators, and women who engaged in public speaking were immoral.[17] One leading Italian Renaissance humanist remarked to wide agreement, 'An eloquent woman is never chaste.' A minority, however, claimed that literacy would increase women's morality via access to religious texts.[18]

Early modern Europe further rigidified the traditional pattern of excluding girls and women from education; most school attendees were male, town-based, and wealthy. Formal education began with the learning of Latin, and such schools catered only to boys.[19] Because the state barred women from any employment that required formal education, most parents were unwilling to pay for even basic schooling for a daughter.[20] And in the Muslim world, the authorities restricted education to mosque schools that trained boys to read the Quran and taught Islamic law, science, and medicine at advanced levels.[21]

Girls' schools began to appear across Europe in the sixteenth century. In central Germany, half of all parishes had boys' schools, compared to one in ten for girls. In Venice, fifteen times as many boys as girls attended school.[22] In Wales and England, out of a population of four or five million, no girls and fewer than twenty thousand boys attended grammar schools. And in France, for every one girl, thirty-three boys attended school.[23]

The state permitted girls to learn to read (typically at home) in order to learn Christian morals, but writing was out of the question. Writing materials were expensive and, besides, women's ideas, by definition, were either trivial or subversive.[24] Like their European counterparts, girls in America learnt basic reading, usually from their mothers, but not writing; sewing was the equivalent skill for girls. Furthermore, the future method of teaching reading in America as in Britain would follow the 'feminine benevolence' model of evangelical author Hannah More rather than the egalitarian model of suffragist Mary Wollestonecraft.[25]

In the early modern period, then, there were few school places for girls; the girls school curriculum was limited; most people assumed that education did not matter or was not healthy for girls; schooling cost money; and, for girls, matching employment prospects were nil. In addition, the state increasingly restricted women's economic activities. By the end of the eighteenth century in England, for example, trades that earlier in the century had accepted girls as apprentices now shut them out.[26]

Then the question arose as to whether the state might benefit from schooling girls.[27] 'Our ladies,' said American politician Benjamin Rush in 1787, 'should be qualified to a certain degree, by a peculiar and suitable education, to concur in instructing their sons in the principles of liberty and government.'[28] When institutionalized education for girls began in the mid-nineteenth century, the state often required teachers of girls to be church based.

In Europe, few girls entered the grammar-style classics education that was a prerequisite for university entrance. America's Oberlin College, on granting entry to a certain Lucy Stone in 1847, barred her from presenting her commencement address orally (which she then refused to write).[29] At the same time, American law prohibited most African-American women from reading and writing.[30]

The first girls' college in London opened in 1848.[31] It took another twenty years for Cambridge to begin admitting women students in 1869 and another seven for British public universities to follow suit. By mid-century, some British art academies had begun admitting women, but only to their 'Female Schools' in which life-drawing models comprised men in armour. By the end of the century, however, art schools permitted women students to attend life classes with next-to-nude models.

For most of women's history, says Lerner, Church and state systematically barred women from the enterprise of creating symbol systems, that is, theory-formation.[32] And when women writers did appear, rather than enjoying a continuous, connected tradition, they often remained isolated only to disappear once more.[33]

And just as women were beginning to access education, an Academy-initiated movement was mobilizing to stop them. The rest of this chapter examines the Academy's case. The education of women, said the Academy, threatened to destroy civilization and the very species itself.

SEX IN MIND AND IN EDUCATION

Did the justification for keeping women disenfranchised and uneducated change over time? When the notion of bodily humors ruled medicine, Woman was too cold, damp, and spongy for her own good. When mechanical models ruled, the laws of energy conservation—as applied to women's bodies—meant they were unable, safely, to use their brains and look after children at the same time: 'What Nature spends in one direction, she must economize in another direction,' says Maudsley in his popular 1874 treatise, "Sex in Mind and in Education." Those who advocate higher education for women are trying to transform women into men, and biology will not allow it: 'Let come what may in the way of assimilation of female and male education and labor, it will not be possible to transform a woman into a man.' Furthermore, says Maudsley, woman herself does not wish to have the freedom of the man: 'If she has been a slave, she has been a slave content with her bondage. . . . In that lies the very pith of the matter—that she is not free and does not care to be free; that she is a slave, and does not know or feel it.'[34]

Maudsley was right that she was a slave; whether she knew it or not was a different matter. The British law of coverture deemed slaves and wives 'civilly dead'; a wife became the person of her husband:[35] 'By marriage, the very being or legal existence of a woman is suspended' or 'consolidated into that of the husband, under whose wing, protection and cover she performs everything.'[36]

Under coverture, a husband owned his wife's earnings and their children. A husband could sell his wife at public auction. Often required to stand with a halter around her neck, the wife for sale was cheaper to buy than a slave or a corpse.[37] Until 1884, the authorities could jail a wife for refusing her husband his conjugal rights and, until 1891, a husband could imprison his wife in their home to obtain those rights.[38] Up until the 1970s, the doctrine of 'marital rape exception' 'was premised in part on medical accounts that defined non-consensual sexual intercourse with a conscious healthy woman to be an impossibility.'[39]

Maudsley berates those who, 'foremost in their zeal for raising the education and social status of woman,' have not given 'proper consideration to the nature of her organization, and to the demands which its special functions make upon its strength.' He documents the physiological dangers peculiar to women: 'Their nerve-centres being in a state of greater instability, they will be the more easily and seriously deranged.' To pit education against biology, says Maudsley, is to cause harm to women, men, and the race itself: the 'advantages of a quantity of female intellectual work' would come at the cost of 'a puny, enfeebled, and sickly race.' 'Women do not and cannot stand on the same level as men.'[40]

We may inquire, Maudsley continues, 'whether any ill effects have been observed from subjecting women to the same kind of training as men.' Although the idea of educating women is comparatively recent in Britain, he said, the American experiment reported dire consequences. Girls who 'have gone with boys through the same curriculum' do so 'at a cost to their strength and health which entails life-long suffering, and even incapacitates them for the adequate performance of the natural functions of their sex.' If these women subsequently married, said Maudsley, they were 'unfit for the best discharge of maternal functions' and 'apt to suffer from a variety of troublesome and serious disorders in connection with them.' 'The brain and the nervous system testify to the exhaustive efforts of undue labor, nervous and even mental disorders declaring themselves . . . even such serious disorders as chorea, epilepsy, insanity.' 'You may hide Nature,' says Maudsley, 'but you cannot extinguish it.' 'If the attempt to do so be seriously and persistently made, the result may be a monstrosity—something which having ceased to be woman is yet not man.'[41]

Maudsley had three additional reasons for keeping women unschooled. First, the breasts of educated girls did not develop properly: 'There is an imperfect development of the structure which Nature has provided in the female for nursing her offspring.' What might evolution be thinking, wonders Maudsley; perhaps it is planning a 'race of sexless beings who . . . shall carry on the intellectual work of the world,' much as 'the sexless ants do the work and the fighting of the community.'

Second, uneducated women make better wives: 'The female qualities of mind adapt her, as her sex does, to be the helpmate and companion of man.' Third, given that educated women make inadequate mothers, Maudsley fears that 'if the nursing of babies were given over to men for a generation or two, they would abandon the task in despair or in disgust, and conclude it to be not worthwhile that Mankind should continue on earth.'[42]

The question of mothers and mothering weighed heavily on Maudsley. His own mother died when he was six and, by all accounts, henceforth his father barely spoke to his sons.[43] In addition, it seems that Maudsley's father considered him insufficiently masculine; on the young Maudsley resigning an appointment, his father described him as 'like a woman, belonging to a *varium et mutabile genus*.[44] At thirty-one, Maudsley married Ann Conolly, five years his senior and the daughter of psychiatrist John Conolly. Conolly then handed over the running of his private hospital to Maudsley and died not long afterward.[45]

Apart from Maudsley's academic brilliance, his colleagues knew him for his cynicism, acerbic criticism, and degenerationalist notions of humankind (i.e., in evolutionary decline rather than progression). Delivering a psychiatric lecture on the link between masturbation and insanity, thirty-three-year-old Maudsley decried the miserable sinners whose minds suffer by reason of self-abuse: 'cunning, deceitful, liars, selfish, in fact, morally insane.' Of one such 'degenerate being,' Maudsley remarked that 'the sooner he sinks to his degraded rest the better for himself and the better for the world which is well rid of him.'[46]

After the death of his wife, the seventy-six-year-old Maudsley, embittered and isolated, 'openly longed for death.'[47] The couple had no children. Maudsley had written: 'Were man robbed of the instinct of procreation, and of all that spiritually springs therefrom, [at] that moment would all poetry, and perhaps also his own moral sense be obliterated from his life.'[48]

To Maudsley's list of the dire consequences of educating women, G. Stanley Hall added that not only does education make for increased difficulties and dangers for the woman when she gives birth, but it also damages the boys and men in the schooling system. 'The more scholastic the education of women,' says Hall, 'the fewer children and the harder, more dangerous, and more dreaded is parturition.'[49] 'The progressive feminization of secondary education works its subtle demoralization on the male teachers who remain. . . . From this deterioration both boys and teachers suffer.'[50] As Möbius had warned: 'All progress is due to man.' 'The woman is like a dead weight on him.' While the woman prevents much of the man's 'restlessness and meddlesome inquisitiveness,' says Möbius, she 'also restrains him from noble actions, for she is unable to distinguish good from evil.'[51]

Möbius's 1900 pamphlet 'On the Physiological Feeblemindedness of Women' depicts Woman as a new Eve heading for a new Fall. Möbius's key assumption, says his translator, is that instinct drives women; thus, their duty is motherhood. Women who fail to attend to motherhood inevitably are struck down with illness; their unwillingness to accept marriage and motherhood leads to disintegration and death.[52]

It was Möbius himself who prematurely met 'disintegration' in the form of cancer of the jaw and died at age fifty-three. His marriage—to a woman eleven years his senior—had been unhappy, and they separated when Möbius was thirty-seven. Möbius had no children.[53]

Möbius's contemporary Otto Weininger shared Möbius's attitudes toward women. Apropos women's emancipation, Weininger concluded that it is only the male (M) aspect of Woman, not the woman herself, which craves emancipation.[54] While the notion of emancipation, says Weininger, can apply to the masculine woman (whose masculine aspect cannot exceed 50 percent), the idea is fundamentally detrimental to the majority of women because it induces 'an artificial aspiration that runs counter to nature and is fundamentally false.' '*Away with the whole woman's movement*,' says Weininger.[55] Emancipation in the sense of the will to be '*internally equal* to a man, attain his intellectual and moral freedom, his interests and creative power' is impossible in that the feminine element has no need and '*no capacity for this kind of emancipation*.' The woman who is truly interested in intellectual matters, says Weininger, is of a sexually intermediate form. Hysteria drives her interests.

Weininger's project, says biographer Chandak Sengoopta, is inseparably intellectual, political, and cultural, aiming at nothing less than the construction of an 'autonomous Aryan male subject' and the denial of autonomy to 'Woman and the Jew.'[56] As Weininger has it, only Aryan males possess the Kantian autonomous self. Given that Woman is 'infinitely porous, infinitely malleable, and infinitely open to external influences,' she is Nothing.[57]

Woman's existence, says Weininger, '*revolves entirely around her sexual life, the sphere of copulation and reproduction*.' In contrast, Man 'knows a dozen other things: fighting and playing, socializing and feasting, discussions and learning, business and politics, religion and art.' Woman takes on extra-sexual matters 'only for the sake of the man she loves' and has no interest in these matters as such. Woman's sexuality 'would be too strong to admit any other serious occupation.'[58] While Woman is 'constantly' sexual, Man is 'only *intermittently sexual.*' Furthermore, Woman cannot become conscious of her sexuality because '*she is nothing but sexuality.*' And because Woman is only sexual, Weininger concludes, 'she lacks the *duality* which is a prerequisite of *noticing* sexuality, or indeed noticing anything.' 'To put it bluntly, Man has the penis, but the vagina has Woman.'[59]

If Weininger's biological construction of Woman reduced her to sexuality alone, says Sengoopta, in an 'emphatically *non*-biological construction' he elevated Man into Genius, a move much less discussed in the critical literature yet inextricable from Weininger's debasement of Woman.[60]

Given contemporary knowledge about the enduring effects of childhood loss and emotional neglect, it is tempting to privilege such psychosocial circumstances—Maudsley's and Freud's lifelong bereavement, anxiety, and depression, for example, and Möbius's undoubted bitterness and depression—as formative of their unsympathetic attitudes.[61] There is little doubt that Weininger suffered from depressive episodes and possible occasional bouts of psychosis. A friend of Weininger remarked that happiness was foreign to Weininger's nature. Another claimed that Weininger had an 'instinctual deviation' that 'infected' his psyche and destroyed it.[62] A previous biographer of Weininger wondered if he was sexually abused in childhood. Weininger had an aggressive sexual urge that he, Weininger, found repulsive. 'I am a murderer!' Weininger said, 'Therefore, I must kill myself.'[63]

From biocultural perspectives, dialectical relationships between factors such as early loss and genetically inherited proclivities (such as anxiety or depression) may produce defensive and aggressive personality characteristics and behaviors in individual cases. And it must be said that to thus contextualize such behaviors is not to condone or excuse them. For example, placing Weininger's theories in context, says Sengoopta, exculpates neither Weininger nor his contemporaries from the charge of being prejudiced. Rather, contextualization shows that 'their views were components of complex, consistent, and widely shared systems of belief . . . far more dangerous in social terms than private prejudices.'[64] Weininger's 'serious, comprehensive and emotionally charged ideological critique of modernity in general, and of women's emancipation in particular,' was not just one man's personal prejudice; it was rooted in his epoch. 'Turn-of-the-century intellectual responses to modernity and modernism were ambivalent at best and hysterical at worst,' says Sengoopta, and debates about the nature of gender were inextricable from debates about the nature and future of civilization.[65]

Weininger's attitudes toward Jews and women were consistent with contemporaneous Viennese cultural criticism 'in which Jews served as a symbol of Mammon and modernity . . . as the greatest threat to the supposedly spiritual culture of the German Volk.' In this scenario, '"Jew" and "German" were opposed categories locked in battle for the soul of (Germanic) civilization.'[66]

Weininger's *Sex and Character* and Möbius's 'On the Physiological Feeblemindedness of Women' (ten editions of which had been published by 1912) fostered enduring misconceptions about women.[67] Influential physicians carried these fallacies into the twentieth century, Freud in particular, who referred to Möbius as one of the founding fathers of modern psychotherapy.[68]

The year 1912 also saw the formation of the German League for the Prevention of Women's Emancipation. The purpose of the League was to prevent the extension of voting, education, and employment rights to women. Described as 'the representative of bourgeois antifeminism,' the league promoted the notion of Imperial Germany's 'ideal' woman who devoted herself entirely to 'the three *K's*: *Kinder, Kuche,* and *Kirche* (children, kitchen, and church).'[69]

The Academy's notion of Woman as failed Man, lacking in intelligence, moral fiber, and capacity for reasoning profoundly shaped Freud as a scientist and physician. The wider Viennese culture treated women as second-class citizens with few rights or opportunities to enter a profession or join a professional organization. However, informed social criticism abounded in Freud's circles, and Freud disagreed just as vehemently with his psychoanalytic colleagues as he did with J. S. Mill over their proemancipation views.

Also in academic vogue was Carl Vogt. The human family, says Vogt, consists of different species, with the Negro and the German positioned at opposite ends of the excellence spectrum as 'two extreme human types.' In Vogt's hierarchy, the species 'Woman' ranks with the lower races: 'The female skull approaches in many respects that of the infant, and still more that of the lower races.'[70] 'The mind of the woman resembles that of the child,' says Harry Campbell. She is 'less intellectual' than Man and demonstrates 'disordered nervous action.'[71] Cesare Lombroso, too, theorized Woman as equivalent to the so-called 'inferior' races, particularly in the woman's demonstrable decline from pre-pubertal precocity to post-pubertal inferiority. The female in the lowest zoological order, says Lombroso, is superior to the male in size. She then falls to the level of 'the humble slave.' Therefore, in our race, before puberty, woman 'appears equal or superior to man,' but then falls behind, leaving 'a proof of the precocity which is typical of the inferior races.'[72]

Thus ran the Aristotelian philosophical, social-scientific, and medical thinking of Freud's day, and thus spoke the Academy on the question of educating women. Mental and physical energy supposedly being finite and competing, disaster befell women if they sought to rival men through education and employment. Most insidiously, too much physical or intellectual activity would drain women of the vital energy required for childcare and, hence, debilitate the race. This was Aristotelian thinking cloaked as modern biology. What was Freud's position on the question? In the next chapter, I make the case that Freud chose a position from which to pursue, above all else, his lifelong ambition: eternal fame.

NOTES

1. Aristotle, "Politics", bk. 1, ch. 5, 1254b.
2. Henry Maudsley (1835–1918), "Sex in Mind and in Education," *Popular Science Monthly* 5, (1874): 210.
3. Christiane Schonfeld, "Under Construction: Gender and the Representation of Prostitutes in Expressionism," in *Expressionism and Gender,* ed. Frank Krause (Goettingen: VandR Unipress GmbH, 2010), 119.
4. Otto Weininger (1880–1903), *Sex and Character* (Vienna: Braumüller, 1903; Reprinted Munich, 1980), 335–36.
5. Menander, *The Fragments of Attic Comedy,* vol. 3, pt. 2, trans. John Maxwell Edmonds (Brill Archive, 1961), Fragment 702, 827.
6. Merry Wiesner-Hanks, *Gender in History: Global Perspectives,* 2nd ed. (Chichester, UK: Wiley-Blackwell, 2011).
7. Wiesner-Hanks, *Gender in History.*
8. Lerner, *Creation of Patriarchy*, 5.
9. Wiesner-Hanks, *Gender in History*, 174.
10. Merrill, "In Silence My Tongue is Broken," 92.
11. Merrill, "In Silence My Tongue is Broken," 92.
12. E.g., Merrill, "In Silence My Tongue is Broken," 11; Wiesner-Hanks, *Gender in History*, 172.
13. Wiesner-Hanks, *Gender in History*, 176.
14. Pavla Miller, "Gender and Education Before and After Mass Schooling," in *A Companion to Gender History,* eds. Teresa Meade and Merry Wiesner-Hanks (London: Blackwell, 2008); Wiesner-Hanks, *Gender in History.*
15. Rozsika Parker and Griselda Pollock, *Old Mistresses: Women, Art, and Ideology* (Kitchener, ON: Pandora Press, 1981), 20.
16. Parker and Pollock, Old Mistresses, 27-28.
17. Wiesner-Hanks, *Gender in History.*
18. Cited in Wiesner-Hanks, *Gender in History*, 178.
19. Wiesner-Hanks, *Gender in History*, 178; Miller, "Gender and Education," 132.
20. Miller, "Gender and Education," 130.
21. Miller, "Gender and Education," 131.
22. Wiesner-Hanks, *Gender in History*, 181.
23. Miller, "Gender and Education," 132.
24. Wiesner-Hanks, *Gender in History*, 182.
25. Catherine Hobbs, ed. "Culture and Practice of United States Women's Literacy," in *Nineteenth-Century Women Learn to Write* (Charlottesville: University of Virginia Press, 1995), 8.
26. Wiesner-Hanks, *Women and Gender*, 126, 116.
27. Wiesner-Hanks, *Gender in History.*
28. Benjamin Rush, reprint of speech in *Women, the Family and Freedom: 1750–1880,* vol. 1, eds. Susan Bell and Karen Offen (Stanford, CA: Stanford University Press, 1983), 76.
29. Lucy Stone became one of the leaders of the American feminist movement.
30. Hobbs, "Culture and Practice," 2.
31. Patricia Ward D'Itri, *Cross Currents in the International Women's Movement 1838–1948* (Bowling Green, OH: Bowling Green State University Popular Press, 1999), 14.

32. Lerner, *Creation of Patriarchy*.
33. Lerner, *Creation of Patriarchy*, 5, 12.
34. Maudsley, "Sex in Mind and in Education," 198, 210.
35. Carol Pateman, *The Sexual Contract* (Stanford, CA: Stanford University Press, 1988), 119.
36. William Blackstone, *Commentaries on the Laws of England 1* (Oxford: Clarendon Press, 1765), 442.
37. Pateman, *Sexual Contract*, 121.
38. Pateman, *Sexual Contract*, 123.
39. Ellen Waldman and Marybeth Herald, "Eyes Wide Shut: Erasing Women's Experiences from the Clinic to the Courtroom," *Harvard Journal of Law and Gender* 28, (2005): 288–89.
40. Maudsley, "Sex in Mind and in Education," 199, 206, 204.
41. Maudsley, "Sex in Mind and in Education," 205, 209.
42. Maudsley, "Sex in Mind and in Education," 209, 204.
43. Theodore Millon, *Masters of the Mind: Exploring the Story of Mental Illness from Ancient Times to the New Millennium* (Hoboken, NJ: John Wiley and Sons, 2004), 177.
44. Trevor Turner, "Henry Maudsley: Psychiatrist, Philosopher and Entrepreneur," *Psychological Medicine* 18, 3, (1988): 553.
45. Turner, "Henry Maudsley," 551.
46. Turner, "Henry Maudsley," 562.
47. Turner, "Henry Maudsley," 566.
48. Maudsley, in Havelock Ellis, *Studies in the Psychology of Sex,* vol. 6 (Teddington, UK: Echo Library, 2007), 126.
49. G. Stanley Hall (1844–1924), "Adolescent Girls and their Education," in *Adolescence: Its Psychology and its Relations to Physiology, Anthropology, Sociology, Sex, Crime, Religion and Education,* vol. 2 (New York: D. Appleton, 1904), 614.
50. Hall, "Adolescent Girls and their Education," 623–24.
51. Paul Möbius, "The Physiological Mental Weakness of Woman," trans. A. McCorn *Alienist and Neurologist* 22, (1901): 629.
52. Schonfeld, "Under Construction," 119.
53. Holger Steinberg, "Paul Julius Möbius (1853–1907)," *Journal of Neurology* 252, (2005): 624–25.
54. Weininger, *Sex and Character*, 335–36. Quotations in this section are from "Emancipated Woman" in *Sex and Character*. Weininger uses (M) and (W) to denote sexual types, not actual men and women. Women who strive for emancipation, he says, *'always display many male properties'* and approximate *'the physical appearance of a man.'* All emphases within quotations are in the original.
55. Weininger, *Sex and Character*, 62.
56. Chandak Sengoopta, *Otto Weininger: Sex, Science, and Self in Imperial Vienna* (Chicago: University of Chicago Press, 2000), 8.
57. Sengoopta, *Otto Weininger*, 11.
58. Weininger, *Sex and Character*, 79.
59. Weininger, *Sex and Character*, 81.
60. Sengoopta, *Otto Weininger*, 7.
61. Marra, *Dialectical Behavior Therapy*. According to contemporary neuroscience, whenever infants sense threat their emotional-pain pathways in the brain fire. Each firing 'kindles' or increases the likelihood of another firing. The more threat sensed, the more hypersensitive to threat the infant becomes, the less able to tolerate the discomfort, and the slower to recover their balance. This is how anxiety becomes chronic, and chronic anxiety often slides into depression.
62. Sengoopta, *Otto Weininger*, 14, 15–16.
63. David Abrahamsen, *The Mind and Death of a Genius* (New York: Columbia University Press, 1946), 93, 83–85.
64. Sengoopta, *Otto Weininger*, 6.
65. Sengoopta, *Otto Weininger*, 5–6.
66. Sengoopta, *Otto Weininger*, 5–6.

67. Diane Guido, *The German League for the Prevention of Women's Emancipation: Anti-feminism in Germany 1912–1920* (New York: Peter Lang Publishing, 2010), 14, 47.
68. Sigmund Freud and A. Albrecht, "American Interview 1909," *Psychoanalytic Review* 55, 3, (1968).
69. Guido, *The German League*, 5, 1.
70. Carl Vogt (1817–1895), "Lecture III," In *Lectures on Man: His Place in Creation and in the History of the Earth* (London: Longman, Green, Longman and Roberts, 1864), 172, 81.
71. Harry Campbell, *Differences in the Nervous Organisation of Man and Woman: Physiology and Pathology* (London: H. K. Lewis, 1891), 163, 84.
72. Nancy Harrowitz, trans. *Antisemitism, Misogyny and the Logic of Cultural Difference: Cesare Lombroso and Matilde Serao* (Lincoln: University of Nebraska Press, 1994), 25.

Chapter Nine

Freud's Fantasy

The prelude to this chapter is an imagined scene based on Freud's case study of Ida Bauer ('Dora'). At the time, Freud wrote to his friend Wilhelm Fliess: 'The expectation of eternal fame was so beautiful. . . . Everything depended upon whether or not [the theory of] hysteria would come out right.' [1]

The maid ushered the pair through the folding glazed and paneled doors into the waiting room at Berggasse 19. Slanting grey light from the facing windows drained the room of colour. The air was chill—despite the fire crackling in the grate—and tinged with cigar smoke. A velvet-upholstered day bed and three upright chairs skirted an ornate occasional table topped by a huge tome encased in a glass box—beckoning but untouchable. Framed portraits, etchings, and diplomas hung against the pale-yellow-print wallpaper behind the day bed.

In silence, father and daughter seated themselves at opposite ends of the bed. The father's gaze, and then the girl's, fell to tracing the scarlet intricacies of the Persian rug at their feet. The year was 1900, the city, Vienna, and the pair, eighteen-year-old Ida Bauer and the prominent Jewish industrialist Philip Bauer. Both had been there before, Ida once, with her father three years previously when her nervous cough was so bad she had to fight for air. During the weeks following the consultation her symptoms dissipated and Philip decided against taking the doctor's advice to pursue a cure.

Philip, on the other hand, was on familiar terms with the waiting room. The doctor had proven discreet and his prescribed treatment regime was eventually successful. Philip drummed his fingers against his thigh.

'How can I make her see reason?' he thought. 'Dr Freud must make her see reason. Why complain about a mere proposition? She must know that men will be men. She ought to feel flattered; Herr K is still a handsome man.

He must be right; she wants to punish him. She's angry with him, jealous of his lovers, angry with me. But what can I do? I can't break it off with Frau K. What would I do without her? Who else would've nursed me through those damnable bouts? The girl must see reason. Her mother doesn't mind—what with all her fuss about dirt and germs, she probably sees it as a blessing. Herr K understands. And it keeps our families close.'

Ida coughed, shivering against the cool velvet. She pulled the heavy shawl tighter and hunched her shoulders. She did not want to see the doctor.

'If only I could get away from them,' she thought, 'go to university as Otto did. It's so unfair. Boys can do whatever they like, be educated, have a career, join the socialists, travel. Get away! And they take whatever they want. When will father see how much he has hurt me—is hurting me—by putting Frau K first? I'm his payment to Herr K. How all of them hurt me, deceive me, use me! All they care about is themselves. Oh to get away from these liars, all of them. Liars!'

As for the doctor, he was on the verge of the fame he had longed for, planned for, since boyhood. His theory of hysteria would turn the world upside down. All he needed now was a fitting case.

What of seed-and-soil and Freud and his theory of sexuality? 'We know less about the sexual life of little girls than boys,' says Freud. 'But we need not feel too ashamed of this distinction; after all, the sexual life of adult women is a dark continent for psychology.'[2] The 'Dark Continent,' retorts Cixous, 'is neither dark nor unexplorable. It is still unexplored only because we have been made to believe that it was too dark to be explored. Because they want to make us believe that what interests us is the white continent, with its monuments to Lack.'[3]

Regarding the roots of femiphobia, depth psychology has it that the little boy—having been made ashamed of his attachment to his mother and to feminine propensities in general—develops an intense envy of his mother that later extends to all women. Yet, according to Freud, biology makes *women* the envious sex. Where does the weight of evidence lie? Here is Freud's case: 'Little girls do not resort to denial when they see that boys' genitals are formed differently from their own. They are ready to recognize them immediately and are overcome by envy for the penis.'[4] Twenty years later, Freud is just as convinced:

> [Girls] notice the penis of a brother or playmate, strikingly visible and of large proportions, at once recognize it as the superior counterpart of their own small and inconspicuous organ, and from that time fall a victim to envy for the penis. . . . She develops, like a scar, a sense of inferiority. . . . Even after penis envy has abandoned its true object, it continues to exist: by an easy displacement it persists in the character trait of jealousy.[5]

The next step in Freud's reasoning is to posit penis envy—or, more precisely, women's refusal to give it up—as a major cause of mental disturbance. 'The hope of some day obtaining a penis in spite of everything and so of becoming like a man may persist to an incredibly late age and may become a motive for strange and otherwise unaccountable actions. [Or] a process . . . which in an adult would mean the beginning of a psychosis.'[6]

Also essential for female mental health, says Freud, is the developmental transfer of sexual pleasure from the 'masculine' clitoris to the 'feminine' vagina. This, and the pubertal repression that overcomes girls' 'childish masculinity,' 'are the chief determinants of the greater proneness of women to neurosis, and especially to hysteria.'[7] When the female gives up her penis envy, Freud says, 'the appeased wish for a penis is destined to be converted into a wish for a baby and for a husband, who possesses a penis.'[8]

That, in essence, is Freud's case. In contrast, feminist psychoanalysts suggest that each sex may envy the other. If girls and women envy the penis, perhaps boys and men envy women's procreative capacities. Might unresolved, unconscious male envy underpin the exclusion of women from much cultural creativity and their predominant confinement to the private sphere?[9] Male envy of women's distinctive sexual apparatus and childbearing functions, says Eva Feder Kittay, could have explanatory force at both individual and cultural levels.[10]

The term 'Non-Menstrual Syndrome' denotes male insecurity in the face of the biologic and reproductive powers of Woman.[11] Men who are envious of women's reproductive power insist that a 'real man' must be 'not-a-woman.' Hence, the theory goes, men seek to dominate women socially as psychological compensation for what they cannot do biologically. Harold Tarpley names vagina envy, womb envy, breast envy, and parturition envy as manifestations of men's desire for these female biological capabilities, together with breast feeding and the social-role freedom to physically nurture children.[12]

Although several of Freud's psychoanalytic colleagues would have openly agreed with Tarpley, Freud could not be more opposed. 'The assumption that all human beings have the same (male) form of genital,' he says, 'is the first of many remarkable and momentous sexual theories of children'[13]: 'We might lay it down that the sexuality of little girls is of a wholly masculine character. . . . It would even be possible to maintain that libido is invariably and necessarily of a masculine nature, whether it occurs in men or women.'[14]

Freud remained attached to the theory of the primacy of the male throughout his life. He placed it at the center of his theory of sexuality, says Christiane Olivier, in a striking convergence between the earliest analytic theory and the dominant political ideology.[15] Freud's writings on the infantile sexuality of girls, she says, come from the imagination of a man set on establishing a comparative relation between the sexes.[16] He must have been supremly confident of possessing the only worthwhile sex (or needed to persuade himself that he did) in order to assume that envy of the boy's genital is the mainspring of girls' psychological development.[17]

'Why does the term "envy" occur to Freud?' asks Irigaray. 'Envy, jealousy, greed are all correlated to lack, default, absence. All these terms describe female sexuality as merely the *other side* or even the *wrong side* of a male sexualism':[18]

> Penis envy means nothing less than that the little girl, the woman, must despise *her own* pleasure in order to procure a . . . remedy for man's castration anxiety. . . . The desire to have it would confirm man in the assurance that he has it. If it were not so, why not also analyze the 'envy' for the vagina? or the uterus? or the vulva? Etc. The desire felt by each pole of sexual difference to have something like it too.[19]

For feminist psychoanalysts, sexual envy is about the attributes of the other; it is not specifically feminine.[20] All through Freud's disquisitions on penis envy, says Olivier, he mentions his own envy of the breast, femininity, and maternity. But in an astonishing reversal women find themselves saddled with men's envy. Freudian theory equates the clitoris with something masculine and requires women to experience orgasm only in the manner sanctioned by men.[21] Women start by renouncing the desiring part of themselves, move into the area of mime, and feign the sexual pleasure of the other. We cannot but wonder (as do many men), says Olivier, is she really having an orgasm or is she faking it in the prescribed manner?[22] Women are only just starting to show themselves as they are, she says, but they know that for doing so they risk rejection by men: 'very often women are still afraid of the death-bearing wishes directed at them, and prefer silence to death.'[23]

FEMIPHOBIA AND ENVY

It is the 'generative power' of the archaic mother, says Kristeva, 'that patrilineal filiation has the burden of subduing.'[24] And, in Horney's view, male envy of Woman manifests in the defensive construction of her as passive and inferior and the consequent devaluation of women and their achievements. Freud's views on women, Horney says, mistake the age-old consequences of social prejudice as biological.[25]

Thus begins an analytic (and gendered) war of words. For his part, Freud notes that Horney's theories diverge from his own. He then points to the difficulties that confront 'the girl who may refuse to accept the fact of being castrated': 'It may put great difficulties in the way of their regular development towards femininity, if it cannot be got over soon enough. . . . A girl may refuse to accept the fact of being castrated, may harden herself in the conviction that she does possess a penis, and may subsequently be compelled to behave as though she were a man.'[26]

On 'the effect of the castration complex on the creature who is without a penis,' Freud says, 'the female acknowledges the fact of her castration, and with it, too, the superiority of the male and her own inferiority; but she rebels against this unwelcome state of affairs.'[27] Horney reiterates her view on the political and cultural oppression of women and remarks that, from cross-cultural as well as clinical work, 'We observe that men are evidently under a greater necessity to depreciate women than conversely.'[28]

Freud responds, 'Horney is of the opinion that we greatly over-estimate the girl's primary penis-envy. . . . This does not agree with the impression that I myself have formed.'[29] In a personal attack aimed primarily at Horney, Freud says[30]:

> The aim of this brief work is to bring together the tenets of psycho-analysis and to state them, as it were, dogmatically . . . only someone who has repeated those observations on himself and on others is in a position to arrive at a judgement of his own upon it.[31] . . . We shall not be so very surprised if a woman analyst who has not been sufficiently convinced of her own desire for a penis also fails to assign adequate importance to that factor in her patients.[32]

Horney was, of course, in a position to come to an independent judgment on the matter. She later moved to the United States, where the psychoanalytic academy expelled her for 'challenging orthodox doctrines.'[33]

Like Horney, Melanie Klein was a loyal Freudian who criticized Freud's theory of sexuality. The ensuing dispute resulted in a tripartite separation within the British psychoanalytic movement in 1944. The Kleinian object-relations school theorized the centrality of envy in both sexes:

> As in the castration complex of girls, so in the femininity complex of the male, there is at bottom the frustrated desire for a special organ. The tendencies to steal and destroy are concerned with the organs of conception, pregnancy and parturition, which the boy assumes to exist in the mother, and further with the vagina and the breasts, the fountain of milk, which are coveted as organs of receptivity and bounty.[34]

Some critics accuse the contemporary psychoanalytic community, too, of failing to acknowledge the substantial body of work on envy. That work, says Cheryl Eschbach, runs up against the legacy of 'Freud's own envious removal of the womb from psychoanalytic theory' and his own unanalyzed envy of female childbearing capacities.[35]

Thus several theorists point to Freud's femiphobia and envy and ask, does Freud protest too much about women's penis envy?

FREUD: CASE STUDY IN FEMIPHOBIA?

Why are some scholars prepared to throw away everything Freudian while others are loathe to criticize Freud at all? Why is Freud billed on the one hand as the worst thing that ever happened to women and on the other as patron saint of feminism? Does criticizing Freud's theories about women entail rejecting the entire psychoanalytic package? Does criticizing Cartesian dualism entail giving up plane geometry and the GPS?[36] Many of Freud's male colleagues disagreed publicly with his patriarchal construction of Woman, and one must query Dorothy Dinnerstein's assessment that Freud's feminist critics are 'wildly ungrateful.'[37] To contest a particular psychoanalytic story such as Freud's about the centrality of penis envy is not to reject psychoanalysis. Nondogmatic psychoanalysis has long tried to develop a better story, says Toril Moi.[38] We can read 'penis envy,' for example, as valid envy of culturally legitimized and totally naturalized male privilege. Furthermore, Freud's biological theory of femininity cannot be disentangled from his overtly anti-feminist position. At the same time as feminists were arguing for a culturally based understanding of women's history, Freud was constructing an elaborate, pseudoscientific case for an entirely biologically based theory of Woman. Even as late as the 1920s, Freud was still denigrating women's moral capacities and pursuing his case against 'the feminists':

> I cannot evade the notion (though I hesitate to give it expression) that for women the level of what is ethically normal is different from what it is in men. . . . They show less sense of justice than men, that they are less ready to submit to the great exigencies of life, that they are more often influenced in their judgements by feelings of affection or hostility. . . . We must not allow ourselves to be deflected from such conclusions by the denials of the feminists, who are anxious to force us to regard the two sexes as completely equal in position and worth.[39]

While Freud's brilliance and contribution to modern psychology are not in doubt, he acknowledged seriously neither his own shortcomings nor the influence of his cultural milieu on his theorizing. Many of Freud's contemporaries supported women's emancipation and reflected critically on their own conditioning under patriarchal culture. They underwent long periods of analysis, not just self-analysis. Apropos Freud's self-analysis, in the light of his various neuroses (e.g., his intense anxiety about being loved) and how little analytic attention he paid them, Freud may have been the most defensive of all his patients.

Despite being disinherited by Freud, or perhaps because of it, feminist-leaning psychoanalysts took psychoanalysis in radically different directions from the orthodox Freud.[40] History has seen psychoanalysis used for both good and ill; its potency alone warrants ongoing deconstruction and reconstruction. While Freud contributed to normalizing certain sexual behaviors and thereby to reducing punitive societal response, he also bolstered the nineteenth- and twentieth-century Academy's Aristotelian view of women as 'failed men.'[41] On J. S. Mill's support for women's emancipation, Freud wrote to his fiancée Martha: 'It seems a completely unrealistic notion to send women into the struggle for existence in the same way as men. Am I to think of my delicate sweet girl as a competitor?'[42] Freud's construction of the passive, docile girl-woman is in direct contrast to his experience of his mother Amalia as aggressive, insensitive, and egocentric. Freud continues: 'No, in this respect I adhere to the old ways, to my longing for my Martha as she is, and she herself will not want it any different; legislation and custom have to grant to women many rights kept from them, but the position of woman cannot be other than what it is: to be an adored sweetheart in youth, and a beloved wife in maturity.'[43]

Moreover, Freud's letters to Martha poignantly reveal his vulnerability and need to feel in charge in their relationship: 'Woe to you, my Princess, when I come, I will kiss you quite red . . . and if you are forward you shall see who is the stronger, a gentle little girl who doesn't eat enough or a big wild man *who has cocaine in his body*.'[44]

Also poignant are Freud's references to his depression and growing reliance on cocaine (billed as the wonder drug of the age) to alleviate his symptoms: 'In my last severe depression,' he wrote to Martha, 'I took coca again and a small dose lifted me to the heights in a wonderful fashion.'[45]

Early in Freud's career he turned on his mentor and co-author Josef Breuer over Breuer's view that sexuality was only one of many important etiological factors for hysteria. In a lecture presenting his own and Freud's collaborative work on hysteria, Breuer praised Freud's discoveries and begged to differ with him on just one point, namely, 'the overvaluation of sexuality.'[46] Breuer had been a pivotal mentor and father-figure for the fledgling physician and, as mildly as Breuer presented his criticism, Freud could not tolerate it. Soon after the publication of *Studies*, Freud cut Breuer out of his life.[47] Breuer's daughter-in-law recalled how on meeting Freud by chance on the street Breuer instinctively opened his arms to embrace his former protégé but Freud ignored him.[48]

Fifteen years later, Freud accused Alfred Adler, physician and president of the Vienna Psychoanalytic Society, of heresy and forced him to resign. 'It had become impossible to go on with Adler,' Freud wrote to Jung early in 1911; 'I mean to be more careful from now on that heresy does not occupy too much space in the *Zentralblatt*.'[49] 'We inside our circle really ought to come to a decision about Adler,' Freud told Jung, 'before he is held up to us by outsiders. His things strike me as more and more stupid as time goes on.'[50] And, later in 1911: 'I have finally got rid of Adler. . . . As a paranoic of course he is right about many things, though wrong about everything. A few rather useless members will probably follow his example.'[51]

What was Adler's heresy? One could read Adler's writings on child development as interpreting Freud's own psychology.[52] Freud's insistence on sexuality and drives as preeminent in human development, Adler said, overlooks a major determining fact of human existence: 'the individual as a social being':[53]

> If we focus on the smallness and helplessness of the child which continues so long and which brings about the impression that we are hardly equal to life, then we must assume that at the beginning of every psychological life there is a more or less deep inferiority feeling: Difficult questions in life, dangers, emergencies, disappointments, worries, losses, especially those of loved persons, social pressures of all kinds, may always be seen as included within the framework of the inferiority feeling . . . as anxiety, sorrow, despair, shame, shyness, embarrassment, and disgust.[54]

Inadvertently or not, Adler had listed the central themes of Freud's childhood. Chronic impoverishment, repeated maternal loss, and the young Freud's deep disappointment in Jacob, his father, remained essentially unanalyzed to the end.

FREUD'S FORMATIVE YEARS

Freud spent the first three years of his life in a cramped one-room apartment that witnessed immense loss and grief. Not long before Freud was born his extended family had grieved the loss of three members. A year later the loss was compounded when the younger brother of his mother died of tuberculosis. Freud's first sibling, named after that late uncle, also died from an infection within the year. His mother, Amalia, had tuberculosis as well and was often absent on rest cures.[55] Freud's early, repeated witnessing of illness and death among his extended family produced enduring morbid fears.[56] Breuer theorized such witnessing as psychically traumatic—in actuality, not merely in fantasy. In his later self-analysis, Freud recalled a childhood nightmare in which he saw his mother dying. Not long prior, the boy had observed Amalia's father in the process of dying.[57]

After Freud's baby brother died, Amalia bore six children in quick succession. The repeated pattern of Freud's childhood was the loss of his mother's care to new babies.[58] At the age of two-and-a-half, Freud lost a mother-figure to whom he was close, his nursemaid, who was imprisoned for theft.[59] Freud repeatedly dreamed about losing his mother, including that she had been locked up in a box as he believed his nursemaid had in actuality.[60] In addition, Jacob's business failed and the extended family broke up, some members moving overseas. As a result of this dislocation, Freud lost two half-brothers, a sister-in-law (all were parent figures to him) and their children, who had been Freud's playmates.[61] When circumstances forced his immediate family to relocate, Freud developed an intense fear of travelling (specifically of being left behind), a phobia that induced travel-related fainting spells and gastric symptoms throughout his life.[62]

Freud wrote at length about the traumatic impact of repeated maternal loss on children. Significantly, though, that writing was in the abstract and concerned only the Girl. Despite this, Freud pointedly describes experiences that could mirror his own:

> But what the child grudges the unwanted intruder and rival is not only the suckling but all the other signs of maternal care. It . . . casts a jealous hatred upon the new baby and develops a grievance against the faithless mother. . . . Especially as this jealousy is constantly receiving fresh nourishment in the later years of childhood and the whole shock is repeated with the birth of each new brother or sister.[63]

According to Freud biographer Louis Breger, Freud's blind spots about anxiety, sexuality, and women were the most striking areas where Freud's disavowed traumas and conflicts intruded into his theories.[64] 'The underlying meaning of sexuality for Freud becomes clear,' says Breger, 'if the words *longing for early love* are substituted for *sexual instinct*.'[65] Traumatic loss left Freud with a powerful longing for these early, pleasurable, 'infant-sexual' experiences, along with a dread of reexperiencing the helplessness and terror associated with them. Freud mastered this fear by becoming moralistic, dutiful, and hardworking—his mother's 'Golden Sigi.'[66] In keeping with Freud's aversion to strong emotion was his aversion to music; he had his mother remove the piano from the house, thus putting an end to his sisters' opportunity to become musicians.[67] Freud projected his wish for pleasure onto his sisters and later onto women in general, while he remained 'civilized' and in control of his impulses. In his theories, says Breger, Freud transformed his 'dread of giving in to his infantile yearnings into the image of a menacing sexual instinct.'[68]

Out of his self-analysis Freud created his revolutionary oedipal theory, not from mining his childhood losses but by inventing a comforting myth in which a little boy competes heroically with the Father-King for the Queen's hand.[69] As to the childhood nightmare about his mother dying, says Breger, Freud invented a 'complicated and strained interpretation' in which the dream was supposedly driven by his sexual desires.[70] 'I was not anxious because I had dreamt that my mother was dying,' Freud says of the dream. 'The anxiety can be traced back . . . to an obscure and evidently sexual craving.'[71] Perhaps 'obscure' is the operative word. From this displaced beginning, Freud's theory constructs the male as hero, reinscribes Woman as

Dark Continent, and becomes the centerpiece of psychoanalysis. The theory is Freud's own invention achieved via the hero's journey—self-analysis.

Freud's elevation of sexuality above all other aspects of development was arguably a defensive strategy. Raising his interpretation of a childhood image to the status of universal law not only enabled him to avoid confronting his worst fears, but also delivered the singular ambition of his life: to be recognized as a hero.[72]

'I have destroyed all my notes,' twenty-eight-year-old Freud wrote to Martha. 'As for biographers . . . each one will be right in his opinion of "The Development of the Hero" and I am already looking forward to seeing them go astray.'[73] And the following year, after receiving a longed-for travel grant, Freud wrote again to his 'Princess, my little Princess': 'Oh, how wonderful it will be! I . . . go on to Paris and become *a great scholar* and then come back to Vienna with a huge, enormous halo, and then we will soon get married, and I will *cure all the incurable nervous cases* and through you I shall be healthy.'[74]

As to his theory of hysteria, Freud wrote to his friend Fliess: 'The expectation of eternal fame was so beautiful. . . . Everything depended upon whether or not hysteria would come out right.'[75]

In the foregoing three paragraphs there is enough material for years of analysis. It is the stuff of adolescent male fantasy yet, like Hollywood, it became a Freudian norm for twentieth-century thought. Freud's early disappointment in his own father, his hero worship of Hannibal,[76] Moses, Napoleon, Alexander, and other military greats (he named his son Oliver after Cromwell),[77] and his self-deprivation all skewed psychoanalysis toward the masculinist notion of heroic struggle.[78] Freud's rules for therapy came out of his intuitive theories about children's needs, theories based on recollections of his own childhood amid precarious family circumstances, including seeing his father permanently cowed. 'I cannot think of any need in childhood,' he says, 'as strong as the need for a father's protection.'[79] The statement is extraordinary not just for its transparency but also for suggesting a lack of understanding of child development compared to that of many of Freud's colleagues.

Freud's obsession with heroic father-figures carries through into a highly paternalistic therapeutic approach compared to the more respectful attitudes of colleagues such as Adler, Jung, Horney, Otto Rank, and Sándor Ferenczi. For Freud, the doctor always knows best: 'Where there is a dispute with the patient as to whether or how he has said some particular thing, the doctor is usually in the right (the doctor can assure him with a quiet feeling of superiority).'[80]

And on the patient's capacity to sublimate sexual instinct, Freud remarks: 'Not every neurotic has a high talent for sublimation; one can assume of many of them that they would not have fallen ill at all if they had possessed the art of sublimating their instincts.'[81] Most of Freud's patients were women and, by Freud's reckoning, women's capacity to sublimate and engage in productive achievement was inferior to men's. Unlike boys, says Freud, girls (having realized they are already castrated) do not fear the threat of castration by the father. Consequently they fail to internalize paternal authority, their super-ego development is stunted, and with it the capacity to sublimate. In addition, shame at their inferior genital organs causes repression rather than sublimation of their sexuality.[82]

Not only psychoanalytic theory but also its practice—according to Freud's rules—was a heroic endeavour. Paternal strength mattered more than maternal support. Freud warned his colleagues against showing support to patients, which, in his view, was dangerously gratifying and harmfully colluded with their infantile-sexual wishes.[83] Kindness, says Freud, must be subordinated to the surgical task of analysis:

> I cannot advise my colleagues too urgently to model themselves during psycho-analytic treatment on the surgeon, who puts aside all his feelings, even his human sympathy, and concentrates his mental forces on the single aim of performing the operation as skilfully as possible. . . . The justification for requiring this emotional coldness in the analyst is that it creates the most advantageous conditions for both parties: for the doctor a desirable protection for his own emotional life and for the patient the largest amount of help that we can give him today.[84]

Given Freud's childhood losses and his lack of insight into their psychic legacy, no doubt his emotional life needed a great deal of protection. Psychoanalytic treatment, says Freud, ought to be carried through under privation, and the analyst 'should be opaque to his patients and, like a mirror, should show them nothing but what is shown to him.'[85] By his own account, Freud began 'from the assumption that my patients knew everything that was of any pathogenic significance and that it was only a question of obliging them to communicate it.'[86] In psychotherapy, 'we force our way into the internal strata, overcoming resistances all the time.'[87] 'Often in my own mind,' says Freud, 'I have compared cathartic psychotherapy with surgical intervention. I have described my treatments as psychotherapeutic operations, and I have brought out their analogy with the opening up of a cavity filled with pus, the scraping out of a carious region.'[88]

Freud's surgical metaphors are disturbing but also moving given the painful repeated surgeries he endured in his final two decades. In 1923, Freud contracted cancer of the palate, but, despite his physician's advice that smoking had caused his cancer (evidence was then available), he could only briefly interrupt his twenty-cigar-a-day habit.[89]

KEEPING THE FAITH

'You know, of course,' Jung wrote to Freud, 'how far a patient gets with self-analysis: not out of his neurosis—just like you. . . . I shall continue to stand by you publicly, while maintaining my own views, but privately shall start telling you in my letters what I really think of you. I consider this procedure only decent. No doubt you will be enraged by this peculiar token of friendship, but it may do you good all the same.'[90]

As Freud's close colleagues began to question aspects of his orthodoxy, Freud branded them traitors, expelled them, and boycotted the publication of their work in the major psychoanalytic journals.[91] Colleagues such as Adler, Rank, and Ferenczi, whose ideas stirred up Freud's unconscious fears about maternal loss, received particularly rough treatment.[92]

In one sense at least, Freud saved some of the roughest treatment for himself. As Freud's letters show, he felt constantly bound to suppress libidinal desire that, in his terms, implied 'feminine' weakness and longing for love, which were unmanly.[93] Throughout his life, Freud's erotic attachments were to charismatic, intelligent, and physically attractive male friends. However, later Freud labelled their enhanced capacity for intimacy as 'feminine' passivity and weakness. The great love of his adult life was Wilhelm Fliess, Freud's physician and prescriber of cocaine.[94] Freud often spoke of the unruly homosexual libido he felt toward Fliess and other esteemed friends,[95] libido he felt bound to suppress along with any masturbatory desires, because yielding to either would indicate weakness.[96] Cigars and cocaine were the (self-medicatory) pleasures he permitted himself, together with submission to the 'tyrant' work.[97]

On his attachment to Fliess, Freud wrote to his confidante Jones,[98] 'This town seems to have acquired a strong connection with my relation to that man. There is some piece of unruly homosexual feeling at the root of the matter.'[99] To his then close friend Ferenczi, Freud wrote in 1910, 'Not only have you noticed that I no longer have any need for that full opening of my personality, but you have understood it correctly and returned it to its traumatic cause. . . . This need has been extinguished in me since Fliess's case. . . . A piece of homosexual investment has been withdrawn and utilized for the enlargement of my own ego.'[100] And later, 'I feel myself to be a match for anything and approve of the overcoming of my homosexuality, with the result being greater independence.'[101] Two years before he died, Freud reprised the dangers of a man's 'passive or feminine attitude to another male':[102] 'A man's heterosexuality will not put up with any homosexuality, and *vice versa*. . . . There is no greater danger for a man's heterosexual function than its being disturbed by his latent homosexuality. . . . Conflict of this sort . . . [is] an expression of the destructive or aggressive [death] instinct.'[103]

Echoing his childhood heroes, Freud described to Martha how his self-imposed emotional deprivation was in the service of sublimation and refinement: 'The mob gives vent to its appetities, and we deprive ourselves. We deprive ourselves in order to maintain our integrity. . . . and this habit of constant suppression of natural instincts gives us the quality of refinement.'[104]

As a young man Freud had shown little interest in women; rather, he disparaged them. 'Aren't young ladies crystallized poison of boredom,' he wrote to Ernst Silberstein, an early intimate.[105] As Freud later reflected, 'probably no male human being is spared the fright of castration at the sight of a female genital. Why the great majority surmounts it, we are frankly not able to explain.'[106] Freud's attitudes toward women remained essentially unchanged, says Breger, his underlying suspicions and distaste for women persisting throughout his life,[107] alternating between expressions of bewilderment on the one hand and theoretical certainty on the other.[108]

Throughout his adolescence, Freud suppressed his anxieties and yearnings for love and became a prudish, censorious, and inexpressive young man. He was a virgin until he married at thirty, and his sexual activity after marriage was by his own admission 'minimal.'[109] In 1882, twenty-six-year-old Freud had fallen in love at first sight with twenty-one-year-old Martha Bernays, the petite, pale, obedient, and dutiful daughter of an Orthodox Jewish widow. Financial constraints in both families delayed the marriage, and Freud and Martha spent only a few months of their three-and-a-half-year-long engagement in each other's physical presence.[110]

Freud's sexual interest in Martha declined early in their marriage. In letters to several friends Freud alluded to his impotence and lack of sexual activity and pleasure. In a letter to Fliess in August 1893, thirty-seven-year-old Freud said of himself and Martha, 'We are now living in abstinence; and you know the reason for this as well.'[111] And again, 'Sexual excitement, too, is no longer of use for someone like me.'[112] He had written to Martha before their marriage, 'I always think that once one is married one no longer—in most cases—lives for each other as one used to. One lives rather with each other for some third thing, and for the husband dangerous rivals soon appear: household and nursery. Then, despite all love and unity, the help each person had found in the other ceases.'[113] And, as Freud wrote later about the first-born of the 'faithless mother,' 'the whole shock is repeated with the birth of each new brother or sister.' In August 1901, Freud wrote to Fliess: 'In my life, as you know, woman has never replaced the comrade, the friend.'[114] Despite, or because of, his own lack of sexual experience, Freud placed sexuality at the center of his theory of neurosis.[115]

In contrast to orthodox Freudianism, Breuer, Jung, Rank, Ferenczi, and, of course, Horney and Klein attributed equal status to mothers and stressed the significance of the mother-infant bond and the therapist's role as including both paternal and maternal functions.[116] Rank, for example, located the primary source of anxiety in separation from the mother.[117] Ferenczi continually challenged his own therapeutic methods and resurrected Breuer's emphasis on actual as well as fantasized trauma.[118] Psychical trauma, says Breuer, may be induced by the loss of a loved person, anxiety, fright, shame, physical pain, moral disgust, tormenting secrets, and so forth: 'In traumatic neuroses, the operative cause of the illness is not the trifling physical injury but the effect of fright, the psychical trauma.'[119]

Ferenczi argued that Freud's stated therapeutic model, which called for emotional privation and distancing strategies (e.g., mirroring, remaining out of view, maintaining lengthy silences), caused more harm than good.[120] Ferenczi's revivifying of Breuer's theories on actual trauma may have rekindled Freud's fear that Breuer's more inclusive theory would overshadow his own.[121] And Freud would have felt particularly threatened by Adler's feminist formulations, including that of the 'masculine protest,' defined by Adler as 'a reaction by either sex to the prejudices of our society about masculinity and femininity. A man's behavior may constitute a protest against the demands made on him by the myths of male superiority.'[122]

Freud would have none of Adler's sacrilege. Adler's 'masculine protest,' says Freud, 'is in fact nothing else than castration anxiety' and 'repudiation of the feminine.'[123] And as Freud had it, Adler harboured even more dangerous heresies: 'Recently, [Adler] expressed the opinion that the motivation even of coitus was not exclusively sexual, but also included the individual's desire to *seem* masculine to himself. So far it hasn't occurred to him that with such a theory there can be no explanation for the sufferings of neurotics.'[124]

For obvious reasons it hadn't occurred to Freud (nor did he want to hear of it from Adler) that the demands made on men by the masculinist enterprise may produce neurosis. In this regard, Freud's feminist colleagues were operating at a level of psychological consciousness beyond his. Also for obvious reasons their theories did not end up on top. Freud's theory left Aristotelian constructions of Woman intact and, hence, was readily assimilable by the Academy.

Freud's construction as a child and man of patriarchy places him in a direct line from Plato, says Rachel Jones: 'Insofar as Plato and Freud can be seen as the beginning and culmination of the Western theoretical tradition, woman remains trapped between them as a necessary absence.'[125] Plato's presentation of sexual difference and the female as deficient, says Jones, is repeated and reinforced throughout the history of Western philosophical thought.[126] In *Timaeus*, for example, Plato uses a procreative metaphor to depict the universe and the natural world as produced by a creator god in his own image: the model (the Forms) is the father and the receptacle (onto which the Forms impose their pattern) is the mother.[127] 'The receptacle,' says Plato, 'must itself be devoid of all character.'[128]

In this view, says Irigaray, Man becomes *the* procreator. Sexual production-reproduction is referable to his 'activity' alone. Woman is the matrix—womb, earth, factory, bank—to which seed capital is entrusted so that it may germinate, produce, and grow fruitful. Woman can lay claim to neither capital nor interest as she has only submitted 'passively' to reproduction.[129] Of Aeschylus's *Oresteia* Freud remarked on the momentousness of that patriarchal revolution in the law:

> Under the influence of external conditions . . . it happened that the matriarchal structure of society was replaced by a patriarchal one. This naturally brought with it a revolution in the existing state of the law. An echo of this revolution can still be heard, I think, in the *Oresteia* of Aeschylus. This turning from the mother to the father, however, signifies above all a victory of spirituality over the senses, that is to say, a step forward in culture, since maternity is proved by the senses whereas paternity is a surmise based on a deduction and a premiss. This declaration in favor of the thought process, thereby raising it above sense perception, was proved to be a step charged with serious consequences.[130]

According to Delaney, Freud was particularly interested in a recurring motif in Greco-Roman/Judeo-Christian mythologies, namely, the son's desire to kill his father.[131] But what of the fathers who would do away with their sons—Kronos with Zeus, Laius with Oedipus—allegedly out of fear of usurpation but perhaps as much out of jealousy? And, says Delaney, despite the fact that Abraham, the father who was willing to sacrifice his son on God's command, is considered the father of the three largest monotheistic religions, Freud did not consider him. Freud became obsessed not with the murderous intent of Laius, father of Oedipus, but with the violence of the son in whom such intent is absent.[132] Why does Freud not even entertain the idea of a 'Laius complex,' asks Delaney, or at least the possibility of a relationship between the destructive wishes of both father and son? Why did Freud go to Moses and fabricate a story of the sons killing the father; why not to Abraham whose willingness to sacrifice his son is integral to his position as the father of faith? What blindness prevented Freud from seeing the relation? Had he done so, he might have produced a theory that took account of the father's infanticidal impulses toward the son.[133] Just as Freud's version of Moses vindicated his oedipal theory, so his theory of Woman justified patriarchal ideology. Moreover, both helped to protect Freud from confronting his early losses.

Historically, Freud's construction of Woman—as caught between hysteria and accepting her 'natural' inferiority and subordination to men—conferred legitimacy on the nineteenth- and twentieth-century campaigns against women's suffrage and also on the medicalization of women's responses to their subordination. This cannot be coincidental: Freud publicly took an antifeminist position and simultaneously developed a theory of femininity diametrically opposed to feminist arguments about the historico-cultural roots of their situation. Freud's public and private stance on the Woman question was equivalent to saying that enslaved peoples are biologically determined for slavery (as, of course, *was* claimed by the Academy).

In contrast to the more inclusive theories of many of his colleagues, Freud's psychoanalysis is fundamentally about fathers and sons, but less as Freud imagined them (sons slaying fathers) than the reverse—fathers preemptively devouring the sons who may forget their duty of obedience. Freud's idealization of the Father skewed psychoanalysis and further legitimized and propagated the Abrahamic myth and its values.[134]

FREUD'S HYSTERIA

At least from the time of Hippocrates and Plato, physicians viewed hysteria as a uterine disorder related to abstinence: the parched uterus wanders about the body seeking sexual activity or moisture and causing trouble for other organs, for example, breathlessness and choking when it ventures into the throat.[135] An alternative etiological theory for hysteria and other mental disorders, says Carol Warren, was the retention of menstrual blood or female semen, which physicians purged from the patient's body using techniques ranging from massage to bloodletting.[136]

Given that the women who appeared less prone to hysteria were typically young, married, and reproductive; marriage, marital intercourse, and motherhood, reasoned physicians, must be protective and curative.[137] However, what could physicians do for women who could not or would not marry? The ancient model had it that the symptoms of hysteria would mount to a 'hysterical paroxysm' and then subside. By this reasoning, the remedy recommended by Egyptian, Greek, and Roman physicians was the therapeutic application of genital stimulation. Galen's so-called 'widow's treatment' involved the physician, midwife, or the patient herself massaging her vulval region with scented oils or ointments. Ibn Sina recommended Galen's treatment, as did al-Majusi, who described it as having 'the same effect of intercourse in warming and thinning the seed, so that it can drain away and the woman can find peace.'[138]

Despite theological doubts about the compatibility of Galen's materialistic conception of medicine with creationism,[139] the resurgence of Arabic medicine in Europe had physicians adopting Galen's widow's treatment, which persisted into the nineteenth century and beyond.[140]

Centuries before, however, in witchcraft-obsessed early modern Europe, such theological unease had turned the abstinence theory on its head by declaring that too much sex, not too little, caused hysteria, the symptoms of which resulted from sexual union with Satan. Women's sexual pleasure was the Devil's work, and men caught it only from satanically influenced women.[141] One sixteenth-century French physician suggested that, like Ethiopian women, all Frenchwomen ought to undergo clitoridectomy as 'a way of disciplining unbridled female sexuality.'[142]

Hence, the 'too little' and 'too much' theories co-existed for centuries, the too-muchers opting for gynecological surgery and the Galenic too-littlers treating clients by genital stimulation. These two seemingly opposed approaches, says Warren, were in fact two sides of the same coin: medical control of female sexuality by medical men.[143] While stimulative treatments of the Victorian era were textually genital (as distinct from nervous), unlike the Galeno-Hippocratic tradition, they were rarely textually sexual; such a suggestion was left to the critics.[144]

Like its predecessors, the nineteenth-century Academy taught that virtually every female ill stemmed from a disordered reproductive system, usually a refusal to participate in wifely duties or 'an unwomanly interest in sex.'[145] And if overarousal was the problem, clitoridectomy was the cure. Clitoridectomies performed in England between 1860 and 1866 for epilepsy, sterility, masturbation, 'hysterical manias,' and various menstrual conditions averaged one hundred per year. Perhaps even more frequently, physicians were removing healthy ovaries to cure, for example, the mental disturbance 'Ovarian insanity,' or 'Old Maid's Mania.'[146] David Gilliam (1844–1923), later an honorary fellow of the American Gynecological and Obstetrical Society, in 1896 pleaded for the 'more general adoption of oophorectomy' (ovary removal). 'Why do we alter our colts and calves?' asks Gilliam. 'That we may make them tractable and trustworthy, that we may convert them into faithful, well-disposed servants.' Some men have 'had the courage of their convictions' to operate, he continues; 'they tell us that castration pays; that the moral sense of the patient is elevated, that she becomes tractable, orderly, industrious, and cleanly.'

The title of Gilliam's paper refers to 'the female' but because, according to Gilliam, insanity and epilepsy are hereditary diseases, he wanted male sufferers castrated as well. Gilliam allowed one caveat: the operation ought to be limited 'to those in whom the malady appears in some way to be connected with or dependent on sexual disturbance.'[147] Thus, the discourse collapses and significantly feminizes notions of insanity, epilepsy, and transgressive sexual behavior and opens the way to increased gynecological surgery for mental and moral disorders.

The innovative obstetrician, Isaac Baker Brown, led the push for gynecological surgery in Britain in the 1860s. Widely respected—including by the *British Medical Journal*—for his avant-garde procedures, and whose private hospital enjoyed royal patronage, Brown accepted the theory of the 'Psychology of the Ovary,' namely that women's medical and emotional problems stemmed from their malfunctioning ovaries. Despite recognition by a more scientifically oriented section of the profession that 'not all disorders of the female genitals stemmed from immoral or otherwise irregular conduct,' Brown and his followers offered clitoridectomy as a cure for a wide range of physical, psychological, and 'moral' disturbances.[148] Reportedly, physicians told patients they would merely undergo 'a slight operation on the external parts.'[149] Brown's 1864 case notes about a twenty-year-old patient with epilepsy read as follows: 'Clitoris excised in the usual manner under chloroform, was restless and hysterical for the first six days, when she improved daily, became cheerful, and much more intelligent.'[150]

After the publication of Brown's *On the Curability of Certain Forms of Insanity, Epilepsy, Catalepsy, and Hysteria in Females*, the medical establishment began to censure him.[151] Although elements of the London Medical Society approved of both clitoridectomy and Brown, having elected him president in 1865, they approved neither of his grandstanding nor of the scrutiny it attracted to the Society's less flamboyant use of the same procedures for the same purposes. Under attack, Brown named and shamed colleagues, all of whom denied their past association with him. In 1866, the Society banned Brown from performing gynecological surgery because his methods, not the procedures per se, had come into question.[152]

In fin-de-siècle Vienna, the Academy's attention turned from the physical to the nervous consequences of uterine disquiet. Physicians no longer held that hysteria was demonically induced; rather, they assumed it was endemic to proper womanhood—not to the possession of a womb per se but to the volatile emotionality demanded by motherhood.[153] Enter Freud's oedipal-sexual theory of hysteria and his case study of Ida Bauer ('Dora'), an eighteen-year-old who, at thirteen and again at fifteen, had shunned the sexual advances of an older male friend of her father.[154]

The 'average uncultivated woman,' when led on by a clever seducer, says Freud, finds, as does a child, 'every sort of perversion to her taste.'[155] Freud interpreted Dora's hysterical symptoms as expressions of her forbidden oedipal-sexual desire, not just for her father Philip and for the seducer Herr K, but also for Frau K—who Dora knew was Philip's lover. When Dora balked at Freud's interpretation, Freud said she was resisting therapy. When she ceased treatment, he said she was punishing him. As Freud's theory of resistance goes, a patient disagreeing with an interpretation simply confirms its accuracy.[156]

Freud's fixation on fame had him claiming case-study confirmation of his theory of hysteria where there was little to be had.[157] The Dora study says less about Dora than it does about the hysterical nature of the Academy's view of female sexuality. Critics have meticulously dissected the case and, by all accounts, Freud's treatment of Dora was aggressive, lacking in empathy, and damaging.[158] And hysteria as a diagnosis disappeared in the early twentieth century, due partly to extreme clinical overextension.[159]

Looking at the fathers of late-nineteenth and early-twentieth-century European medicine, one is struck not only by their intellectual brilliance but also by the depth of their prejudices compared, for example, to Freud's more enlightened psychoanalytic colleagues and the likes of J. S. Mill. Were their prejudices shaped by the dialectics of early trauma and patriarchal ideology? While this stellar group did not invent medical sexism, a phallocentric culture constructed them as highly femiphobic. Their fear of the feminine, together with an all-encompassing desire for acceptance, even fame, locked them into playing the role of the hero. In this light, one could read Freud's continuing refusal to be analyzed (except by himself) not as a sign of mastery but of vulnerability.

GENDER NONCONFORMITY AND EMPIRE

There were, of course, suffragist counter-narratives to the pseudoscience of the Academy. In 1868, American suffragettes organized the Working Women's Association with a group of typesetters (trade unions excluded women from membership). In Britain, J. S. Mill took the suffragist cause to Parliament in 1865, the National Union of Women's Suffrage Societies was formed in 1897, and Emmeline Pankhurst founded the Women's Social and Political Union in 1903.[160] Was the Academy seriously destabilized by the possibility that the educated woman might become the norm? There were political considerations at stake certainly. The Academy's construction of women as opposite to men, for example, provided a sociopolitical resolution of the tension between the widespread enfranchisement of men and its denial to women.[161] Femininity, said the Academy, was antithetical to everything 'that had warranted widespread male participation in the public sphere.'[162] On what grounds, asked the suffragettes, do you deny women everything men have gained from liberalism? Women, answered the Academy, are biologically opposite to men and, hence, unsuited to, damaged by, and damaging to public life.

To what degree the Academy believed its own rhetoric is impossible to tell. Assuming it harbored real fears about the future of civilization, what were they? According to Abrahamic-patriarchal notions of male sacrifice, Woman is the tradition's greatest enemy. Not only does Woman's feminizing/sexualizing effect undermine men's capacity to do battle, but educated (i.e., masculinized) women will not raise adequately masculine son-soldiers.

What drove the Academy to obsess over sexual ambiguity? Was there something in the air? Birth rates had declined, says Nye, and young men were retreating to their clubs and colonial service rather than marrying, all of which raised questions about the quality of masculinity.[163] Some European statesmen claimed that the increasing prevalence of perversion had lowered birth rates and weakened national defenses.[164] Empires were jostling, and future victory depended on soldiers' willingness and fitness to fight to the death.

'A feminist *Volk*,' said Möbius, 'would be defeated by its neighbors.' 'If we want a woman that completely fulfils her motherly profession, then she cannot have a male brain.' 'The unnatural struggles of the feminists,' he says, 'corrupt the human race and culturally signify the beginning of the end.' One of the essential signs of modern degeneration 'is the blurring of sexual characters: i.e. feminine men and masculine women,' and 'the leaders of the women's movement are degenerates.' According to Möbius, the feminists were the real misogynists because they 'want to destroy the differences of the sexes.'[165]

Such talk of the degeneracy and corrupting effect of gender nonconforming women went hand in hand with the increasing trend since the early modern period toward criminalizing the sexual behavior of both sexes. The newly developing field of sexology, with its origins in European medical criminology, was both forensic and misogynous, its knowledge production geared to the interests of social control rather than social critique.[166] Aberrant sexual behavior, said the sexologists, leads to social breakdown. Lombroso's 1893 treatise *Criminal Woman, the Prostitute and Normal Woman* refined and extended the symbiotic relationship of criminology to the Academy's gender norms.[167] Möbius's pamphlet on women's alleged feeblemindedness referred to Lombroso's phrenological studies. Like Lombroso, Möbius accepted that women's smaller brain size was irrefutable proof of their mental inferiority. Andropometric science, says Ralph Leck, functioned sociologically as corroboration of the theological, legal, and social propriety of women's inferiority.[168] In response to both the perceived crisis in gender roles and the Academy's fears that 'normal' sexual drives were being misdirected, the state criminalized and pathologized same-sex sexuality.

These dividing practices occurred at both societal and personal levels. At the societal level, they functioned to constitute and police the boundaries of heterosexual normality by excluding the 'abnormal' Other. At the personal level, take, for example, Weininger, whose belief that 'Judaism is saturated with femininity' drove him to convert to Protestantism.[169] The feminine traits he saw in himself troubled him.[170] Weininger described the choice that men face: Man has elements of Woman, of matter, in him; thus, he 'can allow that part of his nature to develop, that is, he can deteriorate and degenerate' or he can fight it. That is why Man alone 'can arrive at the truth about Woman,' while Woman 'has no possibility of developing except through Man.'[171]

Is it not remarkable, says Weininger, that agitation for women's emancipation rises and falls at certain intervals in history, markedly the tenth, fifteenth, sixteenth, and now in the nineteenth and twentieth centuries? The possibility of a crucial periodicity must be borne in mind, he says, namely, recurring periods of overproduction of hermaphrodite and sexually intermediate forms, male women and female men. The increase in 'dandified homosexuality,' says Weininger, may be a symptom of such periodic overproduction.

Was Weininger grasping at straws? Perhaps, like many commentators of his time, Weininger was looking for a way of plausibly denying a major political ramification of the escalating visibility of women in public life, namely that women were increasingly displaying their skills and talents to increasing acclaim. In fin-de-siècle France, for example, dominant representations of female pathology, says Mesch, concealed male anxiety about female intellect.[172] In contrast to the eighteenth century, which had fostered women's literary sensibilities, for example, nineteenth-century France reconstructed women writers as both products and producers of the moral depravity of their era and associated female intellectual exertion with venereal disease and infertility.[173] Thus, on the cusp of the twentieth century with its angst about political conflict, gender nonconformity, and depopulation, the Academy reinvented the French woman writer—whose visibility and popular success had risen to the point of challenging male literary hegemony—as a reviled and intimidating figure much like the prostitute and the hysteric, and excluded her from the canon.[174] Male fears that women's creativity and intellect may match men's, says Mesch, saturates the medical, sociological, and literary theories of the time.[175]

A remarkable thing about Weininger's theory is the clarity with which it articulates the nineteenth-century femiphobic attitude: Normal women do not desire emancipation; women who do so are masculine and usually hysterical. Inevitably, overproduction of such intermediate forms of Woman will occur from time to time; hence, not only is working toward women's emancipation unnecessary, it can only harm them.

The nineteenth- and early-twentieth-century Academy's anxiety about gender nonconformity reflected not only the prevailing politico-economic-cultural dynamics, but also its members' ambivalent feelings about their own masculinity, feminine/homoerotic aspects, and fantasies about women's capacity to undermine or usurp male power. Of course, these femiphobic fantasies were emblematic of their entire culture. The Academy's pseudoscientific 'evidence' of women's inferiority and of the necessity for their subjugation served the purpose of shoring up patriarchal privilege and power; it also sustained the reasoning by which femiphobic men justified their fear, hostility, and aversion toward the Feminine.

In 1903, shortly after the publication of *Sex and Character*, Weininger took his own life. He was twenty-three. 'The hatred of women,' he had written in his notebook, 'is always only the not yet overcome hatred of one's own sexuality.'[176] As Weininger's reflections on his own sexual ambiguity suggest, the femiphobic man's internal struggle is with his fear and hatred of the Other—the dehumanized Other—in himself and its emasculating presence of which he tries in vain to purge himself. Weininger, it appears, was another Isaac struggling to live up to a standard of masculinity that was beyond him and is in itself pathological.

NOTES

1. Freud, *Freud/Fliess*, 266.
2. Sigmund Freud, "Femininity," *New Introductory Lectures on Psycho-Analysis* (*SE,* vol. 22, 1932), 212.
3. Cixous, "Sorties: Out and Out."
4. Sigmund Freud, "Three Essays on the Theory of Sexuality," (*SE,* vol. 7, 1905), 195.
5. Sigmund Freud, "Some Psychological Consequences of the Anatomical Distinction between the Sexes," (*SE,* vol. 19, 1925), 252–53.
6. Freud, "Some Psychological Consequences," 4.
7. Freud, "Three Essays on the Theory of Sexuality," 22.
8. Sigmund Freud, "Analysis Terminable and Interminable," (*SE,* vol. 23, 1937), 251.
9. Rosalind Minsky, ed. *Psychoanalysis and Gender* (London: Routledge, 1996), 100.
10. Eva Feder Kittay, "Womb Envy: An Explanatory Concept," in *Mothering: Essays in Feminist Theory* ed. Joyce Trebilcot (Totowa, NJ: Rowman and Allanheld, 1984), 95; Daly *Gyn/Ecology*, ch. 1.
11. Robert McElvaine, *Eve's Seed: Biology, the Sexes and the Course of History* (New York: McGraw Hill, 2002), 74.
12. Harold Tarpley, "Vagina Envy in Men," *Journal of the American Academy of Psychoanalysis and Dynamic Psychiatry* 21, (1993).
13. Freud, "Three Essays on the Theory of Sexuality," 195.
14. Freud, "Three Essays on the Theory of Sexuality," 219.
15. Olivier, *Jocasta's Children*, 11.
16. Olivier, *Jocasta's Children*, 11, 13.
17. Olivier, *Jocasta's Children*, 13.
18. Irigaray, *Speculum*, 51.
19. Irigaray, *Speculum*, 51–52.
20. Olivier, *Jocasta's Children*, 14.
21. Olivier, *Jocasta's Children*, 18.
22. Olivier, *Jocasta's Children*, 19.
23. Olivier, *Jocasta's Children*, 21.
24. Kristeva, *Powers of Horror.*
25. Horney, *Feminine Psychology.*
26. Freud, "Some Psychological Consequences," 4.
27. Sigmund Freud, "Female Sexuality," (*SE,* vol. 21, 1931), 7.
28. Horney, *Feminine Psychology*, 62.
29. Freud, "Female Sexuality," 20.
30. Harold Kelman, ed., introduction to *Feminine Psychology* by Karen Horney (New York: W. W. Norton, 1993), 26.
31. Sigmund Freud, "An Outline of Psychoanalysis," (*SE,* vol. 23, 1938), 9.
32. Freud, "Outline of Psychoanalysis," 9.
33. Louis Breger, *Freud: Darkness in The Midst of Vision* (New York: John Wiley, 2000), 439.
34. Melanie Klein, *Love, Guilt and Reparation and Other Works 1921–1945* (New York: Random House, 2011), 190.
35. Cheryl Eschbach, "Toward an Understanding of Womb Envy: Developmental and Clinical Perspectives," in *Jealousy and Envy: New Views about Two Powerful Feelings,* eds. Léon Wurmser and Heidrun Jarass (New York: The Analytic Press, 2008), 50.
36. GPS stands for Global Positioning System.

37. Dorothy Dinnerstein, *The Mermaid and the Minotaur: Sexual Arrangements and Human Malaise* (New York: Other Press, 1999), preface.

38. Toril Moi, "From Femininity to Finitude: Freud, Lacan and Feminism, Again," *Signs, Journal of Women in Culture and Society* 29, 3, (2004): 847.

39. Freud, "Some Psychological Consequences," 7.

40. Bernard Paris, *Karen Horney: A Psychoanalyst's Search for Self-Understanding* (New Haven, CT: Yale University Press, 1994), xv. These colleagues included Adler and Horney.

41. Although Freud disapproved of masturbation, he disagreed with physicians curing it via castration.

42. Sigmund Freud, *Letters of Sigmund Freud* ed. Ernst Freud, trans. Tania and James Stern (New York: Basic Books, 1960), 75–76.

43. Freud, *Letters of Sigmund Freud*, 76.

44. Sigmund Freud, *Cocaine Papers: Sigmund Freud*, ed. Robert Byck (New York: Stonehill, 1974), 10 (letter dated June 2, 1884).

45. Freud, *Cocaine Papers*, 10 (letter dated June 2, 1884).

46. Freud and Breuer, *Studies on Hysteria.* Breuer's Lecture of 1895 is reported in *The Complete Letters of Sigmund Freud to Wilhelm Fliess 1887–1904*, ed. and trans. Jeffrey Masson (Cambridge, MA: Harvard University Press, 1985), p. 151 n.

47. Breger, *Freud: Darkness in The Midst of Vision*, 118.

48. Breger, *Freud: Darkness in The Midst of Vision*, 125.

49. Sigmund Freud and C. G. Jung, *The Freud/Jung Letters: The Correspondence between Sigmund Freud and C. G. Jung*, ed. William Mcguire, trans. Ralph Manheim and R. F. C. Hull (London: The Hogarth Press and Routledge and Kegan Paul, 1977), 400. The *Zentralblatt* was the Society's monthly periodical co-edited by Adler.

50. Freud, *Freud/Jung Letters*, 409.

51. Freud, *Freud/Jung Letters*, 428.

52. Breger, *Freud: Darkness in The Midst of Vision*, 204.

53. Alfred Adler, *Understanding Human Nature: Alfred Adler on the Psychology of Human Personality*, trans. Colin Brett (Oxford: Oneworld Publications, 1998), 46.

54. Adler, "Inferiority Feeling and Masculine Protest," 115.

55. Breger, *Freud: Darkness in The Midst of Vision*, 11.

56. Breger, *Freud: Darkness in The Midst of Vision*, 12.

57. Sigmund Freud, "Interpretation of Dreams" in *SE*, vol. 4 (London: The Hogarth Press, 1900), 583.

58. Breger, *Freud: Darkness in The Midst of Vision*, 14.

59. Breger, *Freud: Darkness in The Midst of Vision*, 15.

60. Breger, *Freud: Darkness in The Midst of Vision*, 15.

61. Breger, *Freud: Darkness in The Midst of Vision*, 16.

62. Breger, *Freud: Darkness in The Midst of Vision*, 16, 291. As an adult, Freud suffered similar intestinal symptoms that coincided with his family's obligatory attendance at Amalia's weekly Sunday lunches. Apart from the symptoms, which persisted until Amalia died, and which Freud always put down to random causes, he always arrived much later than the others.

63. Sigmund Freud, "Femininity," in *New Introductory Lectures on Psycho-Analysis* (*SE*, vol. 22, 1932), 121–123.

64. Breger, *Freud: Darkness in The Midst of Vision*, 168.

65. Breger, *Freud: Darkness in The Midst of Vision*, 168.

66. Breger, *Freud: Darkness in The Midst of Vision*, 24.

67. Breger, *Freud: Darkness in The Midst of Vision*, 33.

68. Breger, *Freud: Darkness in The Midst of Vision*, 168.

69. Breger, *Freud: Darkness in The Midst of Vision*, 19.

70. Breger, *Freud: Darkness in The Midst of Vision*, 19.

71. Freud, "Interpretation of Dreams," 584.

72. Breger, *Freud: Darkness in The Midst of Vision*, 4.

73. Freud, *Letters of Sigmund Freud*, 140–41.

74. Freud, *Letters of Sigmund Freud*, 154 (letter dated June 20, 1885).

75. Sigmund Freud, *The Complete Letters of Sigmund Freud to Wilhelm Fliess 1887–1904*, ed. and trans. Jeffrey Masson (Cambridge, MA: Harvard University Press, 1985), 266.
76. Freud, *Interpretation of Dreams*, 197. Freud tells the story of his father's humiliation and his, Freud's, turning to Hannibal as a heroic father-figure.
77. Breger, *Freud: Darkness in The Midst of Vision*, 27.
78. Breger, *Freud: Darkness in The Midst of Vision*, 160.
79. Sigmund Freud, "Civilisation and its Discontents," (*SE.* vol. 21, 1930), 72.
80. Sigmund Freud, "Recommendations to Physicians Practising Psycho-Analysis," (*SE*, vol. 12, 1912), 113.
81. Freud, "Recommendations to Physicians," 119.
82. Freud, "Femininity."
83. Breger, *Freud: Darkness in The Midst of Vision*, 338.
84. Freud, "Recommendations to Physicians," 115.
85. Freud, "Recommendations to Physicians," 118.
86. Freud and Breuer, *Studies on Hysteria*, 110.
87. Freud and Breuer, *Studies on Hysteria*, 294.
88. Freud and Breuer, *Studies on Hysteria*, 305.
89. Breger, *Freud: Darkness in The Midst of Vision*, 357.
90. Freud and Jung, *Freud/Jung Letters*, 535. This occurred after Freud and Jung's relationship had soured, but before Freud excommunicated him.
91. Breger, *Freud: Darkness in The Midst of Vision*, 352.
92. Rank and Ferenczi were members of Freud's inner circle.
93. Breger, *Freud: Darkness in The Midst of Vision*, 343, 358.
94. Breger, *Freud: Darkness in The Midst of Vision*, 129.
95. Breger, *Freud: Darkness in The Midst of Vision*, 175.
96. Breger, *Freud: Darkness in The Midst of Vision*, 343, 358.
97. Breger, *Freud: Darkness in The Midst of Vision*, 92.
98. Breger, *Freud: Darkness in The Midst of Vision*, 208–209. In 1912, Jones proposed the formation of a clandestine group to act as palace guard to protect Freud and 'defend the cause,' as Freud put it.
99. Ernest Jones, *Sigmund Freud: The Life and Work: The Young Freud 1856–1900* (London: Hogarth Press, 1953), 348.
100. Sigmund Freud and Sándor Ferenczi, *The Correspondence of Sigmund Freud and Sándor Ferenczi*, vol. 1, *1908–1914*, eds. Eva Brabent and Ernst Falzeder, trans. Peter Hoffer (Cambridge, MA: Harvard University Press, 1993), 221.
101. Freud and Ferenczi, *Sigmund Freud/Sándor Ferenczi*, 227.
102. Freud, "Analysis Terminable and Interminable," 250.
103. Freud, "Analysis Terminable and Interminable," 244.
104. Freud, *Letters of Sigmund Freud*, 50–51.
105. Sigmund Freud, *Letters of Sigmund Freud to Eduard Silberstein 1871–1881*, ed. Walter Broehlich, trans. Arnold Pomeranz (Boston, MA: Harvard College, 1990), 40.
106. Sigmund Freud, "Fetishism," *International Journal of Psychoanalysis* 9, (1928): 164.
107. Breger, *Freud: Darkness in The Midst of Vision*, 46.
108. Breger, *Freud: Darkness in The Midst of Vision*, 331.
109. Breger, *Freud: Darkness in The Midst of Vision*, 34.
110. Breger, *Freud: Darkness in The Midst of Vision*, 55.
111. Freud, *Freud/Fliess*, 54.
112. Freud, *Freud/Fliess*, 54.
113. Jones, *Sigmund Freud: The Life and Work*, 154.
114. Freud, *Freud/Fliess*, 447.
115. Breger, *Freud: Darkness in The Midst of Vision*, 332.
116. Breger, *Freud: Darkness in The Midst of Vision*, 338.
117. Breger, *Freud: Darkness in The Midst of Vision*, 367.
118. Breger, *Freud: Darkness in The Midst of Vision*, 347–48.
119. Freud and Breuer, *Studies on Hysteria*, 5–6.

120. Thierry, Bokanowski, "Negative Transference and Transference Depression," in *Ferenczi's Turn in Psychoanalysis,* eds. Peter Rudnytsky, Antal Bókay, and Patrizia Giampieri-Deutsch (New York: New York University Press, 1996), 129.
121. Bokanowski, "Negative Transference," 353, 368.
122. Adler, *Understanding Human Nature*, 231.
123. Freud, "Analysis Terminable and Interminable," 252, 250.
124. Freud and Jung, *Freud/Jung Letters*, 387.
125. Rachel Jones, *Irigaray: Towards a Sexuate Philosophy* (Cambridge: Polity Press, 2011), 137.
126. Jones, *Irigaray: Towards a Sexuate Philosophy*, 75.
127. Plato, *Timaeus* in *Plato in Twelve Volumes,* vol. 9, trans. W.R.M. Lamb (Cambridge, MA: Harvard University Press, 1925), 69 [50d].
128. Plato, *Timaeus*, 70 [51a–b].
129. Irigaray, *Speculum*, 18.
130. Freud, "Moses and Monotheism," 180.
131. Delaney, *Abraham on Trial,* 229.
132. Delaney, *Abraham on Trial,* 229.
133. Delaney, *Abraham on Trial,* 229.
134. Delaney, *Abraham on Trial*, 13.
135. Carol Warren, "Genital Surgeries and Stimulation in Nineteenth-Century Psychiatry," in *Gendered Perspectives on Reproduction and Sexuality*, eds. Marcia Texler Segal, Vasilikie Demos, and Jennie Jacobs Kronenfeld (London: Elsevier, 2004), 168–69; See, e.g., Plato's *Timaeus*, 91a–d.
136. Warren, "Genital Surgeries and Stimulation," 169.
137. Warren, "Genital Surgeries and Stimulation," 170–71; Rachel Maines, *The Technology of Orgasm: 'Hysteria,' the Vibrator, and Women's Sexual Satisfaction* (Baltimore, MD: John Hopkins University Press, 1999), 2.
138. King, "Once Upon a Text," 52, in Warren, "Genital Surgeries and Stimulation," 172.
139. Alain Touwaide, "Galen," in *Medieval Science, Technology, and Medicine: An Encyclopedia*, eds. Thomas Glick, Steven Liversey, and Faith Wallis (New York: Routlege, 2005), 180.
140. Warren, "Genital Surgeries and Stimulation," 192.
141. Janet Beizer, *Ventriloquised Bodies: Narratives of Hysteria in Nineteenth-Century France* (Ithaca, NY: Cornell University Press, 1994), 5.
142. Warren, "Genital Surgeries and Stimulation," 171.
143. Warren, "Genital Surgeries and Stimulation," 171.
144. Warren, "Genital Surgeries and Stimulation," 184, 187, 190.
145. Susan Kingsley Kent, *Sex and Suffrage in Britain 1860–1914* (London: Routledge, 1990).
146. Kent, *Sex and Suffrage*, 44, 47.
147. Gilliam, "Oophorectomy for the Insanity and Epilepsy of the Female," 559–61.
148. Elizabeth Sheehan, "Victorian Clitoridectomy: Isaac Baker Brown and his Harmless Operative Procedure," in *The Gender/Sexuality Reader: Culture, History, Political Economy,* eds. Roger Lancaster and Micaela Di Leonardo (London: Routledge, 1997).
149. Sheehan, "Victorian Clitoridectomy ," 326, 329.
150. Isaac Baker Brown, *On the Curability of Certain Forms of Insanity, Epilepsy, Catalepsy, and Hysteria in Females* (London: Robert Hardwicke, 1866), 61–62.
151. Sheehan, "Victorian Clitoridectomy. "
152. Sheehan, "Victorian Clitoridectomy ," 330.
153. Warren, "Genital Surgeries and Stimulation."
154. Breger, *Freud: Darkness in The Midst of Vision*, 157. To protect Ida's anonymity, Freud changed personal details about her, including adding a year to her actual age.
155. Sigmund Freud, "Three Essays on the Theory of Sexuality," 191.
156. Sigmund Freud and Josef Breuer, *Studies on Hysteria 1893–1895,* trans. James Strachey (New York: Basic Books and London: Hogarth Press, 1895), 110, 294.
157. Todd Dufresne, *Killing Freud: Twentieth-Century Culture and the Death of Psychoanalysis* (London: Continuum, 2003), 169, 173.

158. Breger, *Freud: Darkness in The Midst of Vision*, 158.

159. Maines, *The Technology of Orgasm*, 2; Mark Micale, *Approaching Hysteria: Disease and its Interpretations* (Princeton, NJ: Princeton University Press, 1995), 220. In 1952, the American Psychiatric Association relinquished the term.

160. Sue Davis, *The Political Thought of Elizabeth Cady Stanton: Women's Rights and the American Political Traditions* (New York: New York University Press, 2008), 163; Sophia van Wingerden, *The Women's Movement in Britain 1866–1928* (New York: Palgrave Macmillan, 1999), 9; June Purvis, *Emmeline Pankhurst: A Biography* (London: Routledge, 2002), 67.

161. Kent, *Sex and Suffrage*.

162. Kent, *Sex and Suffrage*, 30.

163. Nye, "Sexuality," 19.

164. Nye, "Sexuality," 19.

165. Möbius, in Ralph Leck, "Anti-Essentialist Feminism Versus Misogynist Sexology in Fin de Siecle Vienna," *Modern Intellectual History* 9, (2012): 43–45.

166. Leck, "Anti-Essentialist Feminism," 33, 37.

167. Leck, "Anti-Essentialist Feminism," 39.

168. Leck, "Anti-Essentialist Feminism," 43.

169. Weininger, *Sex and Character*, 306.

170. Abrahamsen, *Mind and Death of a Genius,* 132.

171. Weininger, *Sex and Character*, 266.

172. Rachel Mesch, *The Hysteric's Revenge: French Women Writers at the Fin de Siècle* (Nashville, TN: Vanderbilt University Press, 2006), 7.

173. Mesch, *The Hysteric's Revenge,* 2.

174. Mesch, *The Hysteric's Revenge,* 4.

175. Mesch, *The Hysteric's Revenge,* 5.

176. Otto Weininger, *Collected Aphorisms, Notebook and Letters to a Friend,* ed. 1.12, trans. Martin Dudaniec and Kevin Solway, (Kevin Solway website, 2002), 71.

Part III

Seed and Soil: Contemporary Ramifications

Chapter Ten

The Contemporary Legacy

We tend to not share our feelings and thoughts as we feel it appears unmasculine and because we fear being ridiculed by our peers and society. Unfortunately, men are portrayed in the media . . . as powerful, solitary and confident heroes.

Eighteen-year-old 'Taylor.'[1]

When I put my hand up [with PTSD], I was basically told to 'f' off. The second time I did it, I was told I was lying. . . . It does stall your career, if not halt it completely, you're less of a person—especially for the guys you're less of a man for owning up to having issues.

Former Australian soldier.[2]

Come the twenty-first century and young men are still struggling. Imagine living in a country where nearly one out of every five young men think their life is not worth living, where more than two in five are experiencing psychological distress, and where, in 2012, nearly one in ten contemplated committing suicide.[3] Suicide is now the main cause of death of Australian men aged between sixteen and twenty-five, followed by motor vehicle accidents, at least some of which—empty straight road, fine weather, broad daylight, no blame—may be masked suicides. Among sexually diverse young people, on average, more than two out of every five have contemplated suicide. Bullying is a key factor: nearly nine out of ten report having experienced homophobic bullying in their lifetime, and the prevalence is increasing. Of the fourteen- to seventeen-year-olds, the same proportion experienced bullying at school.[4] Compared to heterosexual men, young gay men are at much higher risk (by some reports up to fourteen times) of attempting suicide.[5] Australia is not alone; in many European countries, for example, the leading cause of death of young people is suicide.[6] And the rates of anxiety, depression, and self-harm in girls and young women are twice as high as for boys and young men.[7] Over the past decade, Australia's anti-depressant use has doubled and the country is now the second-highest prescriber in the OECD.[8]

Traditional male ideology can devastate love and family life, not least because it subordinates women and despises 'feminine' wishes and behavior in boys and men. The traditional patriarchal nuclear family—typified by the relative omnipresence of women and absence of men—can engender in its children an enduring fear of abandonment or engulfment.[9] Such anxiety can last well into adulthood, corroding love relationships and manifesting in blame and objectification. This chapter argues that analyses of femiphobia as a social phenomenon must encompass both the psychosocial and macro-structural domains. To clarify, this chapter's critique of the traditional nuclear family does not extend to the various forms of non-nuclear family, all of which face differing challenges.

The chapter also examines the link between femiphobia and sexual violence through a survey of recent findings from several major international studies, contemporary literature on traditional male ideology and hegemonic masculinity, anthropological material on an extremely femiphobic culture, and findings from reviews into an admitted culture of abuse in Australia's defense forces. Cross-culturally, the prevalence of sexual violence correlates highly with gender inequality.[10] Hence, in addition to long-term educational and therapeutic strategies, part of the solution lies in designing, implementing, and, above all, enforcing laws aimed at achieving gender equality. That is not to say that law-making constitutes *the* answer, but it is an essential part of the solution.

FEMIPHOBIA, LOVE, AND FAMILY

Patriarchal structures, says psychoanalyst Christiane Olivier, in particular the traditional nuclear family, make of heterosexual love 'the impossible encounter' and condemn its children to an ingrained mistrust of the opposite sex.[11] These configurations trap women and men alike in dysfunctional work patterns that not only cause family breakdown,[12] but also affect children's psychosexual development.[13] While many couples wish to share parenting equally, men's and women's wages are not comparative, part-time work is scarce, and promotional prospects plummet for men who request it.[14] Less because of family-work preferences, then, and more because of the cultural privileging and attendant pressures of men's work,[15] women end up doing most of the parenting regardless of whether they also engage in paid work. More often than not, men work around the office clock while women carers and children move among mostly women-supervised homes and institutions. The low pay and low status associated with childcare and infant teaching discourage men from entering these so-called 'female' professions.

The stay-at-home parent may enjoy being at home yet feel trapped in domesticity, delight in parenthood yet miss adult company. They may also fear their own resentment and next to unilateral power over their children. Feelings of love and anger, says Rich, can exist concurrently, and anger at the cultural conditions of motherhood can translate into anger at the child and fear that we do not love them enough.[16] The mother's grief at what she cannot do for her children in a society that fails to meet human needs may transmogrify into guilt and self-laceration. This powerless responsibility renders her open to charges of moral culpability if society perceives her as having failed her children. The privatization of the home has increased the mother's powerlessness and loneliness. Moreover, says Rich, the mother must deliver her children over to the patriarchal establishment, having prepared them to enter it without rebelliousness or maladjustment and to perpetuate it in adulthood. Patriarchy, Rich says, 'depends on the mother to act as a conservative influence, imprinting future adults with patriarchal values.'[17]

And the children? Together with this broad patriarchal imprinting, says Olivier, the effects of the traditional nuclear family on children skew and undermine their later attempts to create satisfying love lives.[18] The early omnipresence of even the most caring of women (including women teachers) and their wants and demands may leave boys at risk of feeling afraid of being dominated by women, sexually-emotionally vulnerable, and insecure about their capacity to meet women's needs.[19]

And the little girl? Not seeing enough of her father and/or other caring men leaves the girl at risk of feeling somewhat unsatisfied, ill-equipped to satisfy, and discontented with who she is and what she has, including her later love life.[20] In her adult relationships, the theory goes, the woman yearns for recognition from the man who from infancy has learned that he cannot grant her recognition without feeling endangered at some level.[21] Some claim that the less the boy sees of his male caregivers, the more severe are his fear of women and his conflicts around masculinity.[22] And the development of autonomy in girls requires the active presence of male caregivers and other women as well as their mother.[23]

In addition to setting up deep insecurities in both sexes, patriarchal culture trains the girl to put others' needs and wants before her own. And just as society condemns the little boy's identification with the feminine, so it does women's valid response to their subordination. Whereas traditional male ideology lionizes men who express anger as strong and determined, it maligns women who do so as aggressive, weak, and over-emotional. Unsurprisingly, women's suppressed anger emerges unconsciously as disapproval, which in turn triggers men's anxiety about women's wants.

Given these structural obstacles to heterosexual love, Olivier's term 'the impossible encounter' aptly describes the relationship that often ends in divorce—for one in every two American marriages and one in three in Australia.[24] Even without divorce, mutual disappointment is common enough. In findings from a 2011 survey of Australians' satisfaction with their relationships, only 11 percent of respondents said that 'trust and confidence' figured highly and 17 percent that 'We love each other.'[25] 'Friendship and companionship' were nominated by 42 percent of women but only 30 percent of men. Fifteen percent of men said 'compatibility' figured, compared to 8 percent of women.[26] The latter gender-differentiated findings are consistent with other research showing that women initiate divorce twice as often as men do, at least in America.[27] Factors identified by respondents as detracting from relationship were 'lack of time together' (36 percent), 'lack of understanding' (23 percent), and 'lack of communication' (21 percent). Consistent with these figures was a finding about feelings of loneliness. The data indicate that, whereas 37 percent of respondents *not* in a relationship said they 'frequently felt lonely,' 31 percent of people who *were* in a relationship said the same.[28]

Findings from a global study of sexual attitudes and behaviors show that, compared to men, women's satisfaction with their relationships is lower, and that gender equality is the key to sexual-emotional satisfaction for *both* women and men.[29] Among the more gender-egalitarian nations, men's emotional satisfaction was 71 percent compared to women's 63 percent, whereas in the most male-dominated countries, the corresponding figures were 30 percent and 23 percent.[30] In countries with higher gender equality, men and women were more than twice as satisfied as those in extreme patriarchal societies (although across the board women were less satisfied than men were).

Because patriarchal structures have been naturalized into invisibility, men and women often find themselves turning their fears, frustrations, and disappointments against themselves, their partners, and their children. The high divorce statistics, then, are unsurprising. And in many affluent and welfare-state countries, more and more adults are living alone.[31] More than 50 percent of U.S. adults are single. Single-person households make up between 40 and 50 percent of households in prosperous U.S. cities, 50 percent in Paris, and 60 percent in Stockholm.[32] 35 percent of Australian households contain no more than one adult.[33]

At the interpersonal level, how might the impossible encounter be resolved? When the man transcends his initial mistrust of women and offers his partner recognition and regard as a person in her own right, says Olivier, she will feel less needy and more disposed to ask for what she needs in a trusting rather than demanding manner.[34] When the woman shows her partner she trusts his goodwill, he will feel less threatened and, hence, less driven to control her.[35] And at the familial and societal levels? When mothers step back and fathers step in, their sons, having been less subject to the power of women, will feel less need to defend themselves against women and consequently be more willing to meet their needs.[36]

At the structural level, when men make up 50 percent of the teaching and childcare professions and when women and men share parenting equally, the new generation will be better equipped to trust each other.[37] The developmental problem for boys, says Ducat, is not so much the proximity to the mother as the distance from the father.[38] Hence, the solution is for the boy to develop secure attachments to and identification with both parents (or salient caregivers).[39] Extend the apron strings to fathers.

The handful of more egalitarian cultures of which we know, says Ducat, can teach us much about the raising of children. Among both the Batek of Malaysia and the Aka Pygmies, men and women treat each other as equal partners, and fathers have extensive involvement in the primary care of children.[40] Boys in such circumstances may grow up feeling less destructively envious of women and more able to embrace identification with both parents. They may be less fearful and disparaging of women and the internal Feminine and less inclined to act out in hypermasculine ways.[41] Speaking of which, there is now longitudinal evidence that testosterone levels in men who actively engage in fathering can drop by up to one-third.[42] Other cross-cultural comparisons of pre-state societies show that the level of male childcare is directly related to the status of women and that the causal loops run both ways.[43]

In summary, we must address the harmful effects of traditional gender enculturation at the psychosocial level—including helping to pick up the pieces in the lives of men and women who have suffered. At the structural level, given that relationship satisfaction directly correlates with gender equality, we need a political agenda targeted at achieving gender equality at all levels.

GENDERED VIOLENCE

What is the role of femiphobia in gendered violence? Whereas the literature has much to say about the etiology of femiphobia,[44] the association between femiphobia and gendered violence is less explored. The following sections are concerned with male-on-female and male-on-male violence. Femiphobia can be a major factor in male-on-male violence; the femiphobic perpetrator's motivation is both to attack the internal Feminine—which is projected onto the victim—and to force the victim into what the perpetrator perceives as the feminine position. In gender-political terms, femiphobia is an expression of fear of loss of masculine status, power, and privilege; in depth-psychology terms, it is an expression of unconscious hatred and dread of one's inner feminine aspects. The femiphobic man projects the disowned aspects of his personality out onto the world and experiences them as external threats.

I have argued that traditional male ideology enculturates boys in ways that put them at risk of psychological and physical harm and, hence, so, too, the women around them and the societies in which they live. As femiphobia is readily transformed into anger at women and the desire to punish them, some femiphobic men become perpetrators of sexual violence. Physical and sexual violence against women (which is predominantly committed by intimate partners), says the World Health Organization (WHO), is a global public health problem of epidemic proportions requiring urgent action.[45] Regarding women's lifetime exposure to physical or sexual violence, the WHO's regional prevalence figures are: 33 percent in high-income countries, 36 percent in the Eastern Mediterranean, 40 percent in South-East Asia, 46 percent in Africa, and between 60 and 68 percent in the Western Pacific.[46] Nearly 40 percent of all murders of women are committed by intimate partners.[47]

What drives men's violence against women, particularly their intimate partners? Can we study its etiology and dynamics in the interests of prevention and cessation without excusing or ignoring male violence? Many believe so. The UN and the WHO's preventive strategies, for example, engage perpetrators directly in dialogue.[48] Traditional male ideology constructs many men to experience an intimate partner's desire to separate from them as a shameful feminization and an assault on their entire self-worth. Because these men project the internal threat externally onto women, they may retaliate against their partner in an attack driven by humiliated fury.[49] Among the most chilling findings about intimate-partner violence is that, although leaving a violent relationship is the only way for women to stop the abuse, it is also the riskiest thing they can do. For 80 percent of the intimate partners of Canadian men who injure them, for example, the men inflicted the harm *after* their partner had left.[50]

Kimmel's analysis of pornography and male sexuality targets the dialectical relationship between men's sexual violence and their feelings of power and powerlessness.[51] Kimmel homes in on a central paradox: men control virtually all the economic, political, and social institutions of society and yet many feel powerless. Many, in addition, are angry at women, who they perceive as having the power to arouse them and to give or withhold sex.[52] Men can perceive women's beauty and sexuality, says Kimmel, as so injurious that they can fantasize about raping the women in retaliation.[53] Kimmel's hypothesis is consistent with findings from Nicholas Groth's forensic studies of rape perpetrators. While there was no denying the sexual element in rape, after interviewing more than five hundred offenders, Groth concluded that rape is the sexual expression of power and anger, more retaliatory and compensatory than sexual.[54]

One way to think about rape is as the perpetrator's attempt to vanquish his internal Feminine by projecting it onto women or enemy males. Through unconscious fantasies of overpowering and destroying the other, the perpetrator can imagine he is subjugating and annihilating the internal Feminine.[55] Another way to think about sexual violence is via Horney's 1932 account—certainly dated and somewhat over-reaching—of the wounding of the little boy's masculine self-regard when he understands that his mother would reject and deride him as inadequate as a sexual partner.[56] In Horney's view, the boy's phallic impulses to penetrate can merge with his anger and frustration and result in these impulses taking on a sadistic tinge.[57] If the destructive impulses are considerable, the mother's genital becomes an object of direct anxiety.[58] Horney grapples with the likely consequences. This anxiety, says Horney, neither occurs in all men nor is a distinctive characteristic of men's relation to women, yet it leaves traces in every man that either do not exist in women's attitude to men or, if they do, are acquired secondarily. In men analysands, she says, the dread of being rejected and derided is a typical ingredient no matter what the man's mentality or neurosis.[59] Horney's theory, while untestable, is consistent with Kimmel's and Groth's findings.

Findings from two recent UN studies also echo Groth's assessment of motivations for rape. A 2012 household survey of two thousand women and one thousand men aged sixteen to forty-nine across New Delhi found that 51 percent of men admitted to perpetrating sexual harassment or violence against women in public spaces.[60] Blaming women and wanting to punish them featured prominently in the men's thinking; for example, 40 percent agreed that 'Women moving around at night deserve to be sexually harassed,' and 75 percent that 'Women provoke men by the way they dress.' Ninety-five percent of women said they felt at risk of sexual violence in public spaces and 73 percent said they felt unsafe all of the time even in their own surroundings.[61]

A much larger quantitative epidemiological study used standardized household surveys to measure the prevalence of sexual violence in nine sites across six Asia-Pacific countries: Bangladesh, Cambodia, China, Indonesia, Papua New Guinea, and Sri Lanka.[62] Based on theories of masculinities, the United Nations Change Project aimed to better understand the connections between men, gender, power, and violence.[63] Researchers interviewed 10,178 men aged eighteen to forty-nine across the nine sites and surveyed a total of 3,106 women for validation purposes.[64] To the most sensitive questions, the men entered their answers anonymously on audio-enhanced personal digital assistants.[65] Under-reporting appeared to be relatively low (the women's reports generally were consistent with the men's).[66] For similar reasons, over-reporting is probably minimal, although it is not mentioned in the report.

Fifty percent of the men reported having used violence against an intimate partner; the figures varied across sites from 26 percent in Indonesia-rural to 80 percent in PNG-Bougainville, with most sites reporting 30 to 57 percent. One in every four men admitted to having raped an intimate partner; figures varied from 10 percent in Bangladesh-urban to 62 percent in PNG-Bougainville.[67] Of the men who admitted to raping girls or women, more than half said they started raping in their teens.[68] On average, one in every twenty-five men surveyed said they had perpetrated gang rape.

Unsurprisingly, the Change Project found that the prevalence of sexual violence was inversely related to levels of gender equality. As to the perpetrators' motivations, nearly 50 percent said they rape women 'for entertainment' or out of 'boredom.' Around one-third said they rape out of anger and the desire to punish. Most disturbingly, almost three-quarters claimed rape as men's prerogative.[69] Men, they said, are entitled to take what is rightfully theirs—women's bodies—regardless of consent. The respective figures for China and rural Bangladesh were 86 and 82 percent.[70] Raping women, the perpetrators said, showed other men you are man enough to take what you are entitled to. There are resonances here with a separate finding that Australian men who kill their partners do so mainly out of possessiveness.[71]

The idea that men 'prove their manhood' by raping women is central to the problem and also points to its source: the way boys are enculturated into manhood. Nearly 90 percent of the entire sample said that 'to be a man you need to be tough.' The finding supports theories of hegemonic masculinity, as does the evidence that impunity is a major problem:[72] 72 to 97 percent of perpetrators reported experiencing no legal consequences. In addition, marital rape is not criminalized in many countries.[73]

The findings bear out the project's main hypothesis that violence against women is a manifestation of unequal gender relations and harmful aspects of hegemonic masculinity associated with patriarchal beliefs, institutions, and systems.[74] Gendered violence, the researchers found, is associated with a complex interplay of intrapersonal, interpersonal, community, and broader societal factors to do with pervasive gender inequality.[75] For example, one interviewee described a 'good marriage' as 'a loving, caring, devoted and dedicated wife who can always put the husband's family interest on top of anything else. . . . If I am angry, I prefer to teach her an instant lesson. Although I sometimes feel bad about my conduct, it's not a big deal. If she disobeys, she must be punished. That is not wrong at all.'[76]

One critical factor in sexual violence is the effect on men of other men's ridicule and hostility.[77] Many of the peer-pressured men who refused to join in gang rape said they were mocked for their reluctance. 'You are not a real man; you are gay,' is the usual taunt, said one anti-violence educator. 'This is the part that pushes the man [into violence].'[78]

'Above all else,' said the primary investigator, the men who raped 'wanted to fit in with their friends even if it was at the expense of women': 'We are good friends,' said one man she interviewed, 'so we want to have sex together with one girl.' They show off to each other, said another researcher, about who can have sex the longest.[79] Men who perpetrate gang rape are more likely than other men to have been victimized in the past for 'failing' one or other test of manhood, for example, masculine heterosexual performance, dominating women, or perfecting emotional disconnection from sex.[80] In perpetrating gang rape the men involved are bonding with each other, collectively reassuring themselves that they are 'real men.'[81]

In a major European Union project on physical and sexual violence, female interviewers conducted face-to-face interviews with a random sample of 42,000 women aged eighteen to seventy-four years in the general population across the twenty-eight EU member states. One in three women reported experiencing some form of sexual violence since the age of fifteen, and one in twenty said they had been raped. More than 20 percent reported experiencing physical and/or sexual violence from either a current or previous partner, yet less than 15 percent reported their most serious incident to police.[82]

FEMIPHOBIA AND TRADITIONAL MALE IDEOLOGY

Phallocentric cultures, says Bourdieu, use gang rape as one of several testing grounds for 'manliness.' Such practices, he says, test men to prove their virility in the presence of other men, which demonstrates their dependence on the judgment of the male group.[83]

Of his 1960s study of the Kabyle, Bourdieu noted that the Kabyle man's principal duty toward himself is to invest in the social games that make him a 'real' man, that enable him to live up to, in his own eyes, a certain idea of manhood. A 'real' man must rise to the challenge of any opportunity 'to increase his honor by pursuing glory and distinction in the public sphere.' Manliness 'must be validated by other men, in its reality as actual or potential violence, and certified by recognition of membership of the group of "real men."'[84] While some scholars have criticized Bourdieu's model for implying a self-replicating system rather than one open to transformation, the model does highlight the centrality of men's policing of other men's behavior.[85] On the same theme, in the hypermasculinist sphere of military training, says Ben Wadham, the fratriarchal process constructs the cadet's military subjectivity as in opposition to his prior civilian subjectivity. This dividing practice, says Wadham, constructs the 'real man' as self-reliant, stoic, and strong, but also detached from others and untouched by vulnerability.[86]

Men learn that to 'fail' one or other tests of manhood is to feel feminized—in one's own and other men's eyes—and hence excruciatingly vulnerable to exclusion from the group with which they identify. Understandably, then, most men are femiphobic insofar as they fear that other men will judge them as less than a 'real man.' What men call 'courage,' says Bourdieu, is often rooted in a kind of 'cowardice' that arises from 'the fear of losing the respect or admiration of the group, of "losing face" in front of one's "mates" and being relegated to the typically female category of "wimps", "girlies", "fairies."' In circumstances when 'losing face' may invite exclusion and/or violence, however, the word 'cowardice' seems unfair.[87] Manliness, says Bourdieu, 'is an eminently relational notion constructed in front of and for other men and constructed against femininity in a kind of *fear* of the female, firstly in oneself.'[88]

In terms of depth psychology, what the femiphobic man fears most is the internal Feminine,[89] which he imagines could engulf him and destroy what is left of his masculinity, his very identity. It is the fear of ego dissolution that drives the flight from the Feminine, says Klaus Theweleit.[90] I would add that the justifiable fear of marginalization and violation could be equally potent. Unconsciously, the man projects these threats onto women, objectifying them as a group and framing them as harboring an extraordinary desire and capacity to harm men. 'That other sex, the feminine,' says Kristeva, 'becomes synonymous with a radical evil that is to be suppressed.'[91] In compensation, the man must also see women as docile, weak, and inferior. Bourdieu describes this paradox as the exultation of masculine values shadowed by the fears and anxiety aroused by femininity. In this view, women embody weakness, the vulnerability of honor, and exposure to offence. Yet women are also strong, 'armed with the weapons of weakness such as devilish cunning . . . and magic.'[92]

Abuse of women occurs in all male-dominated cultures, but the Change Project demonstrated unequivocably that sexual violence correlates with gender inequality. And military conflict and its aftermath appear to exacerbate the problem. The rates of nonpartner rape, for example, were much higher in the two postconflict nations studied. In the worst case, two out of every five men reported having perpetrated nonpartner rape, which is more than four times the averages in the nonconflict countries.[93]

How are femiphobia and the militaristic mindset related? The political aim of ritual defeminization and masculinization has been to transform boys into fighters as much as into fathers.[94] And as human beings typically are reluctant to take the lives of fellow humans, military training aims to break the individual along with his moral sensibilities and rebuild him as an effective cog in a killing machine.[95] Contemporary Western military training employs femiphobia—among other hypermasculine attributes—to effect this transformation in a matter of months. For the Sambia of eastern Papua New Guinea, however, the process takes more than a decade and is particularly bloody.[96]

As in imperial Greece, Rome, and Britain, warfare is the central organizing force of Sambia culture. Given the temporal, geographical, and cultural chasms between these societies, is any meaningful comparison of their supposed gender norms possible? Without denying the differences within and between them, given the global prevalence of institutionalized patriarchy, militarism, and fear and envy of women, informed comparison is essential. But why include the Sambia? While contemporary Western femiphobia is largely disguised and denied, says Ducat, the developmental and cultural roots of femiphobia among the Sambia are 'compellingly transparent.'[97] Across Papua New Guinea, the prevalence figures for wife-beating range from 60 percent of urban men to 100 percent of men in some rural areas. In more than half of all cases, the men used weapons such as axes, spears, and machetes to disfigure their wives' faces or to amputate their ears, hands, feet, or entire limbs.[98]

Sambia society is male-dominated in most respects. It constructs Woman as polluting, untrustworthy, and dangerous. Fear of women's pollution—which supposedly also infects children—means that children have little contact with their own father or with older boys and men. The Sambia construct menstruation as having toxic effects; hence cohabiting men and women live in separate areas of the house, neither eat nor sleep together, and segregate their eating utensils.[99] There are separate men's and women's pathways through the forest. The men blame all misfortune on women and often beat them—sometimes to death. Suicides among women outnumber those among men by three to one, a remarkable inversion of figures in the West. Kinship is patrilineal, marriages are either prearranged or with women abducted from other tribes, and married couples live with the husband's family.

Just surviving the Sambia's physical environment is grueling, says Robert Stoller.[100] It takes a lot of luck, fierce masculinity, and a steady heterosexual impulse to survive constant war, treacherous terrain, awful weather, severe protein deficiency, strong potential for starvation, and acceptance of killing other humans.[101] Yet the route to this survival is via overt, obligatory, institutionalized homoeroticism. The most secret Sambia male myth about the founding of the cosmos is that originally humans were hermaphroditic and they required repeated insemination to achieve the cultural state of female and male.[102]

Whereas girls develop into women naturally, say the Sambia, boys require extraordinary intervention to assist their development into men. When boys are between seven and ten years of age, the men take them from their mothers, by force if necessary, and subject them to a series of painful initiation rituals that continue for ten to fifteen years, and forbid them, on pain of death, from contact with females including their mother and sisters. Tradition requires the coercion of all males into exclusively homosexual—homosocial and homoerotic—activities from childhood until they are married.[103] If boys fail to injest semen from an outside source—in this case, ritual fellation of adult men—they won't mature physically or be capable of sexual reproduction.[104] The ritualized homosexuality of the Sambia is by no means unique among Melanesian societies; anal penetration of boys by adult men is the preferred method of insemination in some communities.[105]

Sambia males react to the premature, violent, and traumatic rupture of their maternal cocoon and the demand to repudiate everything feminine with an unconscious identification with the feminine. They refer to semen as 'male milk.' The men say that after intercourse the ejaculate travels to the woman's breasts where it becomes milk for babies.[106] After intercourse with women, during initiation, and every month when their wives menstruate, the men insert razor-sharp leaves into their nostrils, bleeding profusely in order to flush out the dangerous female poisons they have absorbed. They refer to this bleeding as 'male menstruation.'[107] According to the Sambia, the fluids of the body—mother's milk, semen, and menstrual fluid in particular—have the power to restore life or, if wrongly displaced, to destroy it.[108]

Arguably, Sambia gender norms have elements in common with the ancient Western tradition whereby no man who has achieved manhood may properly adopt the so-called passive, inferior, feminine sexual role. The community ridicules adult men who do so and treats them as aberrant.[109] Among the Sambia and to some degree in the contemporary West, the psychological (and often physical) absence of fathers contributes to boys' sense of masculinity as a fragile and threatened category. In both cultures, the hypermasculinist acting out of anxious males includes subordinating women and dominating other men.[110] In addition, Sambia customs of warfare resonate with the ancient Athenian military tradition of maintaining morale and motivation to fight via homosexual pair-bonding, which, says Kathleen Kuiperp, maintained the normal fratriarchal bonds at 'ferocious intensity.'[111]

Just as the Sambia appropriate women's reproductive capacity via 'male menstruation,' some Australian Aboriginal tribes do so via penile introcision; they speak of the introcized penis as having a 'vulva' or 'slit.'[112] In addition, among U.S. fathers-to-be, researchers identified the 'couvade' phenomenon (by which the husbands of pregnant women appropriate their wives' symptoms of pregnancy, labor, and parturition). The men who suffered pregnancy-like symptoms, said the researchers, were more likely than the matched asymptomatic group to be more consciously hypermasculine, less egalitarian in the home, and to have had fathers who were absent or emotionally distant.[113]

Numerous cross-cultural studies show that the existence of male initiation rites correlates with gender inequality and segregation, the disparagement of women, and the physical and psychological absence of fathers from children's care.[114] Echoing Sambia boys' masculinization via painful and humiliating ritual, the West has its own more or less forceful traditions, some of which—military training, for example—typically include physically painful and degrading practices of hazing or bastardization.[115] All such traditions have it that boys must be defeminized and demothered (the apron strings cut) as well as intentionally masculinized in order to become men and warriors—whether for King and country or for God. These traditions marginalize women and use the abjected feminine as a tool in the forging of bonds—both tender and punitive—among men.

Official inquiry into an alleged culture of abuse in the Australian Defense Forces (ADF) gained traction in 2011 when Stephen Smith, then defense minister, instigated no less than six reviews. By June 2013, the number of serious complaints of abuse was nearing 2,500.[116] Before the 2011 reviews, the ADF had claimed that the alleged abuse was no worse than in any other organization,[117] and that the training academy of the ADF had 'a remarkably lower rate of sexual harassment and sex crimes than any other university in Australia.'[118] However, a 2005 Australian Senate review of military justice had identified an ADF culture of silence about abuse.[119] *The Age* newspaper reported that it was business as usual for male cadets in the Australian Navy to refer to their women colleagues as 'Abortion' and 'Squid': 'Abortion' meaning 'someone of no value to human life' and 'Squid,' that women 'are flabby, smell of fish, are easy to get into, enfold you with tentacles, and squeeze the moral life out of you.'[120]

'I was called things like ho, front bum, whore, slut, useless sheila and the word starting with "c,"' said one woman of her Australian Army male colleagues serving in East Timor. 'There were times when I was threatened with physical violence.' Another woman soldier still fears reprisal for speaking out: 'I went in to the army as a trusting person . . . I found men there who were woman-haters.' She feared she might be shot in the back. 'I really felt that hated. . . . Most of it was just because I was a girl.'[121]

Some male cadets spied on their fellow women cadets by drilling holes in the ceilings of the women's bedrooms and bathrooms or installing hidden cameras.[122] In addition, serving and former members of the ADF posted hundreds of sexist, racist, and homophobic messages on a closed site used by more than a thousand members.[123] Among the reportedly less odious comments were:

> All women are filthy, lying whores.
> Fuck fodder for the enemy . . . leave female soldier behind wearing bikini.
> Vote 1, ugly chicks in every digger's gunpit!!
> Don't worry about what the enemy will do to them. After a few weeks in the bush, most blokes'd fuck a black snake with a festered arse!!
> If she's half good looking with big tits give's ya something to keep you awake on piquet duty, just hope she's not a noisy fuck or the enemy will know where you are.

After on-air criticism of the ADF Facebook group, an anonymous member sent a text message to the woman critic: 'Just a quick txt to let you know what an aenemic [*sic*] looking asshole you are. cheers.' Other members posted threats against the whistleblower: 'Mystery leak man, pray we never find out who you are you gutless fuck. Enjoy the rest of your life looking over your shoulder wondering if any of us have worked out who you are.' 'Whoever the jack Cunt is that ratted out this site, you'd better hope we don't find out who you are!!'[124] On discovering the identity of the whistleblower, members subjected him to threats and intimidation. 'I had committed treason of the highest order,' said the whistleblower. 'I had betrayed the rule of brothers and broken the principal law of male groups—silence.'[125]

In 2008, a former army officer presented top brass with a statement saying a large number of cadets and senior ADFA officers were engaging in assaults on cadets. The officer said he regretted not trying to stop an attack on 'X,' a fellow cadet who had raised concerns about bastardization at the academy, but that he was 'too afraid of being subjected to the same treatment.' A group of cadets bashed and sexually assaulted X. 'If you fuck with us we will fuck you up the arse you jack faggot,' they said as they shoved a broom handle into his rectum. At the academy, said X, 'it was like running with a pack of wolves. As soon as you are sick or injured they turn on you and eat you.'[126]

However, in the face of dozens of similar reports, the Defense Association director adopted a reassuring posture, not to mention one that afforded a certain degree of plausible deniability: 'I had several golden rules and one was you never did anything to anyone else that wasn't done to you. There had to be a military point to it and the person who was on the receiving end had to see the humour of it. And if you couldn't fulfil those three golden rules, you didn't do anything to anyone. Not everyone followed, ah, those type of golden rules.'[127]

The problems identified by the media, he said, are 'just the odd isolated incident, *not* the result of a sick culture.'[128] And the army chief, Lieutenant-General David Morrison, continued to blame 'human nature' for the ongoing sex-abuse scandals. Will the ADF's plan to bring in more women and recruits from minority groups 'negate any unfortunate incidents in the future?' he said. 'Unlikely; human nature is human nature, irrespective of whether it's lived in uniform or without.'[129]

But twelve months later, Morrison had changed his tune. In response to yet another spate of allegations of misogynous online activity among long-serving army men, Morrison adopted a harder line. 'The leadership of the ADF,' he said, 'no longer accepts the "bad apple" argument. . . . These behaviors are symptoms of a systemic problem.'[130] Morrison addressed his colleagues in a televised message:

> Those who think that it is OK to behave in a way that demeans or exploits their colleagues have no place in this army. . . . On all operations female soldiers and officers have proven themselves worthy of the best traditions of the Australian Army. They are vital to us maintaining our capability now and into the future. If that does not suit you, then get out. . . . The same goes for those who think that toughness is built on humiliating others.[131]

A Defense review headed by Australian law firm DLA Piper documented allegations of vicious beatings and pack rapes in the ADF that had received the supervisory nod.[132] Not only do the allegations 'probably represent a relatively small proportion' of past abuse in Defense, says the review, but Defense's military justice system has been 'largely ineffective' when it comes to serious cases of abuse, particularly sexual abuse, and 'has directly contributed to under-reporting.'[133] The earliest documented allegation is of the rape of a thirteen-year-old naval recruit in 1951. Problems have recurred, says DLA Piper, because in Defense there are 'no adverse consequences for abusive behavior'; the 'culture [discourages] reporting of abuse'; there is an 'absence of positive support for people who report'; and 'the chain of command structure lends itself to superiors abusing juniors with impunity.'[134] Young abused men often became abusers themselves: 'The choice was to join in bashing and assaults on other boys or young males or to continue to be the target of such abuse.'[135] Two years after the publication of the review, the head of the review team reported Defense's continuing reluctance to publicly encourage cadets to report abuse. Up to one-third of women cadets are likely to have been raped or abused, he said, but many are disinclined to report it.[136]

Australia is not the only nation with an endemic problem of military abuse, but in the United States, for example, it is the subject of a considerable body of research and scholarly critique.[137] As to the likely higher prevalence of sexual and other abuse in nations involved in conflict,[138] the U.S. military presence in Iraq and Afghanistan is a case in point. Violent sex crimes committed by U.S. army personnel, says an army report, have nearly doubled since 2006, a rate far above that of the general population.[139] Most of the crimes occurred on U.S. soil. Chronic under-reporting of sexual assault, however, means that the actual number of assaults is reportedly probably more than six times higher.[140] In addition, child abuse cases involving army personnel rose by 43 percent and domestic abuse by roughly one-third. The rise is partly attributed to stress suffered by soldiers undergoing multiple deployments. A person diagnosed with Post-Traumatic Stress Disorder is reportedly three times more likely to participate in some kind of intimate-partner aggression.[141]

PATRIARCHAL TRADITIONS AS TRAUMATOGENIC

Some researchers have it that patriarchal traditions per se are traumatogenic.[142] Why traumatogenic rather than traumatizing? Because, says Greg Forter, these institutions 'embed the *potential* for the repetition and transmission of their injuries—not the *necessity* of such.'[143] And under the 'good woman' stereotype, women are expected to patch up the physical and psychological damage sustained by their boys and men in the outer world. For men, says Rotundo, 'the social fabric was torn every day in the world and mended every night at home.' 'Men's sphere depleted virtue, women's sphere renewed it.'[144]

We may consider patriarchy itself to be a normatively traumatogenic institution, says Forter.[145] Patriarchal identity formation can decisively deform the psyche and manifest in compulsive, rigidified ways of relating to others.[146] Such traumas, he says, are emphatically social disturbances; they are chronic, cumulative, and woven into the social fabric. As such, they cannot count in the 'punctual' version of trauma theory used to analyze historical events of such singularity, magnitude, and horror as the Holocaust.[147] Over millennia, says Forter, societies have sublimated the original violence of patriarchal subjugation into systemic practices, the effects of which are thoroughly naturalized. As such, we must estrange them in order to see them *as* social traumas.[148]

In order to understand these systemic traumatizations, Forter uses Freud's early model of trauma. In Freud's view, hysteria is caused by traumatic experiences that have not been fully integrated into the personality and, hence, resurface as bodily symptoms undecipherable by consciousness.[149] In other words, the material remains unconscious and the symptoms give voice to the suffering. But Freud abandoned this early model, says Forter, because it evoked in him a painful awareness of structures of domination and social violence.[150] For Freud, that knowledge 'concerned the misogyny and femiphobia lodged at the heart of his own masculinity.' He escaped it, Forter says, by developing a second theory that cancelled the force of his own critique and absolved him of historical guilt 'by tracing all human misery . . . to a non-historical or *structural* cause,' namely, the death drive.[151]

Research on systematic traumatization suggests that stereotypical gender patterns arise directly and indirectly out of trauma.[152] What is the nature of these traumatic experiences? 80 percent of U.S. boys, for example, reported having experienced sexual abuse (mostly out of the home). 50 percent reported having been violently victimized. One-third of boys said they had suffered aggravated or simple assault; an additional 15 percent said the assailant had targeted their genitals. 20 percent of navy recruits reported having been physically abused as children.[153] More boys than girls reported suffering 'ruptures of attachment' involving feelings of deprivation and fears of isolation. The prevalence of male childhood sexual abuse varies considerably across cultures: for example, averaging one-third of men in Papua New Guinea and urban Bangladesh and ten to fifteen percent in Australia (compared to 25 percent of women). In one Australian state, 25 percent of the male prison population aged eighteen to twenty-five reported having been sexually abused during their incarceration. Boys are much more likely than girls to be abused by clergy. In another Australian state, for example, 87 percent of more than 500 victims of investigated offences by clergy and church workers were boys aged ten to eleven.[154] Furthermore, compared to men who report not having been abused as boys, abused boys are four times at risk of using injectible drugs.[155]

Findings from a study of American victims of sexual abuse showed that, compared with civilian victims, military victims suffered more serious long-term effects including dissociative disorder, personality disorders, eating disorders, anxiety and alcohol disorders, and symptoms of post-traumatic stress disorder. Rates of self-harm and attempted suicide were double those among nonabused veterans.[156]

Men have long been disproportionately represented among problem populations such as perpetrators of violence, victims of suicide and fatal automobile accidents, sex addicts and sex offenders, the homeless, substance abusers, parents estranged from their children, fatal victims of stress-related illness, and more.[157] Researchers studying men and masculinity in the 1980s formulated social constructionist theories around 'gender ideology,' that is, the internalization of cultural belief systems regarding gender role norms, and 'gender role strain,' the idea that certain traditionally prescribed gender role traits (such as male aggression) are dysfunctional. Violation of gender norms, said the researchers, has more severe consequences for males than for females.[158] Dominant male gender role ideology, said Pleck and O'Neil, is learned from patriarchal structures, including families and schools, and is experienced unconsciously as fear of femininity.[159] While some scholars criticized studies based on instruments such as the Male Role Norms Inventory for its use of trait-oriented terminology,[160] the results generally are consistent with predictions based on hierarchical-masculinity models such as Connell's.[161]

O'Neil, for example, defined 'gender role conflict' as a psychological state in which socialized gender roles cause restriction, devaluation, or violation of others or self via problems in the cognitive, affective, behavioral, and unconscious domains.[162] Conflict-inducing male-gender-role patterns include restrictive emotionality, health care problems, obsession with achievement and success, restrictive sexual and emotional behavior, socialized control, power and competition issues, and homophobia.[163] A review of more than 230 empirical studies to 2007 showed that such patterns significantly correlate with a vast range of intrapersonal and interpersonal problems, among them depression, anxiety, low self-esteem, shame, alexithymia (inability to articulate feelings), substance abuse, suicidality, dysfunctional relationship patterns (including notions of male entitlement), hostile sexism, and self-reported sexual violence. Not only is men's gender role conflict associated with a broad-ranging series of harms to the men themselves, but also with abusive and violent thoughts, attitudes, and behaviors toward women.[164] Findings from the Change Project reinforce O'Neil's conclusions.[165] And the WHO recently concluded that the problem of violence against women is globally pervasive and contributes significantly to women's ill-health.[166]

The WHO's global report and the Change Project are firsts of their kind.[167] The Change Project, for example, asked men *why* they commit sexual violence. Typically, the mainstream media construct rape as primarily sexually motivated (aided by alcohol) rather than, as Groth demonstrated in the 1970s, motivated by both hostility toward women and a traditional male ideology of dominance and entitlement.[168] Did Groth's findings receive inadequate recognition because of commentators' misguided belief that trying to make sense of criminal behavior is the same as condoning it? Some say that involving men in analyzing the problem either excuses the perpetrator or blames the victim, but neither is the case. Part of resolving the problem is in working toward perpetrators accepting responsibility for their crimes.

Furthermore, the WHO noted that studies prior to its 2013 global report had not differentiated between violence committed by intimate partners and by nonpartners[169]: 'Many countries had not collected population-based data on intimate-partner violence or nonpartner violence.'[170] Such omissions are consistent with the policing and judicial tradition of treating intimate-partner violence as relatively trivial. In Australia, for example, one judge proclaimed that 'no' can mean 'yes' and that husbands have the right to apply 'rougher than usual handling' to noncomplying wives.[171]

A related problem in U.S. law enforcement systems in the recent past was the separation of homicide data from family-violence data, which stymied investigations into whether a given killing had a family-violence history.[172] And while some counties are now keeping count of family-violence charges, the figures are underestimates due to inadequate record systems and to underreporting—particularly by women wary of retaliation. In some Australian jurisdictions, the crime statistics include only *prosecuted* cases,[173] hence excluding murder-suicides, of which more than 90 percent are committed by men.[174]

Are there promising directions for change? At the interpersonal and social level, say prominent anti-violence experts, the prevalence of rape, family violence, and sexual harassment is likely to decrease markedly when abusers start to lose status among their peers.[175] At the structural level, even small changes may have far-reaching effects, for example, mandating the collection of gender-relevant data about wages paid to employees, hence exposing hitherto shrouded gendered pay gaps.[176]

At the military-institutional level in Australia, the Piper Review is unequivocal about what it will take after decades of futile attempts to overhaul the ADF's culture of abuse: no less than 'proactive support to those in the ADF who have the courage to stand up for what is right when others in the ADF do wrong.'[177] A noteworthy shift appears to have occurred—during a year or so in the sights of a remarkably persevering defense minister and associated inquiries—in the heart and mind of army chief Morrison ('human nature is human nature, irrespective of whether it's lived in uniform or without'). In the aftermath of yet another scandal about army men degrading and exploiting women, the apparently reconstructed Morrison addressed his colleagues:

> Every one of us is responsible for the culture and reputation of our army and the environment in which we work. . . . If you become aware of any individual degrading another, then show moral courage and take a stand against it. . . . The standard you walk past is the standard you accept. That goes for all of us, but especially those who by their rank have a leadership role. . . . It is up to us to make a difference. If you're not up to it, find something else to do with your life. There is no place for you among this band of brothers and sisters.[178]

What prompted Morrison's about-turn on the gender question? We may never know the inside story, but the public one is telling. Defense Minister Smith placed Morrison, together with other military leaders, under sustained pressure. The government-instigated inquiries found evidence of decades of endemic serious abuse in the ADF,[179] and the typically quiescent mainstream media aired allegations of the ADF's lack of care for the victims. The military establishment struck back at Smith, who 'suffered a concerted attack on his reputation and competence for standing toe-to-toe with Defense.'[180] Although civil society binds the relationship between the military and the state, says Wadham, that is not to say the military complies gracefully. Civil society is, after all, 'the military's feminine.'[181]

Understandably, some analysts are less than sanguine about the new inclusive military rhetoric and the potential for real change. Given the phallocentric logic that to contact the feminine is to risk ego dissolution, says Wadham, what are the chances of the ADF confronting the salient issues? Wouldn't delving into the core of the matter radically destabilize 'the institution itself and the cultural expressions of militarism'?[182] While I agree with Wadham's assessment, I think there is room for hope regarding the individual Morrisons of the world—not that a change in behavior necessarily signifies a change in mind.[183] I return to Morrison's putative reconstruction in the concluding chapter.

Governments are only now starting to investigate gendered violence, at least in Australia, because the assumption that it is both natural and inevitable is becoming increasingly difficult to sustain. In some places in the world, however, girls are being shot for going to school.[184] And, as I write, the Iraqi cabinet has voted for a proposal to change the existing law enshrining women's rights regarding marriage, inheritance, and child custody in order to permit the marriage of nine-year-old girls, grant automatic child custody to fathers, and condone a husband's right to insist on sex at his wish.[185] The enormity of the global problem doesn't alter the direction for change, namely, toward gender equality. Cross-culturally, says the Change Project, the more equal the genders, the lower is the prevalence of sexual violence. One finding that inspires hope is the vast inter-country disparity in its prevalence (ranging from 25 percent to 80 percent). While sexual violence is pervasive, there is hope that it can be altered.[186]

POSTSCRIPT

'The government is not dying of shame. My father did not die of shame.'[187] These words were spoken in Parliament by Australia's first ever woman prime minister. She gave the address a few months before she lost the leadership of her party after a relentless, often personal, sexist, and misogynist campaign against her. The address came to be known as Julia Gillard's 'misogyny speech.' In it she repeatedly accused the then leader of the opposition, Tony Abbott, of sexism and misogyny.

Gillard's 'My father . . .' statement was in response to a public remark by influential broadcaster Alan Jones that Gillard's father had 'died a few weeks ago of shame.'[188] 'Women are destroying the joint,' Jones had said earlier. And of Gillard, 'I'm putting her in a chaff bag and hoisting her into the Tasman Sea.'[189] Gillard's statement also referred to Abbott's comment that the Parliament 'should already have died of shame.' And in a later radio interview, Abbott called on Gillard 'to, politically speaking, make an honest woman of herself.'[190]

The expression 'honest woman' served a double purpose—as a coded reminder of Gillard's unmarried, de facto, and intentionally childless status and to reinforce the branding of Gillard as a liar.[191] What was her alleged lie? Gillard's government had changed its carbon-tax policy as part of a coalition deal with the Australian Greens (under the then leadership of Bob Brown). In this, of course, its actions were no different from the political norm, but only Gillard was subjected to an unprecedented campaign of demonization. On delivering a speech at an anti-carbon-tax rally in 2011, Abbott positioned himself in front of two placards; on one was printed, 'Ditch the Witch,' and on the other was, 'Juliar, Bob Brown's Bitch,' depicted in flames.[192] And for two years prior to Gillard's leadership loss, the very air was saturated with the terms 'liar,' 'Juliar,' and 'the lying bitch.' Throughout Gillard's parliamentary speeches, Abbott would, *sotto voce*, repeatedly call her 'liar.'[193] Other prominent politicians labeled Gillard 'Lady Macbeth,' 'crone,' and 'harridan,' and one strategist suggested on air that Australians 'ought to be kicking her to death.'[194]

Jones claimed that Gillard's carbon-tax policy 'borders on the treasonous' and agreed with a caller, 'Yeah, that's it. Bring back the guillotine!'[195] One cartoonist repeatedly caricatured Gillard, much as Hillary Clinton was in the 1990s, as wearing a huge strap-on dildo, about to assault various male colleagues.[196] Recall that, for millennia, Western culture has labelled the penetrating woman (along with the penetrated man) 'monstrous,' the lowest of the low.[197] At least one analyst branded the anti-Gillard campaign 'political persecution.'[198]

Typically, the media responded to Gillard's misogyny speech by accusing her of 'playing the gender card.' A former government minister said of Gillard that it was 'disgraceful to go around calling people sexist and misogynist.' Others accused her of being 'the driver of the politics of hate in Australia.'[199] A week or so before Gillard lost the leadership, she appeared (as had others before her) as an eponymous menu item: 'Julia Gillard Kentucky Fried Quail—Small Breasts, Huge Thighs and A Big Red Box.'[200] In the same week, a broadcaster put it to her on air that her partner was gay, as if to satisfy the broadcaster's primal suspicion that the penetrating woman and the penetrated man are in it together, dismantling civilization from below.

Several leading observers concluded that Gillard was deliberately sabotaged by political enemies, including from her own camp, using gendered and blatantly sexualized personal denigration to undermine her. 'I never want a young woman to look at the treatment that the prime minister receives,' said a woman cabinet member, 'and think, I don't want to do that job . . . if I'm going to be a target like that.'[201] What does a girl learn from this? 'That if she steps into the public sphere, she is asking to be undressed by the electorate.'[202] 'Female bodily shame is everywhere. . . . The message is clear and it is everywhere: shame is to be her daily companion.'[203] Of Jones's accusation that Gillard's father had died of shame, one commentator made some historical links: 'Paternal shame at the wayward daughter is a loaded concept. Its darkest manifestation is the honor killing.' And, closer to home, 'it speaks to the ancient disgrace of having a daughter on the stage: the shame of female visibility.'[204]

Furnishing a gender nonconforming woman with the power to act on a political stage is one step too far, at least in Australia. And if she begins to interfere with the real command center—the military, industrial, and church traditions—her nonconformity slides into the 'monstrous.' Abbott, a former seminarian, became prime minister at the subsequent election. Gillard had

had five women in her parliamentary cabinet; in Abbott's there is one—one who can be counted on not to covet the leadership.

As for the defense minister who stood 'toe-to-toe with Defense,' Smith resigned when Gillard lost the party leadership. By joining forces against an abusive culture in the ADF, Smith and Gillard had accomplished the hitherto impossible. It seems that a witch hunt with the notion of woman's shame at its core can still be relied on to punish the meddlers even if not entirely restore patriarchal equilibrium.

NOTES

1. Jane Burns et al., "Game On: Exploring the Impact of Technologies on Young Men's Mental Health and Wellbeing," (Melbourne: Young and Well Cooperative Research Centre, 2013).
2. Rachael Brown, "Soldier Suicide," ABC Radio AM program (Australian Broadcasting Commission, April 22 and 23, 2014).
3. Australian Bureau of Statistics, 2012; Burns, "Game On."
4. Australian Human Rights Commission, "Lesbian, Gay, Bisexual, Trans and Intersex Equality," (Australian Human Rights Commission, April 14, 2014).
5. Australian Government, "Tackling Suicide in LGBTI Community," (Australian Government Department of Health and Aging, Canberra, July 20, 2011); Kairi Kõlves, Eeva-Katri Kumpula, and Diego De Leo eds., *Suicidal Behaviour in Men: Determinants and Prevention in Australia* (Brisbane: Australian Institute for Suicide Research and Prevention, 2013), 83.
6. Matt Muijen, Foreword in *Routledge International Handbook of Clinical Suicide Research,* eds. John Cutcliffe et al. (New York: Routledge, 2014), xxxvii.
7. George Patton et al., "The Progression of Common Mental Disorders in Adolescents: A 14-Year Prospective Cohort Study," *Lancet* (January 16, 2014); George Patton et al., "Global Patterns of Mortality in Young People," *Lancet* 374, (2009); P. Moran et al., "The Natural History of Self-Harm from Adolescence to Adulthood," *Lancet* 379, (2012).
8. Lucy Carter, "OECD Snapshot Ranks Australia Second in World in Anti-Depressant Prescriptions," ABC News (Australian Broadcasting Corporation, November 22, 2013).
9. Olivier, *Jocasta's Children.*
10. Fulu, "Why Do Some Men Use Violence."
11. Olivier, *Jocasta's Children*, 50, 90.
12. Paul Shepanski and Michael Diamond, *An Unexpected Tragedy: Evidence for the Connection between Working Patterns and Family Breakdown in Australia* (Sydney: Relationships Forum Australia, 2007).
13. Olivier, *Jocasta's Children;* Dinnerstein, *Mermaid and Minotaur*, preface.
14. E.g., Tom Nightingale, "Pay Gap between Men and Women Wider Now than Twenty Years Ago," ABC News (Australian Broadcasting Corporation, September 4, 2013); Nonee Walsh "Bonuses Paid to Men in the Finance Sector are Worth Triple Those to Women," ABC News (Australian Broadcasting Commission, September 2, 2013).
15. Norman Hermant, "COAG Equity Report Finds Australian Women Still Lag Behind Men in Pay," ABC News (Australian Broadcasting Corporation, November 19, 2013).
16. Rich, *Of Woman Born*, 52.
17. Rich, *Of Woman Born*, 53, 52, 53, 61.
18. Olivier, *Jocasta's Children.*
19. Olivier, *Jocasta's Children*, 38–42.
20. Olivier, *Jocasta's Children.*
21. Olivier, *Jocasta's Children*, 44, 50.
22. Chodorow, *Reproduction of Mothering*, 212.
23. Chodorow, *Reproduction of Mothering*, 213.

24. Anne Minahan, ed., *Encyclopedia of Social Work,* 20 [th] ed., vol. 1 (New York: Oxford University Press, 2008) (re USA) ; Australian Bureau of Statistics, "Parental Divorce or Death during Childhood," *Australian Social Trends 2010* (Sydney: Australian Bureau of Statistics, 2010) (re Australia).

25. Robyn Parker, "Family Statistics and Trends: Relationships Indicators Survey 2006," *Family Relationships Quarterly* 4, (2011) .

26. Parker, "Family Statistics and Trends."

27. Ronald Levant, "The Crisis in Connection between Men and Women," *The Journal of Men's Studies* 5, 1, (1996); Xenia Montenegro, "The Divorce Experience: A Study of Divorce at Midlife and Beyond" (American Association of Retired People, 2004), 14.

28. Parker, "Family Statistics and Trends." A gender-based breakdown for this figure was not provided.

29. Edward Laumann, Anthony Paik, Dale Glasser, et al., "A Cross-National Study of Subjective Sexual Well-Being among Older Women and Men: Findings from the Global Study of Sexual Attitudes and Behaviors," *Archives of Sexual Behavior* 35, 2, (2006).

30. Laumann et al., "Study of Subjective Sexual Well-Being," 14, 22.

31. Eric Klinenberg, *Going Solo: The Extraordinary Rise and Surprising Appeal of Living Alone* (New York: Penguin, 2013).

32. Klinenberg, *Going Solo.*

33. Edith Gray, Anna Reimondos, and Liz Allen. "Australian Census: One in Ten Live Alone," *The Conversation* (June 22, 2012).

34. Olivier, *Jocasta's Children.*

35. Olivier, *Jocasta's Children*, 93.

36. Olivier, *Jocasta's Children.*

37. Olivier, *Jocasta's Children*, 141.

38. Ducat, *Wimp Factor*, 57.

39. Dinnerstein, *The Mermaid and the Minotaur*, preface.

40. Ducat, *Wimp Factor*, 57.

41. Ducat, *Wimp Factor*, 57–58.

42. Lee Gettler et al., "Longitudinal Evidence that Fatherhood Decreases Testosterone in Human Males," *Proceedings of the National Academy of Sciences of the United States of America* 108, 39, (2011), 161–94.

43. Collins et al., "Toward an Integrated Theory of Gender Stratification," 195.

44. E.g., Horney, *Feminine Psychology*; Olivier, *Jocasta's Children*; Gilmore, *Manhood in the Making*; Bergman, "Men's Psychological Development"; Levant, "Reconstruction of Masculinity"; Pollack, "No Man is an Island"; Krugman, "Male Development"; Ducat, *Wimp Factor*; Diamond, "Masculinity Unravelled."

45. World Health Organization, "Global and Regional Estimates of Violence against Women," (Geneva: Department of Reproductive Health and Research, World Health Organization, 2013), 36.

46. World Health Organization, "Global and Regional Estimates of Violence," 36, 20.

47. World Health Organization, "Global and Regional Estimates of Violence," 36, 31.

48. Fulu, "Why Do Some Men Use Violence."

49. Ducat, *Wimp Factor*, 51.

50. Ontario Domestic Violence Death Review Committee, *Annual Report to the Chief Coroner* (Toronto, ON: Office of the Chief Coroner, 2008).

51. Michael Kimmel, *The Gender of Desire: Essays on Male Sexuality* (New York: State University of New York Press, 2005).

52. Kimmel, *Gender of Desire*, 86.

53. Kimmel, *Gender of Desire*, 87.

54. Nicholas Groth, *Men Who Rape: The Psychology of the Offender* (Perseus Books, 2001), 9, 2.

55. Ducat, *Wimp Factor.*

56. Horney , *Feminine Psychology*, 142.

57. Horney , *Feminine Psychology*, 142–43.

58. Horney , *Feminine Psychology*, 143.

59. Horney , *Feminine Psychology*, 143.
60. UN Women, "UN-Women-Supported Survey in Delhi Shows 95 percent of Women and Girls Feel Unsafe in Public Spaces," (UN Women, February 20, 2013). The data were collected between October and November 2012, that is, a month or so before the gang rape in New Delhi of a twenty-three-year-old student who later died from her injuries.
61. UN Women, "UN-Women-Supported Survey in Delhi."
62. Fulu, "Why Do Some Men Use Violence," 12. R. W. Connell is a senior advisor to the Change Project.
63. Fulu, "Why Do Some Men Use Violence," 12.
64. Fulu, "Why Do Some Men Use Violence," 21. Researchers used multivariate logistical regression modeling to analyze the data, a method of particular use in informing multi-factorial interventions aimed at prevention (albeit not for establishing causal links). Researchers selected 10,178 men comprising site-representative samples (at least 1,000 per site) through a multi-stage cluster sampling strategy.
65. Fulu, "Why Do Some Men Use Violence," 21.
66. Fulu, "Why Do Some Men Use Violence," 21.
67. Fulu, "Why Do Some Men Use Violence," 2.
68. Fulu, "Why Do Some Men Use Violence," 3.
69. Fulu, "Why Do Some Men Use Violence," 3.
70. Fulu, "Why Do Some Men Use Violence," 44.
71. Marilyn Gregory, "Masculinity and Homicide," *International Journal of Law, Crime and Justice* 40, 2, (2011).
72. Connell and Messerschmidt, "Hegemonic Masculinity"
73. Fulu, "Why Do Some Men Use Violence," 15.
74. Fulu, "Why Do Some Men Use Violence," 1.
75. Fulu, "Why Do Some Men Use Violence," 16.
76. Fulu, "Why Do Some Men Use Violence", 76.
77. E.g., Ravi Verma, United Nations CSW Side Event "Why Do Men Use Violence and How Do We Stop It?" (UN, March 8, 2013); Bourdieu, *Masculine Domination.*
78. Aela Callan, "It's a Man's World," *101 East* Program (Aljazeera, March 8, 2013).
79. Callan, "It's a Man's World."
80. Verma, United Nations CSW Side Event.
81. Ben Wadham, "Brotherhood: Homosociality, Totality and Military Subjectivity," *Australian Feminist Studies* 28, 76, (2013); Peggy Sanday, *Fraternity Gang Rape: Sex, Brotherhood and Privilege on Campus,* 2nd ed. (New York: New York University Press, 2007), 7; Joshua Goldstein, *War and Gender: How Gender Shapes the War System and Vice Versa* (New York: Cambridge University Press, 2003), 365.
82. European Union Agency for Fundamental Rights, "Main Results Report," In *Violence Against Women: An EU-wide Survey* (Luxembourg: Publications Office of the European Union, 2014), 3. The survey used multistage random sampling stratified by geographical region and urban/rural character. The study did not include women in institutionalized settings and, hence, may have underestimated the actual prevalence of violence (16, 21).
83. Bourdieu, *Masculine Domination*, 52.
84. Bourdieu, *Masculine Domination*, 51–52.
85. E.g., Connell and Messerschmidt, "Hegemonic Masculinity."
86. Wadham, "Brotherhood: Homosociality, Totality," 224.
87. Bourdieu, *Masculine Domination*, 52.
88. Bourdieu, *Masculine Domination*, 53.
89. Ducat, *Wimp Factor*, 51.
90. Klaus Theweleit, *Male Fantasies* (Minneapolis: University of Minnesota Press, 1989).
91. Kristeva, *Powers of Horror*, 70.
92. Bourdieu, *Masculine Domination*, 51.
93. Fulu, "Why Do Some Men Use Violence."
94. Ducat, *Wimp Factor.*

95. Erin Steuter and Deborah Wills, *At War with Metaphor: Media, Propaganda, and Racism in the War on Terror* (Lanham, MD: Lexington Books, 2009), 58; Theodore Nadelson, *Trained to Kill: Soldiers at War* (Baltimore, MD: Johns Hopkins University Press, 2005), 43.
96. Robert Stoller, *Observing the Erotic Imagination* (New Haven, CT: Yale University, 1985).
97. Ducat, *Wimp Factor*, 38.
98. Christine Bradley, "Final Report on Domestic Violence Report No. 14," (Papua New Guinea Law Reform Commission, 1992), 25.
99. Gilbert Herdt, *Rituals of Manhood: Male Initiation in New Guinea* (New Brunswick, NJ: Transaction Publishers, 1982), 52.
100. Stoller, *Observing the Erotic Imagination.*
101. Stoller, *Observing the Erotic Imagination*, 108.
102. Herdt, *Rituals of Manhood,* xiv.
103. Stoller, *Observing the Erotic Imagination*, 108.
104. Ducat, *Wimp Factor*, 39.
105. Stoller *Observing the Erotic Imagination*, 133.
106. Ducat, *Wimp Factor*, 40.
107. Ducat, *Wimp Factor*, 40–41.
108. Herdt, *Rituals of Manhood,* xiv–xv.
109. Dover, *Greek Homosexuality*; Winkler, *Constraints of Desire.*
110. Ducat, *Wimp Factor*, 40.
111. Kathleen Kuiper, ed., *Ancient Greece: From the Archaic Period to the Death of Alexander the Great* (New York: Britannia Educational Publishing, 2011), 40.
112. Ashley Montagu, *Coming into Being among the Australian Aborigines* (London: Routledge, 1974), 312; Warren Shapiro and Uli Linke, *Denying Biology: Essays on Gender and Pseudo-Creation* (Lanham, MD: University Press of America, 1996); Robert Munroe and Ruth Munroe, "Male Pregnancy Symptoms and Cross-Sex Identity in Three Societies," *Journal of Social Psychology* 84, (1971); Robert Munroe and Ruth Munroe, "Psychological Interpretation of Male Initiation Rites: The Case of Male Pregnancy Symptoms," *Ethos* 1, (1973).
113. Munroe and Munroe, "Male Pregnancy Symptoms"; "Psychological Interpretation of Male Initiation Rites."
114. Ducat, *Wimp Factor.*
115. DLA Piper, *Report of the Review of Allegations of Sexual and Other Abuse in Defense,* vol. 1 (June 2012).
116. Minister for Defense, Australian Government, "Paper Presented on the Defense Abuse Response Taskforce" (Canberra: Australian Government, June 20, 2013).
117. Wadham, "Brotherhood: Homosociality, Totality."
118. Geoff Thompson and Clay Hichens, "Culture of Silence," ABC *Four Corners* Program (Australian Broadcasting Corporation, June 13, 2011).
119. Thompson and Hichins, "Culture of Silence."
120. Jo Chandler, "One Woman's War," *The Age* (May 27, 2006).
121. Deborah Snow, "There Will Be Women in Foxholes," *Sydney Morning Herald* (Jan 6, 2012).
122. Snow, "There Will Be Women in Foxholes."
123. Wadham, "Brotherhood: Homosociality, Totality"; ABC, "Facebook Group Reveals Ugly Side of Defense Force Culture," 7.30 Program (Australian Broadcasting Corporation, Feb 29, 2012).
124. ABC, "Facebook Expose Prompts Further Questions," 7.30 Program (Australian Broadcasting Corporation, Mar 1, 2012).
125. Wadham, "Brotherhood: Homosociality, Totality," 225.
126. "Ian McPhedran, "Top Brass Told of Rapist in the Defense Force Ranks," *The Daily Telegraph* (April 12, 2011).
127. Thompson and Hichins, "Culture of Silence."
128. Thompson and Hichins, "Culture of Silence."
129. ABC, "Facebook Expose."
130. David Wroe, "Army Shamed by Sex Videos," *Sydney Morning Herald* (June 14, 2013).

131. Chief of Army, Lieutenant General David Morrison, "Chief of Army Message Regarding Unacceptable Behavior," (YouTube, June 13, 2013).
132. DLA Piper, *Review of Allegations.*
133. DLA Piper, *Review of Allegations*, xxxi.
134. DLA Piper, *Review of Allegations*, xxiii.
135. DLA Piper, *Review of Allegations*, xxix.
136. Mark Schliebs and Sarah Elks, "DART Alone Won't Uncover Abuse in Defence, Garry Rumble Warns." *The Australian*, April 28, 2014.
137. E.g., Antonio Thompson and Christos Frentzos eds., *The Routledge Handbook of American Military and Diplomatic History 1865–Present* (New York: Routledge, 2013), 329; Mic Hunter, *Honor Betrayed: Sexual Abuse in America's Military* (Fort Lee, NJ: Barricade Books, 2007), 174; Marita Gronnvoll, *Media Representations of Gender and Torture Post-9/11* (New York: Routledge, 2010).
138. Fulu, "Why Do Some Men Use Violence."
139. John Rudolf, "U.S. Army: Sex Crimes by Soldiers up 97 Percent in Five Years," *Huffington Post Crime* (January 21, 2012).
140. Rudolf, "U.S. Army: Sex Crimes."
141. Rudolf, "U.S. Army: Sex Crimes."
142. E.g., Greg Forter, *Gender, Race, and Mourning in American Modernism* (Cambridge, UK: Cambridge University Press, 2011); Elizabeth Howell, "'Good Girls,' Sexy 'Bad Girls' and Warriors: The Role of Trauma and Dissociation in the Creation and Reproduction of Gender," in *Trauma and Sexuality: The Effects of Childhood Sexual, Physical and Emotional Abuse on Sexual Identity and Behavior,* eds. James Chu and Elizabeth Bowman (Binghamton, NY: Haworth Medical Press, 2002) ; Connell, *Gender and Power.*
143. Greg Forter, "Absolom! Absolom! The Politics of Traumatic Form," (Speech delivered at Faulkner Society, 2006), 1.
144. E. Anthony Rotundo, *American Manhood: Transformations in Masculinity from the Revolution to the Modern Era* (New York: Basic Books, 1993), 23, in Forter, *Gender, Race.*
145. Forter, *Gender, Race*, 100.
146. Forter, *Gender, Race*, 100.
147. Forter, "Absolom!" 259; Cathy Caruth, *Unclaimed Experience: Trauma, Narrative and History* (Baltimore, MD: John Hopkins University Press, 1996).
148. Forter, "Absolom!" 260.
149. Forter, "Absolom!" 262; Sigmund Freud and Josef Breuer, *Studies on Hysteria 1893–1895,* trans. James Strachey (New York: Basic Books and London: Hogarth Press, 1895).
150. Forter, "Absolom!" 261.
151. Forter, "Absolom!" 261.
152. Howell, "'Good Girls,' Sexy 'Bad Girls' and Warriors ."
153. Howell, "'Good Girls,' Sexy 'Bad Girls' and Warriors," 19–20.
154. Williams, "Facts and Figures Relating to the Childhood Sexual Abuse."
155. Dorais, *Don't Tell,* 172.
156. Ryan Kelty, Meredith Kleykamp, and David Segal, "The Military and the Transition to Adulthood," *The Future of Children* 20, 1, (2010): 197–98.
157. Ronald Levant, "Research in the Psychology of Men and Masculinity Using the Gender Role Strain Paradigm as a Framework," *American Psychologist* 66, (2011): 765–76; Gary Brooks and Louise Silverstein, "Understanding the Dark Side of Masculinity: An Interactive Systems Model," in *A New Psychology of Men,* ed. Ronald Levant and William Pollack (New York: Basic Books, 1995).
158. Pleck, *Myth of Masculinity* ; O'Neil, "Summarizing 25 Years"; Levant, "Research in the Psychology of Men ," 767.
159. E.g., O'Neil, "Summarizing 25 Years," 382.
160. Connell and Messerschmidt, "Hegemonic Masculinity"; Ronald Levant and Katherine Richmond, "A Review of Research on Masculine Ideologies Using the Male Role Norms Inventory," *Journal of Men's Studies* 15, (2007): 132. The Inventory measures seven norms of traditional masculinity ideology: avoidance of femininity, homophobia, self-reliance, aggression, achievement/status, nonrelational attitudes toward sex, and restrictive emotionality.

161. Levant and Richmond, "A Review of Research ," 141.
162. O'Neil, "Summarizing 25 Years," 362.
163. O'Neil, "Summarizing 25 Years," 361.
164. O'Neil, "Summarizing 25 Years," 393.
165. O'Neil, "Summarizing 25 Years," 424.
166. World Health Organization, "Global and Regional Estimates of Violence," 36. The organization endorses prevention programs that challenge social and cultural norms around masculinity, gender power relationships, and violence. It calls for the strengthening of women's economic rights and the elimination of gender inequalities in family law as well as in access to education and formal wage employment.
167. World Health Organization, "Global and Regional Estimates of Violence," 4.
168. Groth, *Men Who Rape.*
169. World Health Organization, "Global and Regional Estimates of Violence," 2.
170. World Health Organization, "Global and Regional Estimates of Violence," 33.
171. R V Johns, Supreme Court, S.A. No. SCCRM/91/452 (August 26, 1992). Judge Derek Bollen in the Supreme Court of South Australia: 'There is, of course, nothing wrong with a husband, faced with his wife's initial refusal to engage in intercourse, in attempting, in an acceptable way, to persuade her to change her mind, and that may involve a measure of rougher than usual handling. It may be that handling and persuasion will persuade the wife to agree. Sometimes it is a fine line between not agreeing, then changing of the mind, and consenting…'
172. Nadia David, "Exploring the Use of Domestic Violence Fatality Review Teams," (Australian Domestic and Family Violence Clearinghouse, 2007).
173. Ruth Pollard, "Shameful Secret of our Family Murder Epidemic," *Sydney Morning Herald* (November 24, 2008).
174. Carlos Carcach and P. Grabosky, "Murder-Suicide in Australia," *Trends and Issues in Crime and Criminal Justice*, No. 82 (Canberra: Australian Institute of Criminology, March, 1998), 4. This statistic is similar in many other nations including Italy (1985–2008), see Paolo Roma et al., "The Epidemiology of Homicide-Suicide in Italy," *Forensic Science International* 214, 1–3, (2012), 1.
175. Jackson Katz, *The Macho Paradox: Why Some Men Hurt Women* (Naperville, IL: Sourcebooks, 2006).
176. Nightingale, "Pay Gap between Men and Women"; Walsh, "Bonuses Paid to Men."
177. DLA Piper, *Review of Allegations*, xl.
178. Chief of Army, "Chief of Army Message."
179. DLA Piper, *Review of Allegations.*
180. Wadham, "Brotherhood: Homosociality, Totality," 217.
181. Wadham, "Brotherhood: Homosociality, Totality," 217.
182. Wadham, "Brotherhood: Homosociality, Totality," 231
183. Hasse Karlsson, et al., "How Psychotherapy Changes the Brain," *Psychiatric Times* (August 11, 2011); M. Lövdén, et al., "Structural Brain Plasticity in Adult Learning and Development," *Neuroscience and Biobehavioral Reviews* (February 28, 2013); Doidge, *Brain that Changes Itself.* Consistent repetition of new behavior patterns gradually restructures one's neurobiology, which literally does change the mind.
184. Malala Yousafzai, Speech to the United Nations (UN, July 13, 2013).
185. ABC, "Iraqi Women Protest Against Proposed Islamic Law to Permit Child Marriage," ABC News (Australian Broadcasting Commission, March 9, 2014).
186. Fulu, "Why Do Some Men Use Violence."
187. Anna Goldsworthy, "Unfinished Business: Sex, Freedom and Misogyny," *Quarterly Essay,* 50, (June, 2013): 1942.
188. Goldsworthy, "Unfinished Business," 1941.
189. Heath Aston, "Gillard's Father Died of Shame: Jones," *Sydney Morning Herald* (September 29, 2012).
190. Malcolm Farr, "Tony Abbott Tells Julia Gillard to 'Make an Honest Woman of Herself' On Carbon Tax," news.com.au (February 25, 2011).
191. Goldsworthy, "Unfinished Business," 1943.
192. Goldsworthy, "Unfinished Business," 1943.

193. Anne Summers, "Her Rights at Work: The Political Persecution of Australia's First Female Prime Minister," (2012 Human Rights and Social Justice Lecture, University of Newcastle, August 31, 2012). Abbott would call Gillard 'liar,' usually *sotto voce*. However, on August 20, 2012, the deputy speaker overheard Abbott and demanded a withdrawal. Abbott's withdrawal was qualified, and he was suspended for an hour.

194. Goldsworthy, "Unfinished Business," 1944.

195. Aston, "Gillard's Father Died of Shame: Jones."

196. Ducat, *Wimp Factor*, 140–44.

197. Carden, "Endangered Ancestress Revisited."

198. Summers, "Her Rights at Work."

199. Goldsworthy, "Unfinished Business," 1945.

200. Kerryn Goldsworthy, "The Unseemly Splatter," Sydney Review of Books (October 22, 2013).

201. Goldsworthy, "Unfinished Business," 1947.

202. Goldsworthy, "Unfinished Business," 1954.

203. Goldsworthy, "Unfinished Business," 1940.

204. Goldsworthy, "Unfinished Business," 1942.

Chapter Eleven

Fair's Fair

In 1950s Australia, I grew up accepting that women were inferior to men. It was not a matter of conscious thought or belief. No one ever said it. Like human blotting paper, I simply soaked up my culture. I knew, with absolute and unconscious certainty, that any woman who hadn't made the house spotless, washed and ironed every drawer-escapee, cooked a three-course breakfast and packed lunches for her family, prepared the evening's roast and baked a cake and, finally, donned the mandatory 'lippy,' hat, and gloves for appearing in public, even on her own front porch . . . yes, any woman who hadn't done all this and more by 8 am . . . was a slut. Some women rose at four simply to forestall attracting that scarlet letter. With unconscious certainty, I also accepted that my working-class family was inferior to those in the ranks above. Hierarchies—in families, schools, churches, and communities—were just how things were and ought to be. And the invisible barriers spoke not as much by word as by body language, tone of voice, and fifty shades of exclusion. I was lucky. In the 1970s, second-wave feminism lifted and carried me into higher education. That wave, however, washed over everyone I knew. It left them floundering. Later, I understood that most of the world remained beyond its reach.[1]

'Most women who have awakened,' says Cixous, 'remember having slept, having been put to sleep.'[2] My aim in this book has been to awaken: to denaturalize the gendered metaphor, seed-and-soil, and its historical role in justifying widespread discrimination against women. Of necessity, the book's historical-snapshots approach has omitted much important and fascinating detail, and I hope the overall project has gained more than it has lost. I have suggested that there is a mutually beneficial tension between big-picture studies and detailed, locally focused analyses, and that mapping continuities and making fine distinctions are vital and complementary processes. This book has examined a handful of historical continuities in the Western natural-philosophical canon that have contributed to the remarkable persistence of gender stratification and women's relative disadvantage. To reiterate a point from the book's introduction, first, the historical counter-narratives are of equal importance as objects of study and, second, the correlations I have outlined throughout the book are for the sake of scholarly interest; I make no claims of universal application or causation.

This chapter summarizes the salient features of seed-and-soil examined in this book, comments on its contemporary legacy, and suggests directions for change. The roots of seed-and-soil lie in the archaic patriarchal tradition of constructing Woman as Man's natural inferior. While the trope's temporal and geographical expression varied, this major continuity, its axiomatic base, remained. This is unsurprising given that it had been around long enough to become part of a tacit worldview. Seed-and-soil was no more questioned than the rising and setting of the sun and, besides, most counter-narratives to the Aristotelian also took for granted the inferiority and thus necessary subordination of women.

In what way did the fathers of the ancient Hebrews and Greco-Romans, the Christians and Muslims, and modern European natural philosophy construct Woman? Even within diverse local conditions and narrative detail, the old story persisted that biology drives Woman. Woman's nature, that is, her sexuality (whether in the form of her wandering womb, nervous excitability, or something else peculiar to her sex), is akin to soil in its less pleasing aspects, namely, dark, damp, cold, passive, profane, and polluted. Several major ramifications of this narrative remain in the Western imagination and still flourish under extremist patriarchal regimes in the Global South and beyond, as well as in many diasporic communities.[3] As I have outlined in this book, certain prejudices about Woman came to the fore during particular historical periods, prejudices that, separately or together, cast her as largely incapable of intellectual, moral, or spiritual development; unwitting and helpless tool of the Devil; inferior partner in procreation; polluted and polluting; and legitimate object of men's anxieties, envy, and hostility. The good woman/bad woman binarism is embedded in each of these constructions; the notion of Woman-as-angel, for example, showed women how they ought to behave. And, no doubt more reasoned and even appreciative counter-narratives existed.

The trope of Woman as privately owned and bounded field is rooted in male monogenetic theory. With the construction of men's role in conception, the primeval notion of procreation as the business of Woman and the divine became one in which Man had the primary role. No one will ever know exactly how the transition occurred, but it was decisive in Greece by the time of Aeschylus. All that remained was to ensure its ongoing enshrinement and justification in the canon. Aeschylus's *Oresteia*, with its depiction of patriarchy's glorious overthrow of Mother right, won first prize in the City Dionysia contest in 458 BCE, and the authorities granted Aeschylus's tragedies alone special permission for restaging in subsequent festivals.[4] By such means—prizes, patronage, publicity, and special privileges—patriarchal elites throughout history have constructed and maintained the canon.

For the Turkish villagers in Delaney's study, as in some other contemporary communities, the authority of the father symbolizes that of God in the world. The goal of marriage is to produce children; wives have no right to refuse sex with their husbands (there is no recognition of marital rape); and fathers alone own their children.

The second major construction of Woman—as unable to resist Satan's wiles —underpinned, for example, the early modern witch hunts. While the hunts unfolded differently according to local conditions and politics, a principal continuity was in the way the prosecutors targeted gender nonconforming women and men. Early modern central Europe reinvented the witch as female; targeted, tortured, and killed tens of thousands of women and thousands of men; terrorized millions more; and reinforced community suspicion toward women and their occupations as potentially diabolical. As a direct consequence, women's employment as physicians, healers, and midwives gradually shifted into the hands of men.[5] Not only did the early modern witch hunts set the scene for the male takeover of women's medicine but also, in their aftermath, the assault on gender nonconformity continued in the form of persecution of so-called perverse behavior.[6] And in some contemporary societies, some women, and to a lesser extent, men are still targeted, tortured, and burned as witches.

Third, the Western tradition constructed Woman as largely incapable of intellectual, moral, or spiritual development due to her biologically based inferiority. The nineteenth- and twentieth-century campaigns against women's suffrage and education turned on this axis. Women's attempts at learning, the Academy said, brought about a kind of masculinization that endangered their health and wellbeing, that of their offspring, and of the species itself. The woman would become a 'monstrosity, something which having ceased to be woman is yet not man.'[7] Medical wisdom had it that women's sexuality was so disabling as to cause nigh every female disease. Consequently, influential sections of the medical Academy argued that treatment options for hysteria, epilepsy, masturbation, 'ovarian insanity,' and associated 'disorders' ought to include genital mutilation and castration. Whereas the last recorded clitoridectomy performed in Britain for an emotional disorder was on a five-year-old girl in the 1940s,[8] such punitive surgery continued in the United States for decades.[9] Throughout history, women have undergone the majority of the most invasive and radical medical treatments, including lobotomy and ECT.[10]

Although Western physicians stopped favoring clitoridectomy, they persisted with other unnecessary and invasive gynecological procedures. Until well into the second half of the twentieth century, for example, a full gynecological examination routinely, but usually unnecessarily, included dilatation and curettage (D&C), the procedure of dilating the cervix and scraping out the uterus with a scalpel.[11] Gynecological operating lists, says Dally, began or ended with a few D&Cs, which mainly the junior staff, even the most junior, performed.[12] Yet there was little or no evidence that the procedure was effective in most conditions, least of all the heavy bleeding for which physicians routinely prescribed it.[13] In the 1970s, if not beyond, some Australian teaching hospitals permitted interns to perform, for training purposes, unnecessary procedures on anesthetized women patients without patients' knowledge or consent. Some women reported that, during uterine surgery, for example, doctors had removed their healthy ovaries without their permission or knowledge.[14] Post-operative complications occasionally revealed the situation. One patient described how she learned that medical staff had subjected her to an unnecessary D&C during an unrelated procedure:

> 'Oh we're awake,' chirrups a hospital mask. 'Bit of a problem. While we were having the D&C, the uterus was perforated. We'll keep you in a bit longer to keep an eye on it. Not to worry.' . . . It takes a while to sink in. A public hospital, a young intern, a young single mum, an unnecessary procedure performed without her consent, an accidental slip of the scalpel, a perforated womb, and a longer stay than anticipated. The supervisor mask is using the passive voice, keeping me in the dark so I can't join the dots.[15]

In the twenty-first century, change is still slow for captive audiences, particularly the anesthetized. In 2001, a storm broke in the United States about the routine training practice by which medical students performed unnecessary pelvic exams on anesthetized women patients without their knowledge or consent.[16] In two studies, 80 to 90 percent of students reported having performed unnecessary pelvic exams on patients.[17] In a third study, one student reported observing 'five or six students doing successive pelvic exams on the same woman.'[18] A decade later, despite the storm of protest, the practice persists,[19] and of course it disproportionately affects women who are poor, uninsured, or from minority groups.[20]

Many physicians justify the nonconsensual procedures by claiming that if they consulted women, too many would withhold permission.[21] The evidence, however, says otherwise. Only 14 percent of one study sample said they would refuse consent (18 percent were unsure),[22] and, in another study, many women were willing to participate as collaborators in the training process, that is, as fully conscious interlocutors with their examining student.[23]

Some American nurse-training institutions require student nurses to perform pelvic examinations on each other in front of male faculty.[24] I do not know if male students perform rectal examinations on each other in front of women faculty, and my point, of course, is not that authorities ought to treat their female and male students equally disrespectfully.

As to other types of genital mutilation, across the globe, tradition inflicts unnecessary and involuntary circumcision on more than 13 million boys and 2 million girls annually.[25] Female circumcision can include clitoridectomy, removal of labia, or infibulation, also known as pharaonic circumcision. To infibulate the girl, the midwife cuts away—often without anesthetic or antiseptics—virtually all of the girl's external genitalia. The girl must remain immobile, legs bound together, for weeks until her vaginal walls fuse around a hollow reed that leaves a tiny hole for urine and menses to pass through.[26] Infibulation makes it impossible for the girl to have intercourse, thus preserving her virginity for a future husband. After the girl's wedding, the midwife cuts open the girl's genital area to enable penetration. In addition, if the husband goes away for a substantial period, the midwife re-infibulates the girl, and she repeats the procedure whenever the girl gives birth. These girls, say researchers, have no legal rights or social status, and they suffer repeated procedures in which infections, hemorrhage, shock, fistulas, and even death, are not uncommon.[27]

In countries such as Somalia, Egypt, and Sudan, on average, more than 90 percent of women undergo pharaonic circumcision, 100 percent in some regions.[28] In contemporary rural Egypt, for example, the girl who remains uncut risks becoming an outcast.[29] In Egypt, the notion of male monogenesis, to which many Egyptians of the urban underclass still subscribe, hails from pre-Islamic times from the teachings of Aristotle and the Pharaonic traditions.[30]

Some contemporary commentators claim, mistakenly in my view, that because tribal women typically carry out the cutting, the tradition itself cannot be patriarchal. On the contrary, countless generations of patriarchal enculturation have had the entire community accepting the tradition as essential for the tribe's survival. And those who disobey tribal law put their very status as 'belonging' at risk. In marginalized communities, not only does that status mitigate disadvantage and aid survival, but also exclusion and isolation can be fatal.[31] In addition, the issue of harm minimization is uppermost for Western surgeons who, while abhorring the practice of female circumcision, are nevertheless responsible for treating clients from diasporic communities that practice female genital mutilation in unsterile, traumatic, and sometimes lethal circumstances.[32]

Finally, while most Westerners would applaud the modern West's abandonment of the practice of female genital mutilation, an important continuity remains. Twenty-first-century enculturation into self-disciplinary practices has increasing numbers of girls and women subjecting themselves to genital cutting in the form of labiaplasty, vaginoplasty, and so forth.[33] And despite the marketing that constructs, for example, vaginal-tightening surgery in the language of feminism and female sexual pleasure, the idea remains grounded in the Freudian prescription that 'mature' feminine pleasure is vaginal rather than clitoral.[34] As to 'vaginal rejuvenation,' says Nikki Sullivan, 'the problematic conflation of tightness with satisfaction clearly raises the question of whose pleasure is really at stake.' What surgeons construe as an imperative—a revolutionary and liberatory practice for women—actually reproduces heteronormative ideals.[35]

The fourth construction of Woman concerns her polluted and polluting nature, which precludes her from approaching the sacred. The ancient practice of barring women from entering sacred areas or touching sacred items is largely to do with fear of contamination by menstrual fluid.[36] In Delaney's Turkish village, menstrual taboos prevented menstruating women from entering the mosque, touching the Quran, or making bread (because wheat is related to seed).[37] In Christianity, unlike the sacrificial blood of Christ, menstrual blood is associated with Eve's sin and ultimately to sexual knowledge and shame.[38] At least until the 1970s, both the Roman Catholic and Greek Orthodox Churches listed the moral impurity of menstruation as a reason for excluding menstruating women from the sanctuary and for resisting the ordination of women priests.[39] Yet, says MacCulloch, Christian history shows that women often filled leadership positions in the Church. There is a pattern of re-emergence of women's leadership, he says, as a survival strategy in times of strife: 'The time for men to take over again is when life has returned to more tranquil patterns . . . and the historical record is adjusted to match.'[40]

In 1878, the *British Medical Journal* contained 'a long and serious correspondence' about 'whether a menstruating woman will contaminate the food she touches.'[41] 'It is undoubtedly the fact,' wrote one contributor: 'that meat will be tainted if cured by women at the [menstrual] period.' Another contributor, an opponent of medical training for women, responded, 'If such bad results accrue from a woman curing dead meat whilst she is menstruating, what would result, under similar conditions, from her attempt to cure living flesh?'[42] This debate, which the Academy undertook in all seriousness, occurred during Freud's formative development as a scientist.

In the twenty-first century, the sparse but definitive literature shows that men and women still view menstrual fluid as odious.[43] At least two major myths predominate, overlap, and are highly gendered. First, men and women still perceive menses as extremely polluting. Perhaps this is unsurprising given that, despite the avalanche of literature demonstrating its benignity, some scientists were still enquiring into the already discredited hypothesis of 'menotoxicity' during the 1970s.[44] Compared to women, men are much more likely to rate menstrual fluid as revolting and polluting. In a survey of men's and women's responses to the mention of various bodily effluvia of a stranger, for example, 80 percent of men, compared with nearly 50 percent of women, gave menstrual fluid a double or triple-R (maximum) rating in revoltingness. In contrast, only 11 percent of both men and women thought of blood from a fresh wound as revolting. Twelve percent of women rated menstrual fluid as not revolting at all, but no man did.[45] A contemporary example reflecting extreme fear of pollution by menstrual fluid is in the diamond district of Manhattan where the Hasidic Jews hire private buses in order to avoid such potential contamination on public transport.[46]

The second major myth about menstruation is that it impairs women's reasoning capacity and, hence, that women ought not to occupy positions of responsibility. Media depictions of Hillary Clinton as potential presidential candidate, for example, included an image of Clinton against the backdrop of a nuclear mushroom cloud with the caption, 'This is what you want once a month?'[47] Women in U.S. Fortune 500 companies hold just 14.3 percent of executive officer positions,[48] and the statistics are even worse for Australian women. Despite recently tightened requirements for gender-diversity reporting, and despite the fact that women comprise 60 percent of university graduates and near 50 percent of workforce participants as well as of those with leadership ambition, men are nine times more likely to advance to senior executive rank.[49]

There is an Australian corporate myth, say researchers, that 'women are biologically incapable of higher office.'[50] That some of the brightest business minds harbor such notions is both disturbing and difficult to read; how does one elucidate tacitly held attitudes, probably forged in childhood, unconsciously reinforced, and hitherto unchallenged? The statistics, however, could not be clearer. Of the top five hundred companies listed on the Australian Stock Exchange, 63 percent have no women executives and less than 10 percent of their CEOs are women.[51] Moreover, the gender pay gap begins at graduation, has been widening since 2006, and the higher one climbs, the wider the gap.[52]

The more one enquires into myths about women's so-called biological incapability, the more paradoxical things become. On the one hand, male leaders rate their female counterparts as only half as skilled as men at problem solving, and on the other hand as on par at delivering corporate outcomes.[53] What does it mean that some male executives perceive men and women leaders as equal corporate performers, yet rate those same women's problem-solving capacities as grossly inadequate? Perhaps the answer is that the media daily provide evidence of the high performance of women executives, whereas people's tacit beliefs about less tangible attributes may not see the light of day.

The twin themes of menstruation-related pollution and cognitive impairment overlap in findings from various studies into men and women's perceptions of women who (the researchers have implied) are menstruating. A mixed sample of European-American college students, for example, compared an ostensible menstruant to a non-menstruant as more 'neurotic,' 'irritable,' 'sad,' and 'angry.' Men students, but not women, perceived the menstruant as less 'clean and fresh,' less 'conscientious,' less 'reasonable,' more 'annoying' and 'spacey.' Another college student sample judged a supposed menstruant as less 'competent' and less 'likeable.'[54] The authors theorize that fear of menstrual fluid is related to human beings' aversion to being reminded of their animal nature and their inevitable death.[55] However, given that people perceive menstrual fluid and venous blood very differently, perhaps the initial fear is of contamination. Sacrificial blood, after all, is masculine and sanctified, not stigmatized.[56]

Among the darkest applications of the construction of women as polluting is in the military abuse of Muslim prisoners at the Guantanamo Bay detention camp.[57] Women interrogators smeared the men with what they said were menses and deprived them of water for washing. The women's defilement of the men rendered the men ritually unclean according to Islamic law and unfit for prayer, and thereby robbed them of a primary source of psychological support.[58]

Historically, political elites have used all these constructions of Woman—as enclosed field, tool of Satan, polluted, and morally and intellectually inferior—to justify barring women from public life, education, trades and professions, and sacred ground, effectively restricting them to the domestic sphere. In today's extremist honor societies, on the rare occasions that a woman's family permits her to leave the house, she wears her portable domestic space in the form of the veil. If women flout the rules—or merely come under such suspicion—they can expect to be punished with violence, even death.

On this punitive base rests a fifth construction of Woman—as legitimate object of men's anxieties, envy, and hostility. As findings from the Change Project show, roughly one in three perpetrators of sexual violence said they rape out of anger and the desire to punish and almost three out of four claimed rape as men's prerogative.[59] Chapter 10 of this book detailed how boys' enculturation into traditional male ideology puts them and those around them in harm's way. The ideology engenders a taboo on male femininity, which the culture enforces through shaming and violence. Femiphobic men often feel angry with women, and some femiphobic men become perpetrators of sexual violence. As discussed previously, for women and gender-nonconforming men, safety has meant appearing to conform, and women's participation in punitive power relations cannot properly be described as the result of free and informed choice. To reiterate, victimhood and agency relate dialectically, no dualistically.

Where to from here? In terms of achieving a feminist future, understanding the historical roots of femiphobia, gender stratification, discrimination, and their legacies is imperative. Recognizing what we are up against, what it means, and where it comes from will better equip us to bring about the changes we seek. Innovative, collaborative cross-cultural research is exemplified in the Change Project, which has pinpointed the key issues without falling prey to pessimism. The Change Project also shows that, across six countries, there were no legal consequences for 72 to 97 percent of perpetrators of rape. Yet in the absence of adequate prosecution, there is no reason to expect a decline in sexual violence. Given that sexual violence correlates with gender inequality, part of the solution lies in designing, implementing, and above all enforcing laws aimed at achieving gender equality.[60] As previously mentioned, laws do not constitute the entire solution but rather a necessary part.

Constructing a feminist future involves freeing ourselves as best we can from the grip of the past. If we do not understand how past events, including the accidental, led to our present situation, we tend to take for granted that the world as we know it is natural, inevitable, and how it ought to be in the future. To understand that the way we see things is because of a chain of historical processes is to release ourselves a little to see more than we saw before, see more clearly, and gain more freedom of action toward the future.[61] In relation to the future of feminism and the dismantling of gender stratification, my rule of thumb is to hope for the best and prepare for the worst; guard what we achieve and offer solidarity wherever it counts.

So wherein lies hope? A case in point is the 2011 governmental inquiries into endemic abuse in the Australia Defense Forces. Perhaps even more potent than the combined pressure of government and media on the defense chiefs appears to be their engagement in dialogue, at the instigation of the sex discrimination commissioner Elizabeth Broderick, with victims and their families. The process, apparently a reconfigured-tailored version of restorative justice, seems to be living up to the latter's reputation for delivering positive change.[62] Painfully aware of innumerable fruitless past efforts, Broderick proposed to enlist Australia's most senior male military and industry leaders in an unprecedented collective commitment to broad-based institutional change. It is early days, but what Broderick has achieved suggests that a powerful source of hope lies in her targeted approach and direct dialogue with top leadership. Broderick's message is that men as well as women must take the message of gender equality to men. Men must take shared ownership with women, she says, to achieve not only improved legislation and more effective enforcement, but also changes in families and informal social structures.[63] Some commentators may argue that the idea of inviting a group of patriarchs to help dismantle patriarchy evokes old stereotypes of women as ineffective and/or easily duped. I suggest a counter-reading: that for women to stand as equal collaborators with men in dismantling what is perhaps the most catastrophic cultural construction in human history rates among the least victim-like behaviors possible. As Kimmel reminds us, the greatest obstacle to change is that privilege is invisible to those who have it,[64] yet some of the most sustainable change occurs when the relatively privileged realize that sharing is better for everyone, including themselves.[65] The main message of this book is not to blame history's patriarchs but to foster awareness that we are still living with their legacy. Blame is futile. What counts is historical insight and commitment to change.

Returning to the pivotal question of what influences seed-and-soil at the center of the Western natural-philosophical canon for more than two millennia, ancient Hebrew, Greco-Roman, Christian, and Muslim elites constructed their canon around the imperative to exclude women from every public sphere. Not only must women be confined to the home, they must also be kept fully occupied with childcare, cleaning, cooking, making, and mending lest their voracious sexuality overtake them and plunge them into prostitution. The coup de grâce was to keep women uneducated and then point to their 'ignorance' as proof that they were ineducable. As Lerner reminds us, the fact that this practice lasted for more than five thousand years and still persists in the Global South is not because of a failing on women's part, but because patriarchy was already well-established before Western civilization had even begun. With the recent passing of Gerda Lerner, I would like to acknowledge my debt to her ground-breaking work in women's history. As human beings, says Lerner, we reflect on the past and look into the future: 'We experience; we give voice to that experience; others reflect on it and give it new form. That new form, in its turn, influences and shapes the way next generations experience their lives. That is why history matters.'[66]

I conclude this book with a recent speech delivered by Ziauddin Yousafzai, Pakistani educator and father of Malala Yousafzai:

In many patriarchal societies and tribal societies, fathers are usually known by their sons, but I'm one of the few fathers who is known by his daughter, and I am proud of it. Malala started her campaign for education and stood for her rights in 2007, and when her efforts were honored in 2011 and she was given the National Youth Peace Prize, before that she was my daughter, but now I am her father.

Ladies and gentlemen, if we glance to human history, the story of women is the story of injustice, inequality, violence, and exploitation. In patriarchal societies, right from the very beginning when a girl is born, her birth is not celebrated. She is not welcomed, neither by father nor by mother. The neighborhood comes and commiserates with the mother, and nobody congratulates the father. And a mother is very uncomfortable for having a girl child. When she gives birth to the first girl child, she is sad. When she gives birth to the second daughter she is shocked and, in the expectation of a son, when she gives birth to a third daughter she feels guilty like a criminal.

Not only the mother suffers but the daughter suffers too. At the age of five, while she should be going to school, she stays at home and her brothers are admitted in a school. Until the age of twelve, somehow, she has a good life. She can play with her friends in the streets and move around in the streets like a butterfly. But when she becomes thirteen years old, she is forbidden to go out of her home without a male escort. She is no more a free individual. She becomes the so-called honor of her father, brothers, and family and, if she transgresses the code of that so-called honor, she could even be killed.

This so-called code of honor does not only affect the life of a girl, it also affects the life of the male members of the family. I know a family of seven sisters and one brother, and that one brother migrated to the Gulf countries to earn a living for his seven sisters and parents because he thinks that it will be humiliating if his seven sisters learn a skill and earn some livelihood. So this brother sacrifices the joys of his life and the happiness of his sisters at the altar of so-called honor.

And there is one more norm of the patriarchal society that is called obedience. A good girl is supposed to be very quiet, very humble, and very submissive. She is supposed to be silent and accept the decisions of her elders even if she does not like them. If she is married to a man she doesn't like or if she is married to an old man, she has to accept, because she does not want to be dubbed as disobedient. If she is married very early, she has to accept. And what happens at the end? She is wedded, bedded, and then she gives birth to more sons and daughters. And it is the irony of the situation that this mother teaches the same lesson of obedience to her daughter and the same lesson of honor to her sons.

And this vicious cycle goes on, goes on.

Ladies and gentlemen, this plight of millions of women could be changed if men and women in the tribal and patriarchal societies in the developing countries can break a few norms of family and society and abolish the discriminatory laws, which go against the basic human rights of the women.

A few days after Malala was born, my cousin brought a family tree of the Yousafzai family tracing back 300 years of our ancestors. But all were men. I drew a line from my name and wrote, 'Malala.' And when she was four-and-a-half-years-old, I admitted her in my school. You will be asking, why should I mention about the admission of a girl in a school? It may be taken for granted in Canada, America, many developed countries, but in poor countries, in patriarchal societies, in tribal societies, it's a big event for the life of girl. Enrollment in a school means recognition of her identity and her name. Admission in a school means that she has entered the world of dreams and aspirations where she can explore her potential for her future life. I have five sisters, and none of them could go to school. When I was filling out the Canadian visa form, I could not recall the surnames of some of my sisters. And the reason was that I have never, never seen the names of my sisters written on any document. That was the reason that I valued my daughter. What my father could not give to his daughters, I thought, 'I must change it.'

I used to appreciate the intelligence and the brilliance of my daughter. I encouraged her to sit with me when my friends used to come. I encouraged her to go with me to different meetings. And all these good values I tried to inculcate in her personality. And I imparted all these good values to my girl students and boy students as well. I used education for emancipation. I taught my girl students to unlearn the lesson of obedience. I taught my boy students to unlearn the lesson of so-called pseudo-honor.

Dear brothers and sisters, we were striving for more rights for women but we came across a new phenomenon lethal to human rights and particularly to women's rights: Talibanization. It means a complete negation of women's participation in all political, economic, and social activities. Hundreds of schools were lost. Girls were prohibited from going to school. Women were forced to wear veils and they were stopped from going to the markets. Musicians were silenced, girls were flogged, and singers were killed. Millions were suffering, but few spoke. And it was the most scary thing when you have all around such people who kill and who flog, and you speak for your rights. It's really the most scary thing.

At the age of ten, Malala stood for the right of education. She wrote a diary for the BBC blog, she volunteered herself for the *New York Times* documentaries, and she spoke from every platform she could. And her voice was the most powerful voice. It spread like a crescendo all around the world. And that was the reason the Taliban could not tolerate her campaign, and, on October 9, 2012, Malala was shot in the head at point blank range.

Dear brothers and sisters, we learned from her how to be resilient in the most difficult times, and I'm glad to share with you that despite being an icon for the rights of children and women, she is like any sixteen-year-old girl. She cries when her homework is incomplete, she quarrels with her brothers, and I am very happy for that.

People ask me, what is special in my mentorship which has made Malala so bold and so courageous and so vocal and poised. I tell them, 'Don't ask me what I did. Ask me what I did not do. I did not clip her wings, and that's all.'[67]

NOTES

1. Jones, "Rougher than Usual."
2. Cixous, "Sorties: Out and Out," 66.

3. The nations of Africa, Central and Latin America, and most of Asia are collectively known as the Global South.
4. Sommerstein, *Aeschylean Tragedy*, 8.
5. Jensen, *Path of the Devil*, 6.
6. Nye, "Sexuality," 18.
7. Maudsley, "Sex in Mind and in Education," 209.
8. Sheehan, "Victorian Clitoridectomy," 333.
9. Ann Dally, *The Trouble with Doctors: Fashions, Motives, and Mistakes* (London: Robson Books, 2003), 129; Daly, *Gyn/Ecology*, ch. 5.
10. Phil Fennell, *Treatment without Consent: Law, Psychiatry, and the Treatment of Mentally Disordered People since 1845* (London: Routledge, 1996), 276.
11. Dally, *The Trouble with Doctors*, 136.
12. Dally, *The Trouble with Doctors*, 136.
13. Dally, *The Trouble with Doctors*, 136.
14. Personal communication, 2000.
15. Jones, *Barking Mad*, 50.
16. Raymond Devettere, *Practical Decision Making in Health Care Ethics: Cases and Concepts*, 3rd ed. (Washington, DC: Georgetown University Press, 2010), 82; Judy Foreman, "Pelvic Exams Performed without Permission," *The Boston Globe* (July 13, 2004), 1.
17. Peter Ubel, Christopher Jepson and Ari Silver-Isenstadt, "Don't Ask Don't Tell: A Change in Medical Student Attitudes of Obstetrics-Gynecology Clerkships toward Seeking Consent for Pelvic Examinations on an Anesthetized Patient, " *American Journal of Obstetrics and Gynecology* 188, 2, (2003): 579; Sara Wainberg, "Teaching Pelvic Examinations under Anaesthesia: What Do Women Think?" *Journal of Obstetrics and Gynecology Canada* 32, 1, (2010).
18. Robin Wilson, "Autonomy Suspended: Using Female Patients to Teach Intimate Examinations without their Knowledge or Consent," *Journal of Health Care Law and Policy* 8, 2, (2005): 245.
19. Shawn Barnes, "Practicing Pelvic Examinations by Medical Students on Women under Anaesthesia: Why Not Ask First?" *Obstetrics and Gynecology* 120, 4, (2012): 941-43.
20. Wilson, "Autonomy Suspended," 245.
21. Wilson, "Autonomy Suspended," 243.
22. Wainberg, "Teaching Pelvic Examinations."
23. Amy Jo Goddard and Julie Carson, "At Your Cervix: A Documentary" (2012).
24. Goddard and Carson, "At Your Cervix."
25. George Denniston and Marilyn Milos, eds. *Sexual Mutilations: A Human Tragedy* (New York: Plenum Press, 1997).
26. Denniston and Milos, *Sexual Mutilations*, 8; Daly *Gyn/Ecology*, ch. 5.
27. Denniston and Milos, *Sexual Mutilations*, 8.
28. Virginia Barnes-Dean, "Clitoridectomy and Infibulation," *Cultural Survival Quarterly* 9.2, (1985); Frances Althaus, "Female Circumcision: Rite of Passage or Violation of Rights," *International Family Planning Perspectives* 23, 3, (1997); M. Mazharul Islam and M. Mosleh Uddin, "Female Circumcision in Sudan," *International Family Planning Perspectives* 27, 2, (2001): 71; Majid Ahmed, "Somalia Leads World in Per Capita Female Circumcision," *Human Rights* (July 31, 2013).
29. Gollaher, *Circumcision*, 193.
30. Inhorn, *Quest for Conception*, 52.
31. Sharon Howard, "Crime, Commonalities, and Authority in Early Modern Wales," (PhD diss., University of Wales, 2003), 258.
32. Ben Mathews, "Female Genital Mutilation: Australian Law, Policy and Practical Challenges for Doctors," *Medical Journal of Australia* 194, 3, (2011).
33. Nikki Sullivan, "Labia," in *Cultural Encyclopedia of the Body*, ed. Victoria Pitts-Taylor (Westport, CT: Greenwood Publishing, 2008), 312.
34. Nikki Sullivan, "Vagina," in *Cultural Encyclopedia of the Body*, ed. Victoria Pitts-Taylor (Westport, CT: Greenwood Publishing, 2008), 539.
35. Sullivan, "Vagina," 539.

36. Grietje, "The Better Blood," 144.
37. Delaney, *Seed and Soil*, 95.
38. Eugenia Georges, *Bodies of Knowledge: The Medicalization of Reproduction in Greece* (Nashville, TN: Vanderbilt University Press, 2008), 92.
39. Albertus McGrath, *Women and the Church* (Image Books, 1976), 135; O'Grady, "The Semantics of Taboo," 12; Ruether, "Women's Body and Blood," 7.
40. Diarmaid MacCulloch, "Women Bishops," *The Observer* (July 7, 2012).
41. Leta Stetter Hollingworth, *Functional Periodicity: An Experimental Study of the Mental and Motor Abilities of Women during Menstruation* (Columbia University, 1914), v.
42. Hollingworth, *Functional Periodicity*, v.
43. E.g., Ingrid Johnston-Robledo and Joan Chrisler, "The Menstrual Mark: Menstruation as Social Stigma," *Sex Roles* 68, 1-2, (2013); Gronnvoll, *Media Representations of Gender and Torture*; M. Stubbs, "Cultural Perceptions and Practices around Menarche and Adolescent Menstruation in the United States," *Annals of the New York Academy of Sciences* 1135, (2008); Keith Allan and Kate Burridge, *Forbidden Words: Taboo and the Censoring of Language* (Cambridge: Cambridge University Press, 2006); G. Forbes et al., "The Role of Hostile and Benevolent Sexism in Women's and Men's Perceptions of the Menstruating Woman," *Psychology of Women Quarterly* 27, (2003); Sophie Laws, *Issues of Blood: The Politics of Menstruation* (London: Macmillan, 1990); Anne Fausto-Sterling, *Myths of Gender* (New York: Basic Books, 1985).
44. Elsimar Coutinho, *Is Menstruation Obsolete?* (Oxford: Oxford University Press, 1999), 32. The term 'menotoxicity' was coined in the 1920s by physician Bela Schick.
45. Allan and Burridge, *Forbidden Words*, 162.
46. Allan and Burridge, *Forbidden Words*, 164.
47. Ryle, *Questioning Gender.*
48. Catalyst, "No Change for Women in Top Leadership."
49. Catherine Fox, "The Higher You Go," 22.
50. Fox, "The Higher You Go," 31.
51. EOWA, "2012 Census of Women in Leadership," 9.
52. CEDA, *Women in Leadership,* 14; Fox, "The Higher You Go."
53. Melanie Sanders et al., "What Stops Women from Reaching the Top: Confronting the Tough Issues," (Bain and Chief Executive Women, 2012), 12.
54. T-A. Roberts, J. Goldenberg et al., "'Feminine Protection': The effects of menstruation on attitudes toward women," *Psychology of Women Quarterly* 26, (2002): 138.
55. Roberts et al., "Feminine Protection."
56. Gronnvoll, *Media Representations of Gender*, 55.
57. Gronnvoll, *Media Representations of Gender*, 55.
58. Gronnvoll, *Media Representations of Gender*, 54.
59. Fulu, "Why Do Some Men Use Violence."
60. Fulu, "Why Do Some Men Use Violence."
61. Harari, *Sapiens.*
62. Elizabeth Broderick, "Men on the Front Line of Gender Equality" (Melbourne: TEDx Lecture, January 12, 2014).
63. Broderick, "Men on the Front Line."
64. Michael Kimmel, *Mars, Venus or Planet Earth? Women and Men in a New Millennium* (DVD, 2008).
65. Richard Wilkinson and Kate Pickett, *Spirit Level: Why Equality is Better for Everyone* (New York: Allen Lane/The Penguin Press, 2009).
66. Lerner, *Why History Matters*, 211.
67. Abridged transcript, Ziauddin Yousafzai, "My Daughter, Malala," (TED.com transcript, Canada: TED2014, March 2014).

Bibliography

ABC. "Iraqi Women Protest Against Proposed Islamic Law to Permit Child Marriage." ABC News, Australian Broadcasting Commission, March 9, 2014.

———. "Facebook Expose Prompts Further Questions." 7.30 program, Australian Broadcasting Corporation, March 1, 2012.

———. "Facebook Group Reveals Ugly Side of Defence Force Culture." 7.30 program, Australian Broadcasting Corporation, February 29, 2012.

Abrahamsen, David. *The Mind and Death of a Genius.* New York: Columbia University Press, 1946.

Abu-Odeh, Lama. "Crimes of Honour and the Construction of Gender in Arab Societies." *Comparative Law Review* 2, (2011).

Adler, Alfred. "Inferiority Feeling and Masculine Protest." In *The Individual Psychology of Alfred Adler: A Systematic Presentation in Selections from his Writings,* edited by Heinz Ansbacher and Rowena Ansbacher, 45–52. New York: Basic Books, 1956.

———. *Understanding Human Nature: Alfred Adler on the Psychology of Human Personality.* Translated by Colin Brett. Oxford: Oneworld Publications, 1998.

———. *The Collected Clinical Works of Alfred Adler.* Vol. 12, edited by Harry Stein. Bellingham, WA: Classical Adlerian Translation Project, 2006.

Aeschylus. *Eumenides.* In *Aeschylus with an English Translation by Herbert Weir Smyth.* Vol. 2. Cambridge, MA: Harvard University Press, 1926.

Ahmed, Leila. "Arab Culture and Writing Women's Bodies." In *Women and Islam: Social Conditions, Obstacles and Prospects*, edited by Haideh Moghissi, 202–215. New York: Routledge, 2005.

Ahmed, Majid. "Somalia Leads World in Per Capita Female Circumcision." *Human Rights*, July 31, 2013.

Akpinar, Aylin. "The Honour/Shame Complex Revisited: Violence against Women in the Migration Context." *Women's Studies International Forum* 26, 5, (2003): 425–442.

Alcoff, Linda Martín. *Visible Identities: Race, Gender and the Self.* New York: Oxford University Press, 2006.

Allan, Keith and Kate Burridge. *Forbidden Words: Taboo and the Censoring of Language.* Cambridge: Cambridge University Press, 2006.

Allen, Danielle. "Burning *The Fable of the Bees*: The Incendiary Authority of Nature." In *The Moral Authority of Nature,* edited by Lorraine Daston and Fernando Vidal. Chicago: University of Chicago Press, 2004.

Allen, Prudence. *The Concept of Woman.* Vol. 2. Cambridge, UK: Wm B Eerdmans Publishing, 2006.

Althaus, Frances. "Female Circumcision: Rite of Passage or Violation of Rights." *International Family Planning Perspectives* 23, 3, (1997).

Anderson, Pamela Sue. *New Topics in Feminist Philosophy of Religion: Contestations and Transcendence Incarnate.* New York: Springer, 2009.

Anlezark, Daniel. "An Ideal Marriage: Abraham and Sarah in Old English Literature." *Medium Aevum* 69, 2, (2000): 187-210.

Appignanesi, Lisa. *Mad, Bad, and Sad: A History of Women and the Mind Doctors from 1800.* Toronto: McCarthur and Co, 2007.

Aquinas, Thomas. *St Thomas Aquinas: Summa Theologica.* Translated by E. Hill. London: Cambridge University Press, 1964.

Aristotle. *Economics*. In *Aristotle in 23 Volumes.* Translated by G. C. Armstrong. Cambridge, MA: Harvard University Press, 1935.

———. *Nicomachean Ethics.* Edited and translated by H. Rackham. Cambridge, MA: Harvard University Press, 1926.

———. *On the Generation of Animals.* Translated by Arthur Platt. Whitefish, MT: Kessinger Publishing, 2004.

———. *Politics*. In *Aristotle in 23 Volumes*. Translated by H. Rackham. Cambridge, MA: Harvard University Press, 1944.

Aristotle and J. A. Smith. *The Works of Aristotle: On the Generation of Animals I and II.* Translated by W. D. Ross. OxfordL Oxford University Press, 1912.

Aristotle, J. A. Smith, and W. D. Ross. *Politics*. In *The Works of Aristotle.* Translated by Benjamin Jowett. Oxford: Oxford University Press, 1921.

Ashbrook-Harvey, Susan. "Women in Early Syrian Christianity." In *Images of Women in Antiquity,* edited by Averil Cameron and Amélie Kuhrt, 288-298. New York: Routledge, 1993.

Aston, Heath. "Gillard's Father Died of Shame: Jones." *Sydney Morning Herald,* September 29, 2012.

Augustine. *The City of God Books I-VII, The Fathers of the Church.* Vol. 8. Translated by Demetrius Zema and Gerald Walsh. Washington, DC: Catholic University of America Press, 1950.

———. *Letters 211-270, 1-29 of St Augustine.* New York: New City Press, 2005.

Austin, Colin Russell. *The Mammalian Egg.* Oxford, UK: Blackwell Scientific Publications, 2011.

———. *Ultrastructure of Fertilization.* New York: Holt, Rinehart and Winston, 1968.

Australian Bureau of Statistics. "Parental Divorce or Death during Childhood." *Australian Social Trends 2010.* Sydney: Australian Bureau of Statistics, 2010.

Australian Government. "Tackling Suicide in LGBTI Community." Australian Government Department of Health and Aging, Canberra, July 20, 2011.

Australian Human Rights Commission. "Lesbian, Gay, Bisexual, Trans and Intersex Equality." Australian Human Rights Commission, April 14, 2014.

Australian Network News. "Indian Man Who Killed Daughter for 'Honour' Has No Regrets." Australian Network News, September 21, 2013.

AYAS (Archives of the Yorkshire Archaeological Society). Leeds, UK, Document DD146/12/2/10, Fawkes of Farnley Collection, 2012.

Aydemir, Murat. *Images of Bliss: Ejaculation, Masculinity, and Meaning.* Minneapolis: University of Minnesota Press, 2007.

Bach, Alice, ed. *Women in The Hebrew Bible.* London: Routledge, 1999.

Baert, Barbara, Liesbet Kusters, and Emma Sidgwick. "An Issue of Blood: The Healing of the Woman with the Haemorrhage (*Mark* 5.24b-34, *Luke* 8.42b-48, *Matthew* 9.19-22)." In *Blood, Sweat and Tears: The Changing Concepts of Physiology from Antiquity into Early Modern Europe,* edited by Manfred Horstmanshoff, Helen King, and Claus Zittel, 307–338. Leiden: Koninklijke Brill NV, 2012.

Bagheri, Elham. "A Qualitative Investigation of Religion, Gender Role Beliefs, and Culture in the Lives of a Select Group of Muslim Men." PhD diss., University Of Iowa, 2012.

Bailey, Michael. "Gender Identity." In *The Lives of Lesbians, Gays, and Bisexuals,* edited by Ritch Savin-Williams and Kenneth Cohen, 71–83. New York: Harcourt Brace, 1996.

———. "The Feminisation of Magic and the Emerging Idea of the Female Witch in the Late Middle Ages." *Essays in Medieval Studies* 19, (2002): 120–134.

Bakan, David. *And They Took Themselves Wives: The Emergence of Patriarchy in Western Civilisation.* New York: Harper Row, 1979.

Baker Brown, Isaac. *On the Curability of Certain Forms of Insanity, Epilepsy, Catalepsy, and Hysteria in Females.* London: Robert Hardwicke, 1866.

Barnes, Shawn. "Practicing Pelvic Examinations by Medical Students on Women under Anaesthesia: Why Not Ask First?" *Obstetrics and Gynaecology* 120, 4, (2012): 941–43.

Barnes-Dean, Virginia. "Clitoridectomy and Infibulation." *Cultural Survival Quarterly* 9, 2, (1985).

Bar On, Bat-Ami, ed. Preface to *Engendering Origins: Critical Feminist Readings in Plato and Aristotle.* Albany, NY: State University of New York Press, 1994.

Barrett, Michèlle and Mary McIntosh. "The 'Family Wage': Some Problems for Socialists and Feminists." *Capital and Class* 4, 2, (1980): 51–72.

Baskin, Judith. "Women in Rabbinic Judaism." In *Judaism from Moses to Muhammad: An Interpretation,* edited by Jacob Neusner, William Scott Green, and Alan J. Avery-Peck, 303–334. Leiden: Koninklijke Brill NV, 2005.

Bataille, Georges. *Death and Sensuality: A Study of Eroticism and the Taboo.* New York: Walker and Co., 1962.

Baumeister, Roy, Karen Dale, and Kristin Sommer. "Freudian Defense Mechanisms and Empirical Findings in Modern Social Psychology: Reaction Formation, Projection, Displacement, Undoing, Isolation, Sublimation and Denial." *Journal of Personality* 66, (1998): 1081–1124.

Beal, John. Review of *Before Dallas: The U.S. Bishops' Response to Clergy Sexual Abuse of Children* by Nicholas Cafardi. *The Catholic Historical Review* 97, 1, (2011): 188–89.

Beckerman, Stephen and Paul Valentine, eds. "The Concept of Partible Paternity among Native South Americans." In *Cultures of Multiple Fathers: The Theory and Practice of Partible Paternity in Lowland South America,* 1–13. Gainesville: University Press of Florida, 2002.

Becking, Bob and Meindert Dijkstra. Introduction. In *Only One God? Monotheism in Ancient Israel and the Veneration of the Goddess Asherah,* edited by Bob Becking, Meindert Dijkstra, Margo Korpel, and Karel Vriezen, 11–16. London: Continuum, 2001.

Behringer, Wolfgang. *Witches and Witch-Hunts: A Global History.* Hoboken, NJ: Wiley-Blackwell, 2004.

———. "Weather, Hunger and Fear: Origins of the European Witch Hunts in Climate, Society and Mentality." In *The Witchcraft Reader,* edited by Darren Oldridge, 69-86. London: Routledge, 2002.

Beizer, Janet. *Ventriloquised Bodies: Narratives of Hysteria in Nineteenth-Century France.* Ithaca, NY: Cornell University Press, 1994.

Bem, Sandra Lipsitz. *The Lenses of Gender: Transforming the Debate on Sexual Inequality.* New Haven, CT: Yale University, 1993.

Benko, Stephen. *Pagan Rome and the Early Christians.* Bloomington: Indiana University Press, 1986.

Bennett, Judith. *History Matters: Patriarchy and the Challenge of Feminism.* Philadelphia: University of Pennsylvania Press, 2006.

———. "Response: Re-Reading *History Matters.*" *Journal of Women's History* 20, 2, (2008): 151–54.

Bergman, Stephen. "Men's Psychological Development: A Relational Perspective." Lecture transcript, Wellesley Centre for Research on Women: Work in Progress, 1991.

Bergmann, M. *In the Shadow of Moloch: The Sacrifice of Children and its Impact on Western Religions.* New York: Columbia University Press, 1992.

Bergren, Theodore. "2 Esdras." In *The New Oxford Annotated Bible,* edited by Michael Coogan, et al. New York: Oxford University Press, 2007.

Berkowitz, Leonard and James Green. "The Stimulus Qualities of the Scapegoat." *Journal of Abnormal and Social Psychology* 7, (1962): 202–207.

Berman, Louis. *The Akedah: The Binding of Isaac.* Lanham, MD: Jason Aronson, 1997.

Berrill, Norman John and Gerald Karp. *Development.* New York: McGraw-Hill, 1976.

Bettelheim, Bruno. *Symbolic Wounds: Puberty Rites and the Envious Male.* New York: Collier Books, 1962.

Bever, Edward. "Old Age and Witchcraft in Early Modern Europe." In *Old Age in Preindustrial Society,* edited by Peter Stearns, 150–90. Teacneck, NJ: Holmes and Meier, 1982.

Billig, Michael. "Freud and Dora: Repressing an Oppressed Identity." *Theory, Culture and Society* 14, (1997): 29–55.

Bilu, Yoram. "From Milah (Circumcision) to Milah (Word): Male Identity and Rituals of Childhood in the Jewish Ultraorthodox Community." *Ethos, Journal of Society for Psychological Anthropology* 31, 2, (June, 2003): 172–203.

Binger, Tilde. *Asherah: Goddesses in Ugarit, Israel and the Old Testament.* Sheffield, UK: Sheffield Academic Press, 1997.

Blackstone, William. *Commentaries on the Laws of England.* Oxford: Clarendon Press, 1765.

Bloch, R. Howard. *Medieval Misogyny and the Invention of Western Romantic Love*. Chicago: University of Chicago Press, 1991.

Blumberg, Rae Lesser. "Toward a Feminist Theory of Development." In *Feminism and Sociological Theory,* edited by Ruth Wallace, 161-199. Beverley Hills, CA: Sage, 1989.

Blumenthal, David. "The Images of Woman in the Hebrew Bible." In *Marriage, Sex, and Family in Judaism,* edited by Michael Broyde and Michael Ausubel, 15-60, Lanham, MD: Rowman and Littlefield, 2005.

Blumoff, Theodore. "*Genesis*, Gender and Community." *Southern California Review of Law and Women's Studies* 5, (1999).

Blundell, Sue. *Women in Ancient Greece.* Cambridge, MA: Harvard University Press, 1995.

Bodin, Jean. *On the Demon-Mania of Witches.* Translated by Randy Scott. Toronto: CRRS University of Toronto, 1995.

Boguet, Henri. *Examen of Witches.* Edited by Montague Summers, translated by E. A. Ashwin. London: Kessinger Publishing, 2003.

Bokanowski, Thierry. "Negative Transference and Transference Depression." In *Ferenczi's Turn in Psychoanalysis,* edited by Peter Rudnytsky, Antal Bókay, and Patrizia Giampieri-Deutsch, 120–44. New York: New York University Press, 1996.

Bolton, Robert. "Definitions of The Soul: *De Anima* ii, 1–3." *Phronesis* 23, 3, (1978): 258–78.

Bossy, John. "Thinking with Clark." *Past and Present* 166, (2000): 242–50.

Bourdieu, Pierre. *Masculine Domination.* Cambridge: Polity Press, 2001.

———. *Practical Reason.* Cambridge: Polity Press, 1998.

———. "The Sentiment of Honour in Kabyle Society." In *Honour and Shame: The Values of Mediterranean Society,* edited by John Peristiany, 192-241. Chicago: Chicago University Press, 1966.

Bourdieu, Pierre and Jean-Claude Passeron. *Reproduction in Education, Society and Culture.* Translated by Richard Nice. London: Sage, 1990.

Bradley, Christine. "Final Report on Domestic Violence Report No. 14." Papua New Guinea Law Reform Commission, 1992.

Brandon, Samuel. *Creation Legends of the Ancient Near East.* London: Hodder and Stoughton, 1963.

Brauner, Sigrid. *Fearless Wives and Frightened Shrews: The Construction of the Witch in Early Modern Germany.* Amherst: University of Massachusetts Press, 2001.

Breger, Louis. *Freud: Darkness in The Midst of Vision.* New York: John Wiley and Sons, 2000.

Bremmer, Jan. *Greek Religion and Culture, the Bible, and the Ancient Near East.* Leiden: Koninklijke Brill NV, 2008.

Britannica Online Encyclopedia. Encyclopedia Britannica Inc, 2012.

Broderick, Elizabeth. "Men on the Front Line of Gender Equality." Melbourne: TEDx Lecture, January 12, 2014. Accessed April 15, 2014. http://www.youtube.com/watch?v=kwQVZWEIXN0

Broedel, Hans Peter, trans. *The Malleus Maleficarum and the Construction of Witchcraft: Theology and Popular Belief.* Manchester: Manchester University Press, 2003.

Brooks, Gary. "Masculinity and the U.S. Presidency." *Psyccritiques* 50, 10, (2005).

Brooks, Gary and Louise Silverstein. "Understanding the Dark Side of Masculinity: An Interactive Systems Model." In *A New Psychology of Men,* edited by Ronald Levant and William Pollack, 280–333. New York: Basic Books, 1995.

Brookes, Rob. "Review *Beyond Human Nature: How Cultural Experience and Experience Shape our Lives.* " *The Conversation*, June, 2012. Accessed April 28, 2014. http://theconversation.com/peer-review-beyond-human-nature-how-culture-and-experience-shape-our-lives-7102.

Brown, Brené, Virginia Rondero Hernandez, and Yolanda Villarreal. "Connections: A 12-Session Psychoeducational Shame Resilience Curriculum." In *Shame in the Therapy Hour,* edited by Ronda Dearing and June Tangney, 355–71. Washington DC: American Psychological Association, 2011.

Brown, Michael. "Biblical Myth and Contemporary Experience: The Akedah in Modern Jewish Literature." *Judaism* 31, (1982): 99–111.

Brown, Rachael. "Soldier Suicide." ABC Radio AM program. Australian Broadcasting Commission, April 22–23, 2014.

Budd, Philip. "Holiness and Cult." In *The World of Ancient Israel,* edited by R. E. Clements. Cambridge: Cambridge University Press, 1989.

Budin, Stephanie. *Images of Woman and Child from the Bronze Age: Reconsidering Fertility, Maternity and Gender in the Ancient World.* New York: Cambridge University Press, 2011.

Buell, Denise. *Making Christians: Clement of Alexandria and the Rhetoric of Legitimacy.* Princeton, NJ: Princeton University Press, 1999.

Burghartz, Susanna. "Ordering Discourse and Society: Moral Politics, Marriage and Fornication during the Reformation and the Confessionalization Process in Germany and Switzerland." In *Social Control in Europe 1500–1800,* edited by Herman Roodenburg and Pieter Spierenburg, 78–98. Columbus: Ohio State University Press, 2004.

———. "The Equation of Women and Witches: A Case Study of Witchcraft Trials in Lucerne and Lausanne in the Fifteenth and Sixteenth Centuries." In *The German Underworld: Deviants and Outcasts in German History,* edited by Richard Evans, 57–74. London: Routledge, 1988.

Burns, Jane et al. "Game On: Exploring the Impact of Technologies on Young Men's Mental Health and Wellbeing." Young and Well Cooperative Research Centre, Melbourne, 2013.

Butler, Judith. "Speaking Up, Talking Back: Joan Scott's Critical Feminism." In *The Question of Gender: Joan W. Scott's Critical Feminism,* edited by Judith Butler and Elizabeth Weed, 11–28. Bloomington: Indiana University Press, 2011.

———. *The Psychic Life of Power.* New York: Routledge, 1999.

———. *Bodies That Matter: On the Discursive Limits of Sex.* New York: Routledge, 1993.

———. *Gender Trouble: Feminism and the Subversion of Identity.* New York: Routledge, 1990.

Butler, Judith and Elizabeth Weed. Introduction to *The Question of Gender: Joan W. Scott's Critical Feminism,* edited by Judith Butler and Elizabeth Weed, 1–10. Bloomington, IN: Indiana University Press, 2011.

Cafardi, Nicholas. *Before Dallas: The U.S. Bishops' Response to Clergy Sexual Abuse of Children.* Mahwah, NJ: Paulist Press, 2008.

Callan, Aela. "It's a Man's World." *101 East* Program, Aljazeera, March 8, 2013.

Campbell, Harry. *Differences in the Nervous Organisation of Man and Woman: Physiology and Pathology.* London: H. K. Lewis, 1891.

Caplan, Paula. *The Myth of Women's Masochism.* Lincoln, NE: iUniverse, 1993.

Caputi, Mary. *Feminism and Power: The Need for Critical Theory.* Lanham, Maryland: Lexington Books, 2013.

Carcach, Carlos and P. Grabosky. "Murder-Suicide in Australia." *Trends and Issues in Crime and Criminal Justice* No. 82. Canberra: Australian Institute of Criminology, March, 1998.

Carden, Michael. "Endangered Ancestress Revisited: Sarah's Miraculous Motherhood and the Restoration of Eden." *The Bible and Critical Theory* 1, 3, (2005): 18.1–18.14

Carter, Lucy. "OECD Snapshot Ranks Australia Second in World in Anti-Depressant Prescriptions." ABC News, Australian Broadcasting Corporation, November 22, 2013.

Caruth, Cathy. *Unclaimed Experience: Trauma, Narrative and History*. Baltimore: John Hopkins University Press, 1996.

Catalyst. "No Change for Women in Top Leadership." In 2012 Catalyst Census: Fortune 500. Catalyst, 2012.

CEDA. *Women in Leadership: Understanding the Gender Gap*. Committee for Economic Development of Australia (CEDA), 2013.

Chafetz, Janet Saltzman, ed. Introduction to *Handbook of the Sociology of Gender*, 3–25. New York: Springer, 2006.

Chandler, Jo. "One Woman's War." *The Age*, May 27, 2006.

Charlton, Sue Ellen, Jana Everett, and Kathleen Staudt. *Women, the State, and Development*. Albany: State University of New York Press, 1989.

Chief of Army, Lieutenant General David Morrison. "Chief of Army Message Regarding Unacceptable Behaviour." YouTube, June 13, 2013. Accessed April 28, 2014. http://www.youtube.com/watch?v=QaqpoeVgr8U.

Chodorow, Nancy. *The Reproduction of Mothering*. Berkeley, CA: University of California Press, 1978.

———. *Feminism and Psychoanalytic Theory*. New Haven, CT: Yale University Press, 1989.

Christian Scriptures. Bible Gateway. Accessed April 28, 2014. http://www.biblegateway.com/.

Chrysostom, John. "Two Exhortations to Theodore after his Fall." In *Nicene and Post-Nicene Fathers, First Series*. Vol. 9, edited by Philip Schaff. Translated by W. R. W. Stephens. Buffalo, NY: Christian Literature Publishing Co., 2009.

Cixous, Hélène. "Sorties: Out and Out: Attacks/Ways Out/Forays." In *The Newly Born Woman* by Hélène Cixous and Catherine Clément, translated by Betsy Wing, 63–128. London: I. B. Tauris and Co, 1996.

Clack, Beverly. *Misogyny in the Western Philosophical Tradition: A Reader*. New York: Routledge, 1999.

Clark, Kelly James. "The Gods of Abraham, Isaiah and Confucius." *Dao, a Journal of Comparative Philosophy* 5, 1, (2005): 109–136.

Clark, Rosemary. *The Sacred Magic of Ancient Egypt: The Spiritual Practice Restored*. Llewellyn Publications, 2003.

Clark, Stuart. *Thinking with Demons*. New York: Oxford University Press, 1999.

Clément, Catherine. "The Guilty One." Translated by Betsy Wing in *The Newly Born Woman*. By Hélène Cixous and Catherine Clément. London: I. B. Tauris and Co, 1996.

Cohen, David. *Law, Sexuality and Society: The Enforcement of Morals in Classical Athens*. New York: Cambridge University Press, 1991.

Cohn, Norman. "The Demonisation of Medieval Heretics." In *The Witchcraft Reader*, edited by Darren Oldridge, 36–52. London: Routledge, 2002.

Coldicott, Yvette. "The Ethics of Intimate Examinations." *British Medical Journal* 97, (2003): 98.

Collins, Randall, et al. "Toward an Integrated Theory of Gender Stratification." *Sociological Perspectives* 36, 3, (1993): 185–216.

Connell, Raewyn. *Confronting Equality: Gender, Knowledge and Global Change*. Cambridge, UK: Polity, 2011.

———. *Gender in World Perspective*, 2nd ed. Cambridge, UK: Polity, 2009.

Connell, R. W. *Masculinities*. Cambridge, UK: Polity, 2005.

———. "Psychoanalysis on Masculinity." In *Theorising Masculinity*, edited by Harry Brod and Michael Kaufman, 11–40. Newbury Park, CA: Sage, 1994.

———. *Gender and Power: Society, the Person and Sexual Politics*. Oxford: Polity Press, 1987.

———. *Which Way Is Up? Essays on Sex, Class and Culture*. Sydney: Allen and Unwin, 1983.

Connell, R. W. and J. W. Messerschmidt. "Hegemonic Masculinity: Rethinking the Concept." *Gender and Society* 19, (2005): 829–859.

Connell, R. W. and Julian Wood. "Globalisation and Business Masculinity." *Men and Masculinities* 7, 4, (2005): 347–64.

Connell, R. W., Lahoucine Ouzgane, and Daniel Coleman. “Cashing Out the Patriarchal Dividend: An Interview with R. W. Connell.” *Jouvert: A Journal of Postcolonial Studies* 2, 1, (1998).

Coombe, Rosemary. “Barren Ground: Re-Conceiving Honour and Shame in the Field of Mediterranean Ethnography.” *Anthropologica* 32, 2, (1990): 221–238.

Coughtry, Sarah. “Patriarchy and the Trap of Masculinity.” MA thesis, Smith College, 2011.

Cornwell, John. *The Dark Box: A Secret History of Confession*. London: Basic Books, 2014.

Coutinho, Elsimar. *Is Menstruation Obsolete?* Oxford: Oxford University Press, 1999.

Currier, Ashley. “Political Homophobia in Postcolonial Namibia.” *Gender and Society* 24, 1, (2010): 110–29.

Dalley, Stephanie, ed. and trans. *Myths from Mesopotamia: Creation, the Flood, Gilgamesh, and Others*. Oxford: Oxford University Press, 2000.

Dally, Ann. *The Trouble with Doctors: Fashions, Motives, and Mistakes*. London: Robson Books, 2003.

Daly, Mary. *Gyn/Ecology*. Boston: Beacon Press, 1978.

Daniels, Cynthia. *Exposing Men: The Science and Politics of Male Reproduction*. New York: Oxford University Press, 2006.

David, Nadia. “Exploring the Use of Domestic Violence Fatality Review Teams.” Australian Domestic and Family Violence Clearinghouse, 2007.

Davidson, James. “Dover, Foucault and Greek Homosexuality: Penetration and the Truth of Sex.” *Past and Present* 170, 1, (2001): 3–51.

———. *The Greeks and Greek Love*. London: Weidenfeld, 2007.

Davis, Sue. *The Political Thought of Elizabeth Cady Stanton: Women's Rights and the American Political Traditions*. New York: New York University Press, 2008.

Dean, Anne. *Teenage Pregnancy: The Interaction of Psyche and Culture*. Hillsdale, NJ: The Analytic Press, 1997.

Dean-Jones, Lesley Ann. “Menstrual Bleeding According to the Hippocratics and Aristotle.” *Transactions of the American Philological Association* 119, (1989): 177–192.

———. *Women's Bodies in Classical Greek Science*. Clarendon Press, 1996.

de Beauvoir, Simone. *The Second Sex*. Translated by Constance Borde and Sheila Malovany Chevallier. London: Vintage, 2010.

Decker, Rainer and H. C. Eric Midelfort. *Witchcraft and the Papacy: An Account Drawing on the Formerly Secret Records of the Roman Inquisition*. Charlottesville: University of Virginia Press, 2010.

Delaney, Carol. “The Abrahamic Theory of Procreation and Spiritual Kinship.” Paper presented at Wenner-Gren Workshop on “The Sacred Social: Investigations of Spiritual Kinship among Abrahamic Faiths.” University of Virginia, March 27–29, 2014.

———. *Abraham on Trial*. Radio interview. Religion for Life program. WEHC-FM (90.7 MHz) Emory, VA. Accessed May 15, 2012. http://religionforlife.me/2012/03/01/carol-delaney-abraham-on-trial-march-8-12/.

———. “Sacrificial heroics: The story of Abraham and its use in the justification of war.” Lecture. New York: Columbia Law School, April 4, 2005. http://cms1.runtime.com/null/heroics?exclusive=filemgr.download&file_id=94783&showthumb=0

———. “Was Abraham Ethical? Should We Admire His Willingness to Sacrifice His Son?” Markkula Ethics Center Lecture, 18 April, 2002. Accessed May 15, 2012. http://www.scu.edu/ethics/publications/submitted/delaney/delaneypaper.html.

———. “Cutting the Ties That Bind.” In *Relative Values: Reconfiguring Kinship* Studies, edited by Sarah Franklin and Susan McKinnon, 445–467. Durham, NC: Duke University Press, 2001.

———. *Abraham on Trial: The Social Legacy of Biblical Myth*. Princeton, NJ: Princeton University Press, 1998.

———. “Abraham and the Seed of the Patriarchs.” In *Genesis: A Feminist Companion to the Bible*, edited by Athalya Brenner, 129–148. Sheffield: Sheffield Academic Press, 1998.

———. *The Seed and the Soil: Gender and Cosmology in Turkish Village Society*. Berkeley: University Of California Press, 1991.

———. "The Legacy of Abraham." In *Anti-Covenant: Counter-reading Women's Lives in the Hebrew Bible*, edited by Mieke Bal, 28–41. Sheffield: Sheffield Academic Press, 1989.

———. "Seeds of Honour. Fields of Shame." In *Honour and Shame and the Unity of the Mediterranean.* Vol. 22, edited by David Gilmore, 35–48. Washington DC: American Anthropological Association Special Publication, 1987.

———. "The Meaning of Paternity and the Virgin Birth Debate." *Man* New Series 21, 3, (1986): 494–513.

———. "The Legacy of Abraham." In *Beyond Androcentrism: New Essays on Women and Religion*, edited by Rita Gross, 217–236. Missoula, MN: Scholars Press, 1977.

Denniston, George and Marilyn Milos, eds. *Sexual Mutilations: A Human Tragedy.* New York: Plenum Press, 1997.

Derrida, Jacques. *Specters of Marx: The State of the Debt, the Work of Mourning, and the New International.* Translated by Peggy Kamuf. New York: Routledge, 1994.

———. *The Gift of Death.* Translated by David Wills. Chicago: University of Chicago Press, 1995.

Deutsch, Morton. "Maintaining Oppression." In *Beyond Intractability*, edited by Guy Burgess and Heidi Burgess. Boulder, CO: Conflict Research Consortium, University of Colorado, 2005.

Devettere, Raymond. *Practical Decision Making in Health Care Ethics: Cases and Concepts* 3rd ed. Washington, DC: Georgetown University Press, 2010

Diamond, Michael. "Masculinity Unravelled: The Roots of Male Gender Identity and the Shifting of Male Ego Ideals throughout Life." *Journal of the American Psychoanalytic Association* 54, 4, (2006): 1099–1130.

Dijkstra, Meindert. "El, the God of Israel—Israel, the People of YHWH: On the Origins of Yahwism in Ancient Israel." In *Only One God? Monotheism in Ancient Israel and the Veneration of the Goddess Asherah,* edited by Bob Becking, Meindert Dijkstra, Margo Korpel, and Karel Vriezen. London: Continuum, 2001.

Dinnerstein, Dorothy. *The Mermaid and the Minotaur: Sexual Arrangements and Human Malaise.* New York: Other Press, 1999.

Diogenes Laertius. *Lives of Eminent Philosophers.* Edited by R. D. Hicks. Cambridge, MA: Harvard University Press, 1972.

Diski, Jenny. "The Will of the Fathers." *London Review of Books, 20*, 24 (December 10, 1998).

Dio Chrystostom. *The Seventh or Euboean Discourse.* Vol. 1. Translated by J. W. Cohoon. Cambridge, MA: Harvard University Press, Loeb Classical Library, 1932.

D'Itri, Patricia Ward. *Cross Currents in the International Women's Movement 1838–1948.* Bowling Green, OH: Bowling Green State University Popular Press, 1999.

DLA Piper. *Report of the Review of Allegations of Sexual and Other Abuse in Defence.* Vol. 1. DLA Piper, June 2012.

Doidge, Norman. *The Brain That Changes Itself.* New York: Viking, 2007.

Dorais, Michel. *Don't Tell: The Sexual Abuse of Boys.* Montreal: McGill-Queens University Press, 2002.

Douglas, Susan. *Enlightened Sexism: The Seductive Message That Feminism's Work Is Done.* New York: Times Books, 2010.

———. "The Rise of Enlightened Sexism." *On The Issues,* (Fall, 2012).

Downing, Francis, ed. *Making Sense in (and of) the First Christian Century.* Sheffield, UK: Sheffield Academic Press, 2000.

Dover, Kenneth. *Greek Homosexuality.* Cambridge, MA: Harvard University Press, 1978.

Dreyfus, Hubert and Paul Rabinow. *Michel Foucault: Beyond Structuralism and Hermeneutics.* Chicago: University of Chicago Press, 1982.

duBois, Page. *Sowing the Body: Psychoanalysis and Ancient Representations of Women.* Chicago: University of Chicago Press, 1991.

Ducat, Stephen. *The Wimp Factor: Gender Gaps, Holy Wars and the Politics of Anxious Masculinity.* Boston: Beacon Press, 2004.

Dufresne, Todd. *Killing Freud: Twentieth-Century Culture and the Death of Psychoanalysis.* London: Continuum, 2003.

Eichman, Peter. "Sex, Blood and Soul: The Transmission of Form in Aristotle's Biology." Accessed September 8, 2013. http://echodin.net/papers/phil515/aristotle.pdf.

Eilberg-Schwartz, Howard. "The Father, the Phallus, and the Seminal Word: Dilemmas of Patrilineality in Ancient Judaism." In *Gender, Kinship, and Power: A Comparative and Interdisciplinary History,* edited by Mary Jo Maynes, et al., 27–42. New York: Routledge, 1996.

Ellis, Havelock. *Studies in the Psychology of Sex.* Vol. 1. F. A. Davis, 1920.

———. *Studies in the Psychology of Sex.* Vol. 6. Teddington, UK: Echo Library, 2007.

el-Zein, Abdul Hamid. "Beyond Ideology and Theology: The Search for the Anthropology of Islam." *Annual Review of Anthropology* 6, (1977): 227–254.

EOWA. "2012 Census of Women in Leadership." Equal Opportunity for Women in the Workforce Agency, 2012.

Engelbrektsson, Ulla-Britt. *The Force of Tradition: Turkish Migrants at Home and Abroad.* Göteborg: Acta Universitatis Gothoburgensis, 1978.

Engels, Friedrich. *Origin of the Family, Private Property and the State.* London: Penguin, 2010.

Eschbach, Cheryl. "Toward an Understanding of Womb Envy: Developmental and Clinical Perspectives." In *Jealousy and Envy: New Views about Two Powerful Feelings,* edited by Léon Wurmser and Heidrun Jarass. New York: The Analytic Press, 2008.

Euripides. *Heraclidae.* In *Euripidea Altera.* Vol. 2, edited by David Kovacs. Leiden: E. J. Brill, 1996.

———. *The Essential Euripides: Dancing in Dark Times.* Edited by Robert Meagher. Wauconda, IL: Bolchaxy-Carducci Publishers, 2002.

European Union Agency for Fundamental Rights. "Main Results Report." In *Violence Against Women: An EU-wide Survey.* Luxembourg: Publications Office of the European Union, 2014. Accessed 4 March, 2014. http://fra.europa.eu/sites/default/files/fra-2014-vaw-survey-main-results_en.pdf.

Evans Grubbs, Judith, Tim Parkin, and Roslynne Bell, eds. *The Oxford Handbook of Childhood and Education in the Classical World.* New York: Oxford University Press, 2013.

Facid, Alda and Martha Morgan. "Equity or Equality for Women?" IWRAW Asia Pacific Occasional Papers No. 14. Kuala Lumpur, Malaysia: International Women's Rights Action Watch Asia Pacific, 2009.

Falk, Avner. *A Psychoanalytic History of the Jews.* London: Associated University Presses, 1996.

Falk, Rabbi P. E. *Modesty: An Adornment for Life: Halachos and Attitudes Concerning Rules of Tznius of Dress and Conduct.* New York: Feldheim Publishers, 1998.

Faludi, Susan. *Backlash: The Undeclared War Against American Women.* New York: Three Rivers Press, 2006.

Farr, Malcolm. "Tony Abbott Tells Julia Gillard to 'Make an Honest Woman of Herself' On Carbon Tax." News.com.au, February 25, 2011. Accessed November 10, 2013. http://www.news.com.au/national/tony-abbott-tells-julia-gillard-to-make-an-honest-womanof-herself-on-carbon-tax/story-e6frfkvr-1226012034629.

Fausto-Sterling, Anne. *Myths of Gender.* New York: Basic Books, 1985.

Federici, Silvia. *Caliban and the Witch: Women, the Body, and Primitive Accumulation.* Brooklyn, NY: Autonomedia, 2004.

Feldman, Yael. *Glory and Agony: Isaac's Sacrifice and National Narrative.* Stanford, CA: Stanford University Press, 2010.

Fennell, Phil. *Treatment without Consent: Law, Psychiatry, and the Treatment of Mentally Disordered People since 1845.* London: Routledge, 1996.

Fernel, Jean. *The Physiologia of Jean Fernel (1567).* Translated by John Forrester. American Philosophical Society, 2003.

Filmer, Robert. *Patriarcha.* 1680. Accessed July 28, 2013. http://www.constitution.org/eng/patriarcha.htm.

Fisher, Elizabeth. *Woman's Creation: Sexual Evolution and the Shaping of Society.* New York: McGraw-Hill, 1979.

Fiske, Shanyn. *Heretical Hellenism: Women Writers, Ancient Greece, and the Victorian Popular Imagination.* Columbus: Ohio University Press, 2008.

Fluehr-Lobban, Carolyn. *Ethics and the Profession of Anthropology: Dialogue for Ethically Conscious Practice,* 2nd ed. Lanham, MD: Rowman and Littlefield, 2003.

———. "Cultural Relativism and Universal Rights." *The Chronicle of Higher Education* (June, 1995): 33–35.

———. *Islamic Values and Social Practice.* Florida: University Press of Florida, 1994.

Forbes, G. et al. "The Role of Hostile and Benevolent Sexism in Women's and Men's Perceptions of the Menstruating Woman." *Psychology of Women Quarterly* 27, (2003): 58–63.

Foreman, Judy. "Pelvic Exams Performed without Permission." *The Boston Globe,* July 13, 2004. Accessed July 3, 2012. http://www.boston.com/yourlife/health/other/articles/2004/07/13/pelvic_exams_performed_without_permission/?page=full.

Forter, Greg. *Gender, Race, and Mourning in American Modernism.* Cambridge: Cambridge University Press, 2011.

———. "Freud, Faulkner, Caruth: Trauma and the Politics of Literary Form." *Narrative* 15, 3, (2007): 259–85

Forter, Greg. "Absolom! Absolom! The Politics of Traumatic Form." Speech delivered at Faulkner Society, 2006. www.faulknersociety.com/mla06forter.doc

Foster, Edgar. "Metaphor and Divine Paternity: The Concept of God's Fatherhood in the Divinae Institutions of Lactantius (250–325 CE)." PhD diss., University of Glasgow, 2008.

Foucault, Michel. *The History of Sexuality.* Vol. 3, *The Care of the Self.* Translated by Robert Hurley. New York: Vintage Books, 1988.

———. *The History of Sexuality.* Vol. 2, *The Use of Pleasure.* Translated by Robert Hurley. New York: Random House, 1986.

———. "The Subject and Power." In *Michel Foucault: Beyond Structuralism and Hermeneutics,* edited by Hubert Dreyfus and Paul Rabinow, 208–226. Chicago: University of Chicago Press, 1982.

———. "Method." In *The History of Sexuality.* Vol. 1, *An Introduction.* Translated by Robert Hurley. New York: Random House, 1979.

———. *The Archaeology of Knowledge.* London: Tavistock, 1972.

Fox, Catherine. "The Higher You Go, the Wider the Gap." In *Women in Leadership: Understanding the Gender Gap.* Committee for Economic Development of Australia (CEDA), 2013.

Frankenberry, Nancy. "Philosophy of Religion in Different Voices." In *Feminist Philosophy of Religion: Critical Readings,* edited by Pamela Sue Anderson and Beverly Clack. New York: Routledge, 2004.

Fraser, Nancy. *Justice Interrupus: Critical Reflections on the 'Postsocialist' Condition.* New York: Routledge, 1997.

Frede, Michael. "Monotheism and Pagan Philosophy in later Antiquity." In *Pagan Monotheism in Late Antiquity,* edited by P. Athanassiadi and M. Frede, 41–68. Oxford: Oxford University Press, 2002.

Frede, Michael. "Aristotle." 1997. Accessed June 2, 2012. http://chemistry.mtu.edu/~pcharles/SCIHISTORY/aristotle.html.

Freud, Sigmund. Note on references: unless otherwise indicated, Freud's works are from James Strachey, ed. and trans. *The Standard Edition of the Complete Psychological Works of Sigmund* Freud. London: The Hogarth Press (abbreviated as *SE*).

———. "Interpretation of Dreams." *SE.* Vol. 4. 1900.

———. "Three Essays on the Theory of Sexuality." *SE.* Vol. 7. 1905.

———. "'Civilised' Sexual Morality and Modern Nervous Illness." In *SE.* Vol. 9. 1908.

———. "The Dynamics of the Transference." In *SE.* Vol. 12. 1912.

———. "Recommendations to Physicians Practising Psycho-Analysis." In *SE.* Vol. 12. 1912.

———. *Totem and Taboo: Resemblances between the Psychic Lives of Savages and Neurotics.* London: George Routledge and Sons, 1919.

———. "Some Psychological Consequences of the Anatomical Distinction between the Sexes." *SE.* Vol. 19. 1925.

———. "Fetishism." *International Journal of Psychoanalysis* 9, (1928): 161–66.

———. "Civilisation and its Discontents." *SE.* Vol. 21. 1930.

———. "Female Sexuality." *SE.* Vol. 21. 1931.

———. "Femininity." *New Introductory Lectures on Psycho-Analysis. SE.* Vol. 22. 1932.

———. "Analysis Terminable and Interminable." *SE.* Vol. 23. 1937.

———. "An Outline of Psychoanalysis." *SE.* Vol. 23. 1938.

———. "Moses and Monotheism." Translated by Katherine Jones. *SE.* Vol. 23. 1939.

———. *Letters of Sigmund Freud.* Edited by Ernst Freud, translated by Tania and James Stern. New York: Basic Books, 1960.

———. *Cocaine Papers: Sigmund Freud.* Edited by Robert Byck. New York: Stonehill, 1974.

———. *The Complete Letters of Sigmund Freud to Wilhelm Fliess 1887–1904.* Edited and translated by Jeffrey Masson. Cambridge, MA: Harvard University Press, 1985.

———. *The Letters of Sigmund Freud to Eduard Silberstein 1871–1881.* Edited by Walter Broehlich. Translated by Arnold Pomeranz. Boston, MA: Harvard College, 1990.

———. *Sigmund Freud, Martha Bernays.* Vol. 1. Edited by Ilse Grubrich-Simitis. S. Fischer Verlag via Bloomberg, 2012.

Freud, Sigmund and A. Albrecht. "American Interview 1909." *Psychoanalytic Review* 55, 3, (1968): 33–39.

Freud, Sigmund and C. G. Jung. *The Freud/Jung Letters: The Correspondence between Sigmund Freud and C. G. Jung.* Edited by William Mcguire, translated by Ralph Manheim and R. F. C. Hull. London: The Hogarth Press and Routledge and Kegan Paul, 1977.

Freud, Sigmund and Ernest Jones. *The Complete Correspondence of Sigmund Freud and Ernest Jones: 1908–1939.* Edited by R. A. Paskauskas. Cambridge, MA: Harvard University Press, 1993.

Freud, Sigmund and Josef Breuer. *Studies on Hysteria 1893–1895.* Translated by James Strachey. Basic Books and the Hogarth Press, 1895.

Freud, Sigmund and Sándor Ferenczi. *The Correspondence of Sigmund Freud and Sándor Ferenczi.* Vol. 1. Edited by Eva Brabent and Ernst Falzeder. Translated by Peter Hoffer. Cambridge, MA: Harvard University Press, 1993.

Fromm, Erich. *The Crisis of Psychoanalysis.* New York: Holt, Rinehart and Winston, 1970.

Fukuyama, Francis. *End of History and the Last Man.* New York: Free Press, 1992.

Fulu, Emma, Xian Warner et al. "Why Do Some Men Use Violence Against Women And How Can We Prevent It? Quantitative Findings from the United Nations Multi-Country Study on Men and Violence in Asia and the Pacific." Bangkok: UNDP, UNFPA, UN Women and UNV, 2013.

Galen. *Galen on the Usefulness of the Parts of the Body: Peri Chreias Moriōn [Romanized Form] De Usu Partium.* Translated by Margaret Tallmadge. Ithaca, NY: Cornell University Press, 1968.

Gearin, Mary et al. "UK Arranged Marriages: Kidnapping, Rape and Murder in the Name of Family Honour." ABC *Foreign Correspondent* program, Australian Broadcasting Corporation, November 12, 2013. Accessed November 12, 2013. http://www.abc.net.au/news/2013-11-12/foreign-correspondent-honour-killings/5082946.

Geertz, Clifford. "Religion as a Cultural System." In *The Interpretation of Cultures: Selected Essays,* 87–125. New York: Basic Books, 1973.

Gellman, Jerome. *Abraham! Abraham! Kierkegaard and the Hasidim on the Binding of Isaac.* Hants, UK: Ashgate Publishing, 2003.

Georges, Eugenia. *Bodies of Knowledge: The Medicalization of Reproduction in Greece.* Nashville, TN: Vanderbilt University Press, 2008.

Gettler, Lee, Thomas McDade, Alan Feranil, and Christopher Kazawa. "Longitudinal Evidence that Fatherhood Decreases Testosterone in Human Males." *Proceedings of the National Academy of Sciences of the United States of America* 108, 39 (2011): 16194–16199.

Gilliam, David. "Oophorectomy (Removal of the Uterine Appendages) for the Insanity and Epilepsy of the Female: A Plea for its More General Adoption." *Transactions of the American Association of Obstetricians, Gynaecologists, and Abdominal Surgeons* 9, (1896): 319.

Gilligan, Carol. Interview by L. Granek (Video Recording). Psychology's Feminist Voices Oral History and Online Archive Project, New York, September 15, 2009.

———. *In a Different Voice: Psychological Theory and Women's Development.* Cambridge, MA: Harvard University Press, 1993.

Gilligan, James. *Preventing Violence.* New York: Thames and Hudson, 2001.

Gilman, Sander. Introduction to *Hysteria beyond Freud,* edited by Sander Gilman, vii–xxiv. San Francisco: University of California Press, 1993.

Gilmore, David. *Manhood in the Making.* New Haven, CT: Yale University Press, 1990.

———. *Misogyny: The Male Malady.* Philadelphia: Pennsylvania University Press, 2001.

———. *Honour and Shame and the Unity of the Mediterranean.* Arlington, VA: American Anthropological Association, 1987.

Girard, René. *Violence and the Sacred.* London: Continuum Publishing, 2005.

Glancy, Jennifer. *Corporal Knowledge: Early Christian Bodies.* New York: Oxford University Press, 2010.

Glancy, Jennifer. "The Law of the Opened Body: Tertullian on the Nativity." *Henoch* 30, (2008): 272.

Goddard, Amy Jo and Julie Carson. "At Your Cervix: A Documentary." 2012. Accessed December 2, 2013. http://www.atyourcervixmovie.com/.

Goddard, Victoria. "Honour and Shame: The Control of Women's Sexuality and Group Identity in Naples." In *The Cultural Construction of Sexuality,* edited by Pat Caplan, Ch 6. London: Tavistock, 1987.

Goff, Barbara. *Citizen Bacchae: Women's Ritual Practice in Ancient Greece.* Berkeley, CA: University of California Press, 2004.

Goldberg, Harvey, ed. *Judaism Viewed from Within and from Without: Anthropological Studies.* Albany: State University of New York Press, 1987.

———. Review of *Abraham on Trial* by Carol Delaney. *American Anthropologist* 102, 1, (2000): 188–89.

Goldenberg, N. R. "A Theory of Gender as a Central Hermeneutic in the Psychoanalysis of Religion." In *Hermeneutical Approaches in Psychology of Religion,* edited by J. A. Belzen, 65–84. Atlanta, GA: Rodopi, 1997.

Goldstein, Joshua. *War and Gender: How Gender Shapes the War System and Vice Versa.* New York: Cambridge University Press, 2003.

Goldsworthy, Anna. "Unfinished Business: Sex, Freedom and Misogyny." *Quarterly Essay, 50,* June, 2013.

Goldsworthy, Kerryn. "The Unseemly Splatter." Sydney Review of Books, October 22, 2013.

Gollaher, David. *Circumcision: A History of the World's Most Controversial Surgery.* New York: Basic Books, 2000.

Gonsoulin, Margaret. "Women's Rights and Women's Rites: Religion at the Historical Root of Gender Stratification." *Electronic Journal of Sociology,* (2005).

Goodman, Alan and Thomas Leatherman. "Traversing the Chasm between Biology and Culture: An Introduction." In *Building a New Sociocultural Synthesis: Political Economic Perspectives on Human Biology,* edited by Alan Goodman and Thomas Leatherman. Ann Arbor: University of Michigan Press, 1998.

Grant, Edward. *The Foundations of Modern Science in the Middle Ages: Their Religious, Institutional and Intellectual Contexts.* Cambridge: Cambridge University Press, 1996.

Gray, Edith, Anna Reimondos, and Liz Allen. "Australian Census: One in Ten Live Alone but That Doesn't Mean They're Lonely." *The Conversation,* June 22, 2012. Accessed June 10, 2013. https://theconversation.com/australian-census-one-in-ten-live-alone-but-that-doesnt-mean-theyre-lonely-7674.

Green, Monica. *Making Women's Medicine Masculine: The Rise of Male Authority in Pre-Modern Gynaecology.* New York: Oxford University Press, 2008.

Greenberg, David. *The Construction of Homosexuality.* Chicago: University of Chicago Press, 1990.

Greenson, Ralph. "Dis-identifying from Mother: Its Special Importance for the Boy." *International Journal of Psychoanalysis* 49, 3, (1968): 70–74.

Gregory, Marilyn. "Masculinity and Homicide." *International Journal of Law, Crime and Justice* 40, 2 (2011).

Grietje, Dresen. "The Better Blood: On Sacrifice and the Churching of New Mothers in the Roman Catholic Tradition." In *Wholly Woman, Holy Blood: A Feminist Critique of Purity and Impurity,* edited by Kristin De Troyer et al., 143–164. Harrisburg, PA: Trinity Press International, 2003.

Gronnvoll, Marita. *Media Representations of Gender and Torture Post-9/11.* New York: Routledge, 2010.

Gross, Paul. "Bashful Eggs, Macho Sperm, and Tonypandy." In *A House Built on Sand,* edited by Noretta Koertge, 59–70. New York: Oxford University Press, 1998.

Grosz, Elizabeth. *Volatile Bodies: Toward a Corporeal Feminism.* Bloomington: Indiana University Press, 1993.

Groth, Nicholas. *Men Who Rape: The Psychology of the Offender.* New York: Perseus Books, 2001.

Guardian. "Suicides At 10-Year High in US Military." *Guardian,* June 8, 2012. http://www.guardian.co.uk/world/2012/jun/08/suicide-rise-us-military

Guido, Diane. *The German League for the Prevention of Women's Emancipation: Antifeminism in Germany 1912–1920.* New York: Peter Lang Publishing, 2010.

Guiley, Rosemary. *The Encyclopedia of Witches, Witchcraft, and Wicca.* New York: Infobase Publishing, 2008.

Hacker, Jeanette and Kandeel Fouad. "Male Reproduction: Evolving Concepts of Procreation and Infertility through the Ages." In *Male Reproductive Dysfunction: Pathophysiology and Treatment,* edited by Kandeel Fouad, 3–8. New York: Informa Healthcare, 2007.

Halberstam, Judith. *Female Masculinity.* Durham, NC: Duke University Press, 1998.

Hales, Barbara. "Woman as Sexual Criminal: Weimar Constructions of the Criminal *Femme Fatale.*" In *Woman in German Yearbook: Feminist Studies in German Literature and Culture.* Vol. 12. 101–121. Lincoln: University of Nebraska Press, 1996.

Halperin, David. *One Hundred Years of Homosexuality and Other Essays on Greek Love.* New York: Routledge, 1990.

Hall, G. Stanley. "Adolescent Girls and their Education." In *Adolescence: Its Psychology and its Relations to Physiology, Anthropology, Sociology, Sex, Crime, Religion and Education.* Vol. 2. New York: D. Appleton, 1904.

Hammurabi. *The Code Of Hammurabi, King Of Babylon: About 2250 BC*, 2nd ed. Edited by Robert Francis Harper. Union, NJ: The Lawbook Exchange, 1999.

Hann, Chris. "Culture and Anti-Culture: The Spectre of Orientalism in New Anthropological Writing on Turkey." *Journal of the Anthropological Society of Oxford* 24, 3, (1993): 223–43.

Hannerz, Ulf. *Exploring the City: Inquiries toward an Urban Anthropology.* New York: Columbia University Press, 1980.

Harari, Yuval Noah. *Sapiens: A Brief History of Humankind.* London: Random House, 2014.

Hare-Mustin, Rachel and Jeanne Merecek. "The Meaning of Difference: Gender Theory, Postmodernism and Psychology." *American Psychologist* 43, (1988): 455–64.

Harlow, Mary. "In the Name of the Father." In *Thinking Men: Masculinity and its Self-Representation in the Classical Tradition,* edited by Lin Foxall and John Salmon, 155–68. London: Routledge, 1998.

Harrowitz, Nancy, trans. *Antisemitism, Misogyny, and the Logic of Cultural Difference: Cesare Lombroso and Matilde Serao.* Lincoln, Nebraska: University of Nebraska Press, 1994.

Hart, George. *The Routledge Dictionary of Egyptian Gods and Goddesses.* New York: Routledge, 2005.

———. *Egyptian Myths.* London: British Museum Press, 1990.

Hart, Paul 't. "Irving L. Janis' Victims of Groupthink." *Political Psychology* 12, 2, (1991): 247–78.

Harwood, Jonathon. *Styles of Scientific Thought: The German Genetics Community: 1900–1930.* Chicago: University of Chicago Press, 1993.

Hassan, Riffat. "Eidal-Adha (Feast of Sacrifice) in Islam: Abraham, Hagar and Ismael." In *Commitment and Commemoration: Jews, Christians, Muslims in Dialogue,* edited by Andre Lacocque, 131–150. Chicago: Exploration Press, 1992.

Hays, Harold. *The Organization of the Pyramid Texts.* Leiden: Koninklijke Brill NV, 2012.

Hebrew Scriptures. Bible Gateway. Accessed April 28, 2014. http://www.biblegateway.com/passage/?search=Genesis+22.

Heesacker, Martin and Shawn Prichard. "In a Different Voice Revisted: Men, Women and Emotion." *Journal of Mental Health Counseling* 14, (1992): 274–90.

Heinsohn, Gunnar. "Population, Conquest and Terror in the 21st Century." WebCitation.com, 2005. Scribd has since deleted this document and suggests another source. Accessed April 28, 2014. http://www.scribd.com/deleted/46528782.

Heinsohn, Gunnar and Otto Steiger. "The Elimination of Medieval Birth Control and the Witch Trials of Modern Times." *International Journal of Women's Studies* 3, 1, (1982): 93–214.

Herdt, Gilbert. *The Sambia: Ritual, Sexuality and Change in Papua New Guinea.* New York: Thomson/Wadsworth, 2006.

———. *The Sambia: Ritual and Gender in New Guinea.* New York: Holt, Rinehart and Winston, 1987.

———. *Rituals of Manhood: Male Initiation in New Guinea.* New Brunswick, NJ: Transaction Publishers, 1982.

———. *Sambia Sexual Culture: Essays from the Field.* Chicago: University of Chicago Press, 1999.

Hermant, Norman. "COAG Equity Report Finds Australian Women Still Lag Behind Men in Pay." ABC News, Australian Broadcasting Corporation, November 19, 2013.

Hesiod. *Works and Days.* In *The Homeric Hymns and Homerica,* translated by Hugh G. Evelyn-White. Cambridge, MA: Harvard University Press, 1914.

———. *Theogony.* In *The Homeric Hymns and Homerica,* translated by Hugh G. Evelyn-White. Cambridge, MA: Harvard University Press, 1914.

Hesiod and Apostolos Athanassakis, trans. *Hesiod: Theogony, Works and Days, Shield* 2nd ed. Baltimore, MD: Johns Hopkins University Press, 2004.

Hiebert, Paul. *Anthropological Insights for Missionaries.* Grand Rapids, MI: Baker Book House, 1985.

Hill, Darryl and Brian Willoughby. "The Development and Validation of the Genderism and Transphobia Scale." *Sex Roles* 53, 7–8, (2005): 531–44.

Hillel, David. *The Natural History of the Bible.* New York: Columbia University Press, 2006.

Hippocrates. *Lex.* In *The Genuine Works of Hippocrates,* edited by Charles Darwin Adams. New York: Dover Books, 1868.

Hobbs, Catherine, ed. "Culture and Practice of United States Women's Literacy." In *Nineteenth-Century Women Learn to Write,* 1–33. University Press of Virginia, 1995.

Hodder, Ian. "New Finds and New Interpretations at Çatalhöyük." *Çatalhöyük 2005 Archive Report.* Catalhoyuk Research Project, Institute Of Archaeology, 2005.

Hollingworth, Leta Stetter. *Functional Periodicity: An Experimental Study of the Mental and Motor Abilities of Women during Menstruation.* New York: Columbia University, 1914.

Holmes, Brooke. *Gender: Antiquity and its Legacy.* London: I.B. Tauris, 2012.

Homer. *Iliad.* Translated by Samuel Butler. Orange Street Press, 1998.

Hooper, Charlotte. *Manly States.* Ithaca, NY: Cornell University Press, 2008.

Horney, Karen. *Feminine Psychology.* Edited by Harold Kelman. New York: Norton, 1993.

Horowitz, Maryanne Cline. "Aristotle and Women." *Journal of History of Biology* 9, 2, (1976): 183–213.

———. *Seeds of Virtue and Knowledge.* Princeton, NJ: Princeton University Press, 1998.

Howard, Sharon. "Crime, Commonalities, and Authority in Early Modern Wales." PhD diss., University of Wales, 2003.

Howell, Elizabeth. *The Dissociative Mind.* New York: Routledge, 2005.

———. "'Good Girls', Sexy 'Bad Girls' and Warriors: The Role of Trauma and Dissociation in the Creation and Reproduction of Gender." In *Trauma and Sexuality: The Effects of Childhood Sexual, Physical and Emotional Abuse on Sexual Identity and Behaviour,* edited by James Chu and Elizabeth Bowman. Binghamton, NY: Haworth Medical Press, 2002.

Huber, Joan. "Comparative Gender Stratification." In *Handbook of the Sociology of Gender,* edited by Janet Saltzman Chafetz, 65–80. New York: Springer, 2006.

Human Rights Watch. "Item 12—Integration of the Human Rights of Women and the Gender Perspective: Violence against Women and 'Honour' Crimes." Human Rights Watch Oral

Intervention at the 57th Session of the UN Commission on Human Rights. April 6, 2001. Accessed December 2, 2013. http://www.hrw.org/news/2001/04/05/item-12-integration-human-rights-women-and-gender-perspective-violence-against-women.

Hunter, Mic. *Honor Betrayed: Sexual Abuse in America's Military*. Fort Lee, NJ: Barricade Books, 2007.

Hunter, Preston. "Major Religions of the World Ranked by Number of Adherents." Accessed December 2, 2013. http://www.adherents.com/Religions_By_Adherents.html.

Hunter, Virginia. "The Origins of Patriarchy: Gender and Class in the Ancient World." *Labour/Le Travail* 22 (1988): 239–246.

Hutton, Ronald. *Witches, Druids and King Arthur*. New York: Hambledon Continuum, 2006.

———. "Counting the Witch Hunt." Accessed December 2, 2013. http://www.summerlands.com/crossroads/remembrance/current.htm.

———. *Triumph of the Moon*. New York: Oxford University Press, 1999.

Hyde, Janet Shibley. "The Gender Similarities Hypothesis." *American Psychologist* 60, 6, (2005): 581–92.

İlkkaracan, Pinar. *Deconstructing Sexuality in the Middle East*. Burlington, VT: Ashgate Publishing, 2008.

———, ed. *Women and Sexuality in Muslim Societies*. Istanbul: Women for Women's Human Rights, Kadinin Insam Haklari Projesi, 2000.

Inckle, Kay. "Cultural History of Menstruation." In *Cultural Encyclopedia of the Body*, edited by Victoria Pitts-Taylor, 407–11. Westport, CT: Greenwood Publishing, 2008.

Inhorn, Marcia. *Quest for Conception: Gender, Infertility and Egyptian Medical Traditions*. Philadelphia: University of Pennsylvania Press, 1994.

———. *Infertility and Patriarchy: The Cultural Politics of Gender and Family Life in Egypt*. Philadelphia: University of Pennsylvania Press, 1996.

———. *Local Babies, Global Science: Gender, Religion, and In Vitro Fertilization*. New York: Routledge, 2003.

Innocent III, Pope. *On the Misery of the Human Condition*. Translated by Margaret Mary Dietz. New York: Bobbs-Merrill, 1969.

Irigaray, Luce. *Speculum of the Other Woman*. Translated by Gillian Gill. Ithaca, NY: Cornell University Press, 1985.

Islam, M. Mazharul and Uddin, M. Mosleh. "Female Circumcision in Sudan." *International Family Planning Perspectives* 27, 2, (2001): 71–76.

Jacobs, Louis and Ari Sagi. *Encyclopedia Judaica*. Vol. 1. 2nd ed. Edited by Michael Berenbaum and Fred Skolnik. Detroit, MI: Macmillan References USA, 2007.

Jaeggi, Stefan. "Lucerne Method of Sorcery and Witchcraft until the Mid-Sixteenth Century." *Swiss Journal of History* 52, (2002): 103–62.

Jain, Shall. "Lifetime Marriage and Divorce Trends." *Australian Social Trends 2007*. Sydney: Australian Bureau of Statistics, 2007.

Janis, Irving. *Groupthink: Psychological Studies of Policy Decisions and Fiascos* 2nd ed. New York: Houghton Mifflin, 1982.

Jantzen, Grace. *Violence to Eternity: Death and the Displacement of Beauty*. Vol. 2. London: Routledge, 2009.

Janzen, David. *The Social Meanings of Sacrifice in the Hebrew Bible: A Study of Four Writings*. Berlin: Walter De Gruyter GmbH and Co., 2004.

Jay, Nancy. *Throughout Your Generations Forever: Sacrifice, Religion and Paternity*. Chicago: Chicago University Press, 1992.

Jensen, Gary. *The Path of the Devil: Early Modern Witch Hunts*. Lanham, Maryland: Rowman and Littlefield Publishers, 2007.

Jerome. "Commentary on Zachariah, 3:13:1." In *Voices of Early Christianity: Documents from the Origins of Christianity*, edited by Kevin Kaatz. Santa Barbara, CA: Greenwood, ABC-CLIO, LLC., 2013.

Johnston-Robledo, Ingrid and Joan Chrisler. "The Menstrual Mark: Menstruation as Social Stigma." *Sex Roles* 68, 1–2, (2013): 9–18.

Jones, A. "Rougher than Usual Handling." In *Joining the Dots: Essays and Opinion*. Melbourne: Publish in Ink, 2014.

———. *Barking Mad.* Melbourne: Publish In Ink, 2011.

Jones, David. *The Soul of the Embryo.* London: Continuum, 2004.

Jones, Ernest. *Sigmund Freud: The Life and Work: The Young Freud 1856–1900.* London: Hogarth Press, 1953.

Jones, Rachel. *Irigaray*: Towards a Sexuate Philosophy. Cambridge: Polity Press, 2011.

Josephus, Flavius. "Antiquities of the Jews. Book 1, Section 232." In *The Works of Flavius Josephus,* translated by William Whiston. Buffalo: John Beardsley, 1895.

Jouanna, Jacques. *Greek Medicine from Hippocrates to Galen.* Leiden: Koninklijke Brill NV, 2012.

Kandiyoti, Deniz. "The Paradoxes of Masculinity: Some Thoughts on Segregated Societies." In *Dislocating Masculinity: Comparative Ethnographies,* edited by Andrea Cornwall and Nancy Lindesfarne, 197–213. London: Routledge, 1994.

———. "Bargaining with Patriarchy." *Gender and Society 2*, 3, (1988): 274–290.

Kanefsky, Yosef. "Witness to Redemption Parshat Vayerah (Genesis 18:1–22:24)." JewishJournal.com. November 9, 2006. Accessed December 2, 2013. http://www.jewishjournal.com/torah_portion/article/witness_to_redemption_20061110.

Kant, Immanuel. *The Conflict of the Faculties.* Translated by Mary Gregor. Norwalk, CT: Abaris Books, 1979.

Karlsson, Hasse et al. "How Psychotherapy Changes the Brain." *Psychiatric Times* August 11, 2011.

Katz, Jackson. *The Macho Paradox: Why Some Men Hurt Women.* Naperville, IL: Sourcebooks, 2006.

Kaufman, Michael. "Differential Participation: Men, Women and Popular Power." In *Community Power and Grassroots Democracy,* edited by Michael Kaufman and Alfonso, Haroldo. London: Zed Books and Ottawa: International Development Research Centre, 1997.

Kelman, Harold, ed. Introduction to *Feminine Psychology* by Karen Horney. New York: Norton, 1993.

Kelty, Ryan, Meredith Kleykamp, and David Segal. "The Military and the Transition to Adulthood." *The Future of Children* 20, 1, (2010): 181–207.

Kent, Susan Kingsley. *Gender and Power in Britain 1640–1990.* London: Routledge, 1999.

———. *Sex and Suffrage in Britain 1860–1914.* London: Routledge, 1990.

Kibler, William. *Medieval France: An Encyclopedia*. London: Routledge, 1995.

Kierkegaard, Søren. *Fear and Trembling.* Translated by C. Evans and S. Walsh. Radford, VA: Wilder Publications, 2008.

Kieckhefer, Richard. "Witch Trials in Mediaeval Europe." In *The Witchcraft Reader,* edited by Darren Oldridge, 25–35. London: Routledge, 2002.

———. *European Witch Trials: Their Foundations in Popular and Learned Culture 1300–1500.* Berkeley: University of California Press, 1976.

Kimmel, Michael. *Mars, Venus or Planet Earth? Women and Men in a New Millennium.* DVD, 2008. www.MediaEd.org. Accessed March 31, 2013. http://www.youtube.com/watch?v=JgaOK74HqiA.

———. *The Gender of Desire: Essays on Male Sexuality.* Albany: State University of New York Press, 2005.

———. *Manhood in America: A Cultural History.* New York: The Free Press, 1996.

———. "Masculinity as Homophobia: Fear, Shame and Silence in the Construction of Gender Identity." In *Theorising Masculinity,* edited by Harry Brod and Michael Kaufman, 119–141. Newbury Park, CA: Sage, 1994.

———. *Changing Men: New Directions in Research on Men and Masculinity.* New York: Sage Publications, 1987.

King, Helen. Introduction to *Blood, Sweat and Tears: The Changing Concepts of Physiology from Antiquity into Early Modern Europe,* edited by Manfred Horstmanshoff, Helen King, and Claus Zittel, 1–24. Leiden: Koninklijke Brill NV, 2012.

———. "Once upon a Text: Hysteria from Hippocrates." In *Hysteria Beyond Freud,* edited by Sander Gilman, 3–90. Berkeley: University of California Press, 1993.

———. "Sacrifical Blood." In *Rescuing Creusa: New Methodological Approaches to Women in Antiquity,* edited by Marilyn Skinner, 117–126. Lubbock, TX: Texas Tech University Press, 1987.

King James I. *Demonology.* Translated by Paul Tice. San Diego, CA: The Book Tree, 2002.

King, Mark, Sam Winter, and Beverley Webster. "Contact Reduces Transprejudice: A Study on Attitudes towards Transgenderism and Transgender Civil Rights in Hong Kong." *International Journal of Sexual Health* 21, 1, (2009): 17–34.

Kirkpatrick, Martha. "Thoughts about the Origins of Femininity." *The Journal of the American Academy of Psychoanalysis and Dynamic Psychiatry* 18, (1990): 554–565

Kilmartin, Christopher and Julie Allison. *Men's Violence against Women: Theory, Research, and Activism.* Mahwah, NJ: Lawrence Erlbaum Associates, 2007.

Kistler, S. I. "Rapid Bloodless Circumcision." *Journal of the American Medical Association* 54, (1910): 1782.

Kittay, Eva Feder. "Womb Envy: An Explanatory Concept." In *Mothering: Essays in Feminist Theory,* edited by Joyce Trebilcot. Totowa, NJ: Rowman and Allanheld, 1984.

Klaits, Joseph. *Servants of Satan: The Age of the Witch Hunts.* Bloomington: Indiana University Press, 1985.

Klare, Michael. *War without End: American Planning for the Next Vietnams.* New York: Knopf, 1972.

Klein, Melanie. *Envy and Gratitude: A Study of Unconscious Sources.* London: Tavistock Publications, 1957.

———. *Love, Guilt, and Reparation and Other Works 1921–1945.* The Melanie Kline Trust. New York: Random House, 2011.

———. "Early Stages of the Oedipus Conflict." *The Writings of Melanie Klein: Envy and Gratitude and Other Works.* London: Hogarth Press, 1975.

Klinenberg, Eric. *Going Solo: The Extraordinary Rise and Surprising Appeal of Living Alone.* New York: Penguin, 2013.

Kofman, Sarah. *The Enigma of Woman: Woman in Freud's Writings.* Translated by Catherine Porter. Ithaca, NY: Cornell University Press, 1985.

Kohn, Risa Levitt and Rebecca Moore. *A Portable God: The Origin of Judaism and Christianity.* Lanham, MD: Rowman and Littlefield, 2007.

Kõlves, Kairi, Eeva-Katri Kumpula, and Diego De Leo, eds. *Suicidal Behaviour in Men: Determinants and Prevention in Australia.* Australian Institute for Suicide Research and Prevention, Brisbane, 2013.

Kors, Alan and Edward Peters, eds. *Witchcraft in Europe 400–1700: A Documentary History* 2nd ed. Philadelphia: University of Pennsylvania Press, 2001.

Kramarae, Cheris and Dale Spender, eds. *Routledge International Encyclopedia of Women: Education: Health to Hypertension.* London: Routledge, 2000.

Kristeva, Julia. *Tales of Love.* Translated by Leon Roudiez. New York: Columbia University Press, 1989.

———. *Powers of Horror: An Essay on Abjection.* Translated by Leon Roudiez. New York: Columbia University Press, 1982.

Krugman, Steven. "Male Development and the Transformation of Shame." In *A New Psychology of Men,* edited by Ronald Levant and William Pollack, 91–128. New York: Basic Books, 1995.

Kuiper p, Kathleen, ed. *Ancient Greece: From the Archaic Period to the Death of Alexander the Great.* New York: Britannia Educational Publishing, 2011.

Lachman, Gary. "Otto Weininger." *The Dedalus Book of Literary Suicide.* Dedalus, 2008.

Lactantius. "On the Workmanship of God." In *Lactantius, the Minor Works, the Fathers of the Church: A New Translation.* Vol. 54. Translated by Mary Francis McDonald. Washington, DC: Catholic University of America Press, 1965.

———. "On the Workmanship of God." In *Ante-Nicene Fathers.* Vol. 7, *Fathers of the 3rd and 4th Century*, edited by Philip Schaff. Translated by William Fletcher. Grand Rapids, MI: Christian Classics Ethereal Library, 2009.

———. *Divine Institutes.* Translated by Anthony Bowen and Peter Garnsey. Liverpool, UK: Liverpool University Press, 2003.

Laqueur, Thomas. *Making Sex: Body and Gender from the Greeks to Freud.* Cambridge, MA: Harvard University Press, 1992.

Lambert, P. M. "Health Versus Fitness: Competing Themes in the Origins and Spread of Agriculture." *Current Anthropology* 50, 5 (2009): 603–8.

Lamont, Corliss. *The Philosophy of Humanism,* 8th ed. New York: Basic Books, 1997.

Larner, Christina. *Enemies of God: The Witch Hunt in Scotland.* Oxford: Basil Blackwell, 1983.

———. "The Crime of Witchcraft in Early Modern Europe." In *The Witchcraft Reader,* edited by Darren Oldridge, 276–88. London: Routledge, 2002.

———. "Was Witch-Hunting Woman-Hunting?" In *The Witchcraft Reader,* edited by Darren Oldridge. London: Routledge, 2002.

Laumann, Edward, Anthony Paik, Dale Glasser, et al. "A Cross-National Study of Subjective Sexual Well-Being among Older Women and Men: Findings from the Global Study of Sexual Attitudes and Behaviors." *Archives of Sexual Behaviour* 35, 2, (2006): 145–161.

Laws, Sophie. *Issues of Blood: The Politics of Menstruation.* London: Macmillan, 1990.

Leatherman, Tom and Alan Goodman. "Critical Biocultural Approaches in Medical Anthropology." In *A Companion to Medical Anthropology,* edited by Merrill Singer and Pamela Erickson. Chichester, UK: Blackwell, 2011.

Leck, Ralph. "Anti-Essentialist Feminism Versus Misogynist Sexology in Fin de Siecle Vienna." *Modern Intellectual History* 9, (2012): 33–60.

Lee, Richard and Richard Daly. "Man's Dominance and Women's Oppression: The Question of Origins." In *Community Power and Grassroots Democracy,* edited by Michael Kaufman and Haroldo Alfonso, 151–170. London: Zed Books and Ottawa: International Development Research Centre, 1997.

Leemhuis, F. "Ibrāhīm's Sacrifice of his Son in the Early Post-Koranic Tradition." In *The Sacrifice of Isaac: The Aqedah (Genesis 22) and its Interpretations,* edited by Ed Noort and Eibert Tigchelaar, 125–39. Boston: Brill, 2002.

Leeming, David. *The Oxford Companion to World Mythology.* New York: Oxford University Press, 2005.

———. *Creation Myths of the World: An Encyclopedia,* 2nd ed. Santa Barbara, CA: ABC-CLIO, LLC, 2010.

Lemay, Helen and St Albertus Magnus. *Women's Secrets: A Translation of Pseudo-Albertus Magnus's* De Secretis Mulierum *with Commentaries.* Albany: State University of New York Press, 1992.

Leo the Great. "Sermons of Leo the Great." In *Nicene and Post-Nicene Fathers:* Second Series. Vol. 12, *Leo the Great*, edited by Philip Schaff and Henry Wallace. New York: Cosimo, 2007.

Lerner, Gerda. "Women and History Part I: The Creation of Patriarchy." Interview with Gerda Lerner by Jeffrey Mishlove, 1999. Accessed April 24, 2012. http://www.intuition.org/txt/lerner1.htm.

———. *Why History Matters.* New York: Oxford University Press, 1997.

———. *The Creation of Patriarchy.* New York: Oxford University Press, 1986.

———. *The Witch Hunt in Early Modern Europe.* London: Pearson Longman, 2006.

Levant, Ronald. "Research in the Psychology of Men and Masculinity Using the Gender Role Strain Paradigm as a Framework." *American Psychologist* 66, (2011): 765–76.

———. "The Crisis in Connection between Men and Women." *The Journal of Men's Studies* 5, 1, (1996): 1–12.

———. "Toward the Reconstruction of Masculinity." In *A New Psychology of Men,* edited by Ronald Levant and William Pollack, 229–251. New York: Basic Books, 1995.

Levant, Ronald and Katherine Richmond. "A Review of Research on Masculine Ideologies Using the Male Role Norms Inventory." *Journal of Men's Studies* 15, (2007): 130–46.

Levant, Ronald and William Pollack. Introduction to *A New Psychology of Men,* edited by Ronald Levant and William Pollack, 1–10. New York: Basic Books, 1995.

Lévi-Strauss, Claude. *The Elementary Structures of Kinship.* Boston: Beacon Press, 1969.

Levins, Richard and Richard Lewontin. *The Dialectical Biologist.* Cambridge, MA: Harvard University Press, 1985.

Lewontin, Richard, Steven Rose and Leon Kamin. "The Determined Patriarchy." In *Family Bonds and Gender Divisions: Readings in the Sociology of the Family*, edited by Bonnie Fox, 209. Toronto: Canadian Scholars' Press, 1988.

Library of Congress. "A finding aid to the papers in the Sigmund Freud Collection in the Library of Congress." April, 2012. Sigmund Freud Collection, Manuscript Division, Library of Congress, Washington, DC. Accessed January 3, 2013. http://lcweb2.loc.gov/service/mss/eadxmlmss/eadpdfmss/2004/ms004017.pdf.

Lightfoot-Klein, Hanny. "Prisoners of Ritual: Some Contemporary Developments in the History of Female Genital Mutilation." Paper Presented at the Second International Symposium on Circumcision, San Francisco, May, 1991.

Linehan, Marsha. *Cognitive-Behavioural Therapy for Borderline Personality Disorder*. New York: Guilford Press, 1993.

Ljungqvist, Fredrik. "Female Shame, Male Honor: The Chastity Code in Juan Luis Vivres' *De Institutione Feminae Christianae*." *Journal of Family History* 37, 2, (2012): 139–154.

Llewellyn-Jones, Lloyd. *Aphrodite's Tortoise*. Swansea, UK: The Classical Press of Wales, 2003.

Loader, William. *The Pseudepigrapha on Sexuality: Attitudes toward Sexuality in Apocalypses, Testaments, Legends, Wisdom and Related Literature*. Cambridge, UK: Wm. B. Eerdmans Publishing, 2011.

Lochrie, Karma. *Covert Operations: The Medieval Uses of Secrecy*. Philadelphia: Pennsylvania University Press, 1999.

Lombroso, Cesare and Guglielmo Ferrero. *Criminal Woman, the Prostitute, and Normal Woman*. Translated by Nicole Hahn Rafter and Mary Gibson. Durham, NC: Duke University Press, 2004.

Lorber, Judith. *Gender Inequality: Feminist Theories and Politics*. Oxford: Oxford University Press, 2011.

Lotto, David. "Review of *The Wimp Factor* by Stephen Ducat." *Journal of Psychohistory* 34, 2, (2006): 171–73.

Lövdén, M., E. Wenger, J. Märtensson, U. Lindenberger, and L. Bäckmann. "Structural Brain Plasticity in Adult Learning and Development." *Neuroscience and Biobehavioural Reviews*, (February 28, 2013).

Luft, David. *Eros and Inwardness in Vienna: Weininger, Musil, Doderer*. Chicago: University of Chicago Press, 2003.

MacCulloch, Diarmaid. *Silence: A Christian History*. London: Allen Lane, 2013.

———. "Women Bishops." *The Observer*, July 7, 2012.

MacKinnon, Catharine. *Are Women Human? And Other International Dialogues*. Cambridge: Belknap Press, 2006.

———. *Toward a Feminist Theory of the State*. Cambridge, MA: Harvard University Press, 1989.

Maclean, Ian. *The Renaissance Notion of Women: A Study in the Fortunes of Scholasticism*. New York: Cambridge University Press, 1999.

Maddison, Sarah and Emma Partridge. *How Well Does Australian Democracy Serve Australian Women? The Democratic Audit of Australia*. Canberra: Australian National University, 2007.

Maines, Rachel. *The Technology of Orgasm: 'Hysteria,' the Vibrator, and Women's Sexual Satisfaction*. Baltimore, MD: John Hopkins University Press, 1999.

Malinowski, Bronislaw. *The Father in Primitive Psychology*. London: Routledge, 2002.

Malinowski, Bronislaw and Robert Redfield. *Magic, Science and Religion and Other Essays 1948*. San Francisco: The Free Press, 1948.

Mann, Michael. *The Sources of Social Power*. Vol. 1, *A History of Power from the Beginning to AD 1760*. New York: Cambridge University Press, 2012.

Manusmriti: "The Laws of Manu. No. 41" (c.1500 BCE.) Translated by G. Buhler. Internet Indian History Sourcebook, 2014. Accessed December 2, 2013. http://hinduism.about.com/library/weekly/extra/bl-lawsofmanu1.htm.

Marra, Thomas. *Dialectical Behaviour Therapy in Private Practice*. Oakland, CA: New Harbinger, 2005.

Mathews, Ben. "Female Genital Mutilation: Australian Law, Policy and Practical Challenges for Doctors." *Medical Journal of Australia* 194, 3, (2011): 139–141.

Mathews, Kenneth. *The New American Commentary: An Exegetical and Theological Exposition of Holy Scripture Genesis 11.27–50.26.* Vol. 1B. Broadman and Holman, 2005.

Maudsley, Henry. *The Physiology and Pathology of Mind.* London: Macmillan, 1868.

———. "Sex in Mind and in Education." *Popular Science Monthly* 5, (1874).

Mayhew, Robert. *The Female in Aristotle's Biology: Reason or Rationalisation.* Chicago: University of Chicago Press, 2004.

McElvaine, Robert. *Eve's Seed: Biology, the Sexes and the Course of History.* McGraw Hill, 2002.

McEwan, Bruce. *The End of Stress as We Know It.* Washington, DC: Joseph Henry Press, 2002.

McGrath, Albertus. *Women and the Church.* Colorado Springs, CO: Image Books, 1976.

McKeen, Catherine. "The Female in Aristotle's Biology." *Philosophical Books* 48, 1, (2007): 60–65.

McKeon, Michael. "Historicizing Patriarchy: The Emergence of Gender Difference in England 1660–1760." *Eighteenth-Century Studies* 28, 3, (1995): 295–322.

McKinley, Ann. "Identities Cubed: Perspectives of Multidimensional Masculinity Theory." Scholarly Works, Paper 761, 2013. Accessed December 2, 2013. http://scholars.law.unlv.edu/facpub/761.

McLaren, Angus. *I mpotence: A Cultural History.* Chicago: University of Chicago Press, 2007.

McPhedran, Ian. "Top Brass Told of Rapist in the Defence Force Ranks." *Daily Telegraph*, April 12, 2011 . Accessed January 9, 2012. http://www.dailytelegraph.com.au/news/national/top-brass-told-of-rapist-in-the-ranks/story-e6freuzr-1226037474092.

Mead, Margaret. "Sex Role." In *Cross-Cultural Human Development,* edited by Ruth Munroe, Robert Munroe, and Beatrice Whiting. Stamford, CT: Brooks/Cole Publishing, 1975.

Medvei, V. C. *The History of Clinical Endocrinology: A Comprehensive Account of Endocrinology from the Earliest Times to the Present Day.* New York: Parthenon Publishing, 1993.

Meeker, Michael. "The Black Sea Turks: A Study of Honour, Descent and Marriage". PhD diss., Chicago: University of Chicago, 1970.

Menander. *The Fragments of Attic Comedy.* Vol. 3, pt. 2. Translated by John Maxwell Edmonds. Brill Archive (Fragment 702, P. 827), 1961.

Merriam-Webster Online Dictionary. Accessed April 28, 2014. http://www.merriam-webster.com/.

Merrill, Yvonne. "'In Silence My Tongue is Broken': The Social Construction of Women's Rhetoric Before 1750." PhD diss., University of Arizona, 1994.

Mesch, Rachel. *The Hysteric's Revenge: French Women Writers at the Fin de Siècle.* Nashville, TN: Vanderbilt University Press, 2006.

Micale, Mark. *Approaching Hysteria: Disease and its Interpretations*. Princeton, NJ: Princeton University Press, 1995.

Midelfort, H. C. Erik. *Witch Hunting in Southwestern Germany 1562–1684: The Social and Intellectual Foundations.* Stanford, CA: Stanford University Press, 1972.

Miller, Alan, Shlomo Riskin, and Sheldon Zimmerman. "The Binding of Isaac: Three Rabbis Study and Teach a Text." *The Jewish Spectator* 31 (1983).

Miller, J. *Toward a New Psychology of Women.* Boston: Beacon Press, 1976.

Miller, Pavla. "Gender and Education before and after Mass Schooling." In *A Companion to Gender History,* edited by Teresa Meade and Merry Wiesner-Hanks. London: Blackwell, 2008.

Miller, Sarah. *Medieval Monstrosity and the Female Body.* New York: Routledge, 2010.

Millett, Kate. *Sexual Politics.* University of Illinois Press, 2000.

Millon, Theodore. *Masters of the Mind: Exploring the Story of Mental Illness from Ancient Times to the New Millennium.* John Wiley and Sons, 2004.

Milman, Yoseph. "The Sacrifice of Isaac and its Subversive Variations in Contemporary Hebrew Protest Poetry." Religion and Literature 23, (1991): 61–83.

Minahan, Anne, ed. *Encyclopedia of Social Work,* 20th ed. Vol. 1. New York: Oxford University Press, 2008.

Minister for Defence, Australian Government. "Paper Presented on the Defence Abuse Response Taskforce." Canberra: Australian Government, June 20, 2013. Accessed June 21, 2013. http://www.minister.defence.gov.au/2013/06/20/minister-for-defence-stephen-smith-paper-presented-on-the-defence-abuse-response-taskforce/.

Minsky, Rosalind, ed. *Psychoanalysis and Culture.* London: Routledge, 1998.

———, ed. *Psychoanalysis and Gender.* London: Routledge. 1996.

Möbius, Paul. *Ueber den Physiologischen Schwachsinn des Weibes* (*On the Physiological Imbecility of Women).* Halle: Marhold, 1903.

———. "The Physiological Mental Weakness of Woman." Translated by A. McCorn. *Alienist and Neurologist* 22, (1901): 624–642.

Moebius, Matthew. "Clerical Conceptions of Magic and the Stereotype of the Female Witch." *Oshkosh Scholar* 6, (2011): 9–24. Accessed April 26, 1012. http://www.uwosh.edu/grants/students/showcase/oshkosh-scholar/volumes/2011-volume-vi/osvol6-8Web.pdf.

Moi, Toril. "From Femininity to Finitude: Freud, Lacan and Feminism, Again." *Signs, Journal of Women in Culture and Society* 29, 3, (2004): 841–78.

———. *Sexual/Textual Politics: Feminist Literary Theory.* London: Routledge, 2002.

Muijen, Matt. Foreword. In *Routledge International Handbook of Clinical Suicide Research,* edited by John Cutcliffe et al. New York: Routledge, 2014.

Munroe, Robert and Ruth Munroe. "Male Pregnancy Symptoms and Cross-Sex Identity in Three Societies." *Journal of Social Psychology* 84, (1971): 11–25.

———. "Psychological Interpretation of Male Initiation Rites: The Case of Male Pregnancy Symptoms." *Ethos* 1, (1973): 490–98.

Montagu, Ashley. *Coming into Being among the Australian Aborigines.* London: Routledge, 1974.

Montaigne, Michel de. "On Affectionate Relationships." In *The Complete Essays,* edited and translated by M. A. Screech. Penguin Books, 1989.

Montenegro, Xenia. "The Divorce Experience: A Study of Divorce at Midlife and Beyond." American Association of Retired People, 2004.

Monter, E. William. "The Sociology of Jura Witchcraft." In *The Witchcraft Reader,* edited by Darren Oldridge. London: Routledge, 2002.

Moran, P., et al. "The Natural History of Self-Harm from Adolescence to Adulthood." *Lancet* 379 (2012): 236–43.

Murphy, Patricia. *In Science's Shadow: Literary Constructions of Late Victorian Women.* Columbia: University of Missouri Press, 2006.

Nadelson, Theodore. *Trained to Kill: Soldiers at War.* Baltimore, MD: Johns Hopkins University Press, 2005.

Nasr, Sayyed Hossein. *Ideals and Realities of Islam.* London: Allen and Unwin, 1985.

Needham, Joseph. *Chemical Embryology.* Vol. 1. Cambridge: Cambridge University Press, 1931.

Neumann, Erich. *The Fear of the Feminine and Other Essays on Feminine Psychology.* Princeton, NJ: Princeton University Press, 1994.

Neusner, J., trans. *Genesis Rabbah: The Judaic Commentary to the Book of Genesis: A New American Translation.* Atlanta, GA: Scholars Press, 1985.

New Interpreter's Bible. Nashville, TN: Abingdon Press, 1998.

New World Encyclopedia contributors. "Atum." *New World Encyclopedia.* Accessed March 1, 2014. www.newworldencyclopedia.org/entry/Atum.

Niehoff, Maren. *Philo on Jewish Identity and Culture.* Tübingen: J. C. B. Mohr, 2001.

Nightingale, Tom. "Pay Gap between Men and Women Wider Now than Twenty Years Ago." ABC News, Australian Broadcasting Corporation, September 4, 2013.

Nitzan, Tal, ed. *With an Iron Pen: Twenty Years of Hebrew Protest Poetry.* Albany: State University of New York Press, 2009.

Nixon, Lucia. "The Cults of Demeter and Kore." In *Women in Antiquity: New Assessments,* edited by Richard Hawley. London: Routledge, 1995.

Nye, Robert. "Sexuality." In *A Companion to Gender History,* edited by Teresa Meade and Merry Wiesner-Hanks. Hoboken, NJ: Wiley, 2008.

O'Donovan, Denis. "Health and Femiphobia." Paper Presented at the 95th Annual Meeting of the American Psychological Association, 1987.

O'Donovan, Denis. "Femiphobia: Unseen Enemy of Intellectual Freedom." *Men's Studies Review* 5, (1988): 5–8

Ogden, Daniel. "Homosexuality." In *A Cultural History of Sexuality in the Classical World,* edited by Mark Golden and Peter Toohey. New York: Oxford, 2011.

O'Grady, Kathleen. "The Semantics of Taboo: Menstrual Prohibitions in the Hebrew Bible." In *Wholly Woman, Holy Blood: A Feminist Critique of Purity and Impurity,* edited by Kristin de Troyer et al. Harrisburg, PA: Trinity Press International, 2003.

Oldridge, Darren, ed. General Introduction to *The Witchcraft Reader.* London: Routledge, 2002.

Olivier, Christiane. *Jocasta's Children: The Imprint of the Mother.* London: Taylor and Francis, 1989.

O'Neil, James. "Summarizing 25 Years of Research on Men's Gender Role Conflict Using the Gender Role Conflict Scale (GRCS)." *The Counseling Psychologist* 36, 3, (2008): 58–445.

Ontario Domestic Violence Death Review Committee. *Annual Report to the Chief Coroner.* Toronto, ON: Office of the Chief Coroner, 2008. February 23, 2012. http://www.makeitourbusiness.com/index.php?option=com_content&view=article&id=39&Itemid=50

Orme, Nicholas. *Medieval Children.* New Haven, CT: Yale University Press, 2003.

Padel, Ruth. "Women: Model for Possession by Greek Daemons." In *Images of Women in Antiquity,* edited by Averil Cameron and Amélie Kuhrt, 3–19. Oxon: Routledge, 1993.

Paris, Bernard. *Karen Horney: A Psychoanalyst's Search for Self-Understanding.* New Haven, CT: Yale University Press, 1994.

Parker, Robyn. "Family Statistics and Trends: Relationships Indicators Survey 2006." *Family Relationships Quarterly* 4, (2011). Accessed March 5, 2013. www.relationships.org.au.

Parker, Rozsika and Griselda Pollock. *Old Mistresses: Women, Art, and Ideology* (Kitchener, ON: Pandora Press, 1981).

Parla, Ayşe. "Honor: Turkey and the Caucasus." In *Encyclopedia of Women and Islamic Cultures: Family, Law and Politics.* Vol. 2, edited by Suad Joseph and Afsaneh Najmabadi. Leiden: Koninklijke Brill NV, 2005.

Pateman, Carol. *The Sexual Contract.* Stanford, CA: Stanford University Press, 1988.

Patton, George et al. "The Progression of Common Mental Disorders in Adolescents: A 14-Year Prospective Cohort Study." *Lancet* (January 16, 2014). doi:10.1016/S0140–6736(13)62116–9.

———. "Global Patterns of Mortality in Young People." *Lancet* 374 (2009): 881–92.

Pavlac, Brian. *Witch Hunts in the Western World: Persecution and Punishment from the Inquisition through the Salem Trials.* Westport, CT: Greenwood Press, 2009.

———. "Ten common errors and myths about the witch hunts." Accessed April 27, 2014. http://departments.kings.edu/womens_history/witch/werror.html.

Pellegrin, Pierre. "Aristotle." In *Greek Thought: A Guide to Classical Knowledge,* edited by Jacques Brunschwig and Geoffrey Lloyd, translated by Rita Guerlac and Anne Slack, 554–574. President and Fellows, Harvard College, 2000.

Peristiany, John, ed. *Honour and Shame: The Values of Mediterranean Society.* Chicago: Chicago University Press, 1965.

Pickthall, Mohammed, trans. *The Meaning of the Glorious Koran.* Hyderabad-Deccan: Government Central Press, 1938.

Pindar. "Pindar's Fourth Pythian Ode." In *Pindar's Victory Songs,* translated by F. J. Nisetich. Johns Hopkins Press, 1980. http://mkatz.web.wesleyan.edu/cciv243/cciv243.Pythian4.html

Pitt-Rivers, Julian. "Honour and Social Status." In *Honour and Shame: The Values of Mediterranean Society,* edited by John Peristiany. Chicago: Chicago University Press, 1965.

Plato. *The Laws of Plato.* Translated by Thomas Pangle. New York: Basic Books, 1980.

———. *Meno.* In *Plato in Twelve Volumes.* Vol. 3. Translated by W.R.M. Lamb. Cambridge, MA: Harvard University Press, 1967.

———. *Timaeus.* In *Plato in Twelve Volumes.* Vol. 9. Translated by W.R.M. Lamb. Cambridge, MA: Harvard University Press, 1925.

———. *Symposium*. In *Collected Works of Plato* 4th ed. Translated by Benjamin Jowett. Oxford: Oxford University Press, 1953.

Pleck, Joseph. *The Myth of Masculinity*. Cambridge, MA: MIT Press, 1981.

———. "The Gender Role Strain Paradigm: An Update." In *A New Psychology of Men*, edited by Ronald Levant and William Pollack, 11–32. New York: Basic Books, 1995.

Pliny the Elder. *Natural History*. Translated by John Bostock. London: Taylor and Francis, 1855. Accessed June 15, 2012. http://www.perseus.tufts.edu/hopper/text?doc=Perseus:text:1999.02.0137:book=7:chapter=13&highlight=woman%2Cstate%2Capproach.

Plutarch. *Conjugalia Praecepta*. Translated by Frank Cole Babbitt. London: William Heinemann, 1928. 2.

Pollard, Ruth. "Shameful Secret of our Family Murder Epidemic." *Sydney Morning Herald*, November 24, 2008. Accessed February 25, 2012. http://www.smh.com.au/news/national/shameful-secret-of-our-family-murder-epidemic/2008/11/23/1227375062534.html.

Pollack, William. "No Man is an Island: Toward a New Psychoanalytic Psychology of Men." In *A New Psychology of Men*, edited by Ronald Levant and William Pollack, 33–67. New York: Basic Books, 1995.

Pollock, Susan. *Ancient Mesopotamia: The Eden That Never Was*. Cambridge: Cambridge University Press, 1999.

Posner, George, Kenneth Strike, Peter Hewson, and William Gertzog. "Accommodation of a Scientific Perception: Toward a Theory of Conceptual Change." *Science Education* 66, 2, (1982): 211–227.

Potts, Malcolm and Martha Campbell. "The Origins and Future of Patriarchy: The Biological Background of Gender Politics." *Journal of Family Planning and Reproductive Health Care* 34, 2, (2008).

Prioreschi, Plinio. *A History of Medicine*. Vol. 3, *Roman Medicine*. Omaha, NE: Horatius Press, 2001.

Purkiss, Diane. *The Witch in History: Early Modern and Twentieth-Century Representations*. London: Routledge, 2012.

Purvis, June. *Emmeline Pankhurst: A Biography*. London: Routledge, 2002.

Quaife, Geoffrey. *Godly Zeal and Furious Rage*. London/Sydney: Croom Helm, 1987.

R v. Johns. Supreme Court, S.A. No. SCCRM/91/452, August 26, 1992.

Rabelais, François. *The Complete Works of François Rabelais*. Translated by Donald Frame. Berkeley: University of California Press, 1991.

Rackham, Oliver. "The Physical Setting." In *The Mediterranean in History*, edited by David Abulafia. London: Thames and Hudson, 2003.

Rathmann, W. G. "Female Circumcision: Indications and a New Technique." *General Practitioner* 20, (1959): 115–20.

Ray, Suranjita. "Understanding Patriarchy." Online paper. University of Delhi, 2006. Accessed December 9, 2013. http://www.du.ac.in/fileadmin/DU/Academics/course_material/hrge_06.pdf.

Relationships Australia (S.A.) Relationships Australia web page, 2012. Accessed June 11, 2012. http://www.rasa.org.au/en/whoweare/history.aspx.

Rémy, Nicolas. *Demonolatry: An Account of the Historical Practice of Witchcraft*. Edited by Montague Summers, translated by E. A. Ashwin. New York: Dover Publications, 2008.

Reynolds, Philip, trans. *Food and the Body: Some Peculiar Questions in High Medieval Theology*. Leiden: Koninklijke Brill, 1999.

Rich, Adrienne. *Of Woman Born: Motherhood as Experience and Institution*. New York: Norton, 1995.

Richlin, Amy. *Garden of Priapus*. New York: Oxford University Press, 1992.

Riddle, John. "The Great Witch-Hunt and the Suppression of Birth Control: Heinsohn and Steiger's Theory from the Perspective of an Historian." In *Witchcraft, Population Catastrophe and Economic Crisis in Renaissance Europe: An Alternative Macroeconomic Explanation* by Gunnar Heinsohn and Otto Steiger. Discussion Paper, University of Bremen, 2004.

Riley, Denise. *Am I That Name? Feminism and the Category of 'Women' in History*. Minneapolis: Minnesota University Press, 1988.

Riskin, Shlomo. "The Akedah: A Lesson in Martyrdom." *Jerusalem Post,* Oct 21, 1994.

Rissel, Chris and Juliet Richters. *Doing it Down Under: The Sexual Lives of Australians.* Sydney: Allen and Unwin, 2005.

Roberts, A., J. Donaldson and A. Coxe. *The Ante-Nicene Fathers.* Vol. 1, *Translations of the Fathers down to AD 325.* Oak Harbor: Logos Research Systems, 1997.

———. *The Ante-Nicene Fathers.* Vol. 4, *Translations of the Fathers down to AD 325.* Oak Harbor, WA: Logos Research Systems, 1997.

Roberts, T-A., J. Goldenberg et al. "'Feminine Protection': The effects of menstruation on attitudes toward women." *Psychology of Women Quarterly* 26, (2002): 131–139.

Rowbotham Sheila. "The Trouble with Patriarchy." In *No Turning Back: Writings from the Women's Liberation Movement 1975–80.* London: The Women's Press, 1981.

Roma, Paolo, Antonella Spacca, et al. "The Epidemiology of Homicide-Suicide in Italy." *Forensic Science International* 214, 1–3, (2012): 1–5.

Romanska, Magda. "Gender and Jewish Culture: Performing the Covenant: Akedah and the Origins of Masculinity." *Gender Forum an Internet Journal for Gender Studies* 21, (2008). http://www.genderforum.org/index.php?id=193

Roots, Peter. "The De Opificio Dei: The Workmanship of God and Lactantius." *Classical Quarterly* 37 (1987): 466–86.

Rosenmeyer, Thomas. *The Art of Aeschylus.* Berkeley, CA: University of California Press, 1982.

Rössler, Augustin and William Fanning. "Woman." In the Catholic Encyclopedia online. New Advent, 2013. Accessed June 5, 2012. http://www.newadvent.org/cathen/index.html.

Rotundo, E. Anthony. *American Manhood: Transformations in Masculinity from the Revolution to the Modern Era.* New York: Basic Books, 1993.

Rowlett, Lori. *Joshua and the Rhetoric of Violence: A New Historicist Analysis.* Sheffield, England: Sheffield Academic Press, 1996.

Rubin, Gabrielle. *Les Sources Inconsciente de la Misogynie*. Paris: R. Laffont, 1977.

Rudolf, John. "U.S. Army: Sex Crimes by Soldiers up 97 Percent in Five Years." *Huffington Post Crime,* January 21, 2012. Accessed June 21, 2013. http://www.huffingtonpost.com/2012/01/20/army-sex-crimes-by-soldiers-sexual-assault-domestic-violence_n_1219246.html.

Ruether, Rosemary. *Women and Redemption: A Theological History* 2nd ed. Minneapolis, MN: Fortress Press, 2012.

———. "Women's Body and Blood: The Sacred and the Impure." In *Through the Devil's Gateway: Women, Religion and Taboo,* edited by Alison Joseph. London: SPCK Publishing and Channel Four Television Company, 1990.

———. "Male Clericalism and the Dread of Women." In *Women and Orders,* edited by Robert Heyer, 1–13. Paulist Press, 1974.

Ruggi, Suzanne. "Commodifying Honour in Female Sexuality: Honour Killings in Palestine." In *Women and Islam: Social Conditions, Obstacles and Prospects*, edited by Haideh Moghissi, 125–30. New York: Routledge, 2005.

Rupp, Leila. "Revisiting Patriarchy." *Journal of Women's History* 20, 2, (2008): 136–140.

Rush, Benjamin. In *Women, the Family and Freedom: 1750–1880.* Vol. 1, edited by Susan Bell and Karen Offen. Stanford, CA: Stanford University Press, 1983.

Ryle, Robyn. *Questioning Gender: A Sociological Exploration.* London: Sage, 2012.

Sadegh-Zadah, Kazem. *Handbook of Analytic Philosophy of Medicine.* New York: Springer, 2011.

Saint Albertus (Magnus). *On Animals: A Medieval Summa Zoologica.* Translated by Kitchell, Kenneth and Irven Resnick. Baltimore, MD: Johns Hopkins University Press, 1999.

Saint Gregory (Thaumaturgus), Saint Dionysius (of Alexandria), Archelaus (Bishop of Charcar) and Stewart Dingwall Fordyce Salmond. *The Works of Gregory Thaumaturgus, Dionysius of Alexandria and Archelaus.* Vol. 20, *Ante-Nicene Christian Library.* Translated by Benjamin Plummer Pratten. London: T&T Clark, 1871.

Saint Paul. "Letter to the Galatians." (Gal 4:21–31) Christian Scriptures, c.60–70 CE. Accessed April 28, 2014. http://www.biblegateway.com/passage/?search=Galatians+1&version=NIRV.

Sanday, Peggy. *Fraternity Gang Rape: Sex, Brotherhood and Privilege on Campus* 2nd ed. New York: New York University Press, 2007.
Sanders, Melanie, Jayne Hrdlicka, Meredith Hellicar, Dale Cottrell, and Joanna Knox. "What Stops Women from Reaching the Top: Confronting the Tough Issues." Bain and Chief Executive Women, 2012. Accessed November 27, 2013. http://www.bain.com/Images/BAIN_BRIEF_What_stops_women_from_reaching_the_top.pdf.
Schatten, Gerald and Daniel Mazia. "The Penetration of the Spermatozoon through the Sea Urchin Egg Surface at Fertilization." *Experimental Cell Research* 98 (1976): 325–337.
Schiefsky, Mark. "Galen's Teleology and Functional Explanation." In *Oxford Studies in Ancient Philosophy 33*, edited by D. Sedley, 369–400. Oxford: Oxford University Press, 2007.
Schlesinger, Arthur. *The disuniting of America: Reflections on a Multicultural America*. New York: W. W. Norton, 1991.
Schliebs, Mark and Sarah Elks. "DART Alone Won't Uncover Abuse in Defence, Garry Rumble Warns." *The Australian*, April 28, 2014.
Schneider, David. "What is Kinship all About?" In *Kinship Studies in the Morgan Centennial Year,* edited by P. Reining. Washington, DC: The Anthropological Society, 1972.
Schonfeld, Christiane. "Under Construction: Gender and the Representation of Prostitutes in Expressionism." In *Expressionism and Gender,* edited by Frank Krause, 117–130. Goettingen: VandR Unipress GmbH, 2010.
Schwartz, Michael. *War without End: The Iraq War in Context.* Chicago: Haymarket Books, 2008.
Scott, Joan W. *Only Paradoxes to Offer: French Feminists and the Rights of Man.* Cambridge, MA: Harvard University Press, 1996.
Sengoopta, Chandak. *Otto Weininger : Sex, Science, and Self in Imperial Vienna.* Chicago: University of Chicago Press, 2000.
Sered, Susan. "Towards a Gendered Typology of Sacrifice: Women and Feasting, Men and Death in an Okinawan Village." In *Sacrifice in Religious Experience,* edited by Albert Baumgarten. 13–38. Boston: Brill, 2002.
Shapiro, Warren and Uli Linke. *Denying Biology: Essays on Gender and Pseudo-Creation.* Lanham, MD: University Press of America, 1996.
Shariati, Ali and Hamid Algar, trans. "The Philosophy of History: Cain and Abel." In *On the Sociology of Islam*, 97–110. Berkeley, CA: Mizan Press, 1979.
Sharples, Robert. *Peripatetic Philosophy 200 BC to 200 AD*. New York: Cambridge University Press, 2010.
Sheehan, Elizabeth. "Victorian Clitoridectomy: Isaac Baker Brown and his Harmless Operative Procedure." In *The Gender/Sexuality Reader: Culture, History, Political Economy,* edited by Roger Lancaster and Micaela Di Leonardo, 325–34. London: Routledge, 1997.
Shepanski, Paul and Michael Diamond. *An Unexpected Tragedy: Evidence for the Connection between Working Patterns and Family Breakdown in Australia.* Sydney: Relationships Forum Australia, 2007.
Sheridan, Mark and Thoman Oden. *Ancient Christian Commentary on Scripture, Old Testament II, Genesis 12–50.* Downers Grove, IL: InterVarsity Press, 2002.
Short, Julian. *An Intelligent Life.* Sydney: Random House, 2005.
Shulman, Dennis. *The Genius of Genesis: A Psychoanalyst and Rabbi Examines the First Book of the Bible.* Lincoln, NE: Iuniverse Inc., 2003.
Singer, Peter. *A Darwinian Left: Politics, Evolution and Cooperation.* New Haven, CT: Yale University Press, 1999.
Siraisi, Nancy. *Medicine and the Italian Universities 1250–1600*. Leiden: Brill, 2001.
Sjolander, Claire and Kathryn Trevenen. "One of the Boys? Gender Disorder in Times of Crisis." *International Feminist Journal of Politics* 12, 2, (2010): 58–76.
Skinner, Marilyn. *Sexuality in Greek and Roman Culture* 2nd ed. Hoboken, NJ: Wiley-Blackwell, 2013.
Smith, Thomas, trans. "The Recognitions of Clement." In *Ante-Nicene Fathers.* Vol. 8, edited by Philip Schaff. Grand Rapids, MI: Christian Classics Ethereal Library, 1886.
Smuts, Barbara. "The Evolutionary Origins of Patriarchy." *Human Nature* 6, 1, (1995): 1–32.

Snow, Deborah. "There Will Be Women in Foxholes." *Sydney Morning Herald,* Jan 6, 2012. http://www.smh.com.au/national/there-will-be-women-in-foxholes-20120106-1po7m.html#ixzz1ijDYt8Uj

Sommerstein, Alan. *Aeschylean Tragedy* 2nd ed. London: Bloomsbury, 2013.

Sokoloff, Michael and Joseph Yahalom. *Jewish Palestinian Aramaic Poetry from Late Antiquity: Critical Edition with Introduction and Commentary.* Jerusalem: Publications of the Israel Academy of Sciences and Humanities, 1999.

Sophocles. *The Complete Sophocles.* Vol. 1, *The Theban Plays.* Edited by Peter Burian, and Alan Shapiro. New York: Oxford University Press, 2010.

———. *Antigone.* In *The Complete Greek Tragedies,* edited by D. Grene and R. Lattimore. Chicago: University of Chicago Press, 1959.

Spade, Paul Vincent. "Medieval Philosophy." In *The Stanford Encyclopedia of Philosophy,* edited by Edward Zalta. Stanford, CA: Stanford University, 2010. Accessed June 5, 2012. http://plato.stanford.edu/archives/sum2010/entries/medieval-philosophy.

Speiser, E. A. *Genesis: Translation and Commentary for the Anchor Bible.* New York: Doubleday, 1964.

Steffensmeier, Darrell and Emilie Allan. "Gender and Crime." *Annu. Rev. Sociol., 22,* (1996): 459–87.

Steinberg, Holger. "Paul Julius Möbius (1853–1907)." *Journal of Neurology* 252, (2005): 624–25.

Steuter, Erin and Deborah Wills. *At War with Metaphor: Media, Propaganda, and Racism in the War on Terror.* Lanham, MD: Lexington Books, 2009.

Stock, Jay and Ron Pinhasi eds. *Introduction to Human Bioarchaeology of the Transition to Agriculture.* Chichester, UK: Wiley-Blackwell, 2011.

Stoller, Robert. *Observing the Erotic Imagination.* New Haven, CT: Yale University, 1985.

Stoller, Robert and Gilbert Herdt. "The Development of Masculinity: A Cross-Cultural Contribution." *Journal of the American Psychoanalytic Association* 30, (1982): 29–59.

Stone, Kenneth. *Queer Commentary and the Hebrew Bible.* Cleveland, OH: Pilgrim Press, 2001.

———. *Practicing Safer Texts: Food, Sex, and Bible in Queer Perspective.* London: T&T Clark International, 2005.

Stringer, Rebecca. *Knowing Victims: Feminism, Agency, and Victim Politics in Neoliberal Times.* New York: Routledge, 2014.

Stubbs, M. "Cultural Perceptions and Practices around Menarche and Adolescent Menstruation in the United States." *Annals of the New York Academy of Sciences* 1135, (2008): 58–66.

Studd, J. "Ovariotomy for Menstrual Madness and Premenstrual Syndrome—19th-Century History and Lessons for Current Practice." *Gynaecological Endocrinology* 22, 8, (2006): 411–15. Accessed July 29, 2012. http://www.ncbi.nlm.nih.gov/pubmed/17012101.

Sullivan, Nikki. "Labia." In *Cultural Encyclopedia of the Body,* edited by Victoria Pitts-Taylor, 311–13. Westport, CT: Greenwood Publishing, 2008.

———. "Vagina." In *Cultural Encyclopedia of the Body,* edited by Victoria Pitts-Taylor, 532–39. Westport, CT: Greenwood Publishing, 2008.

Summers, Anne. "Her Rights at Work: The Political Persecution of Australia's First Female Prime Minister." 2012 Human Rights and Social Justice Lecture, University of Newcastle, 31 August, 2012. Accessed November 10, 2013. http://annesummers.com.au/speeches/her-rights-at-work-r-rated/.

———. *Damned Whores and God's Police.* New York: Penguin, 1975.

Summers, Martin. "This Immoral Practice." In *Gender Nonconformity, Race and Sexuality,* edited by Toni Lester, 21–42. Madison: University of Wisconsin Press, 2002.

Summers, Montague, trans. *Malleus Maleficarum.* London: John Rodker, 1928.

Tarpley, Harold. ' Vagina Envy in Men.' *Journal of the American Academy of Psychoanalysis and Dynamic Psychiatry* 21, (1993): 457–464.

Terman, Lewis and Catherine Miles. *Sex and Personality: Studies in Masculinity and Femininity.* New York: Mcgraw-Hill Book Company, 1936.

Terry, Karen, et al., eds. "The Causes and Context of Sexual Abuse of Minors by Catholic Priests in the United States." A report presented to the United States Conference of Catholic Bishops by the John Jay College Research Team, 2011.

Tertullian (Quinti Septimi Florentis Tertulliani). *On the Flesh of Christ.* Translated by Peter Holmes. Christian Classics Ethereal Library, 1970. Accessed June 29, 2012. www.tertullian.org/works/de_carne_christi.htm.

———. "On the Apparel of Women." In *Ante-Nicene Fathers.* Vol. 4, edited by Alexander Roberts, James Donaldson, and A. Cleveland Coxe. Translated by S. Thelwall. Buffalo, NY: Christian Literature Publishing Co., 1885.

———. *The Writings of Tertullian.* Vol. 1. Translated by S. Thelwall. Edinburgh: T&T Clark, 1869.

———. *De Anima*. Translated by Jan Hendrik Waszink. Leiden: Koninklijke Brill, 2010.

Theophila. "'Banquet of the Ten Virgins." In *Ante-Nicene Fathers.* Vol. 6, *Fathers of the 3rd Century,* edited by Philip Schaff. Grand Rapids, MI: Christian Classics Ethereal Library, 2004.

Theweleit, Klaus. *Male Fantasies.* Minneapolis, MN: University of Minnesota Press, 1989.

Thomasset, Claude. *Une Vision de Monde a la Fin du Xiiie Siècle: Commentaire du Dialogue de* Placides et Timèo. Geneva: Librairie Droz, 1982.

Thompson, Antonio and Christos Frentzos, eds. *The Routledge Handbook of American Military and Diplomatic History 1865–Present*. New York: Routledge, 2013.

Thompson, Geoff and Clay Hichens. "Culture of Silence." ABC *Four Corners* program, Australian Broadcasting Corporation, June 13, 2011. Accessed January 24, 2012. http://www.abc.net.au/4corners/content/2011/s3242652.htm.

Thurston, Herbert. "Witchcraft." *The Catholic Encyclopedia.* Vol. 15. New York: Robert Appleton Co, 1912. Accessed April 27, 2014. http://www.newadvent.org/cathen/15674a.htm.

Tiele, C. P. *Outlines of the History of Religion.* Translated by J. Estlin Carpenter. New Delhi: Logos Press, 2005.

Titus Lucretius Carus. *The Nature of Things: A Didactic Poem in Two Volumes.* Vol. 2. Translated by John Mason Good. London: Longman, Hurst, Rees, and Orme, 1805.

Tjørnhøj-Thomsen, Tine. "'It's a Bit Unmanly in a Way': Men and Infertility in Denmark." In *Reconceiving the Second Sex: Men, Masculinity, and Reproduction,* edited by Marcia Inhorn, et al., 226–52, New York: Berghahn Books, 2009.

Touwaide, Alain. "Galen." In *Medieval Science, Technology, and Medicine: An Encyclopedia*, edited by Thomas Glick, Steven Liversey, and Faith Wallis, 179–181. New York: Routlege, 2005.

Triebs, Carl. *About Celibacy I Have No Instructions from the Lord: An Inquiry into the Origins of Clerical Celibacy.* Victoria, BC: Trafford Publishing, 2004.

Trinkaus, E. and A. Buzhilova. "The Death and Burial of Sunghir 1." *International Journal of Osteoarchaeology* 22, 6, (2012): 655–666.

Turner, Trevor. "Henry Maudsley: Psychiatrist, Philosopher and Entrepreneur." *Psychological Medicine* 18, 3, (1988): 551–74.

Ubel, Peter, Christopher Jepson and Ari Silver-Isenstadt. "Are Patients Willing to Participate in Medical Education?" *Journal of Clinical Ethics* 11, 3, (2000): 230–235.

———. "Don't Ask Don't Tell: A Change in Medical Student Attitudes of Obstetrics-Gynecology Clerkships toward Seeking Consent for Pelvic Examinations on an Anesthetized Patient." *American Journal of Obstetrics and Gynecology* 188, 2, (2003): 575–579.

UN Women. "UN-Women-Supported Survey in Delhi Shows 95 percent of Women and Girls Feel Unsafe in Public Spaces." UN Women, February 20, 2013. Accessed March 6, 2013. http://www.unwomen.org/2013/02/un-women-supported-survey-in-delhi-shows-95-percent-of-women-and-girls-feel-unsafe-in-public-spaces/.

van Balen, Frank. "Infertility and Culture: Explanations, Implications and Dilemmas." In *Marginalized Reproduction: Ethnicity, Infertility and Reproductive Technologies,* edited by Lorraine Culley, Nicky Hudson, and Floor Van Rooij, 34–48. London: Earthscan, 2009.

van Bekkum, Wout. "The Aqedah and its Interpretations in Midrash and Piyyut." In *The Sacrifice of Isaac: The Aqedah (Genesis 22) and Its Interpretations,* edited by Ed Noort and Eibert Tigchelaar. 86–95. Boston: Brill, 1999.

Vandello, Joseph, Jennifer Bosson, Dov Cohen, Rochelle Burnaford, and Jonathan Weaver. "Precarious Manhood." *Journal of Personality and Social Psychology* 95, 6, (2008): 1325–1339.

Van der Horst, P. "Sarah's Seminal Emission: Hebrews 11:11 in the Light of Ancient Embryology." In *Greeks, Romans, and Christians: Essays in Honor of Abraham J. Malherbe*. Minneapolis, MN: Fortress Press, 1990.

van't Land, Karine. "Sperm and Blood, Form and Food: Late Medieval Medical Notions of Male and Female in the Embryology of *Membra*." In *Blood, Sweat, and Tears: The Changing Concepts of Physiology from Antiquity into Early Modern Europe,* edited by Manfred Horstmanshoff, Helen King, and Claus Zittel, 363–392. Leiden: Koninklijke Brill NV, 2012.

van Wingerden, Sophia. *The Women's Movement in Britain 1866–1928.* New York: Palgrave Macmillan, 1999.

Verma, Ravi. United Nations CSW Side Event "Why Do Men Use Violence and How Do We Stop It?" UN, March 8, 2013. Accessed March 22, 2013. UN Webcast http://webtv.un.org/watch/why-do-men-use-violence-and-how-do-we-stop-it:-new-evidence-on-mens-use-of-violence-against-women-and-girls-and-its-uses-for-enhanced-prevention-csw-side-event/2212787287001/.

Vivres, Juan Luis. *De Institutione Feminae Christianae*. Edited and translated by C. Fantazzi and C. Matheeussen. Leiden: E. J. Brill, 1996.

Vogt, Carl. "Lecture III." In *Lectures on Man: His Place in Creation and in the History of the Earth.* London: Longman, Green, Longman and Roberts, 1864.

Vööbus, Arthur. *History of Asceticism in the Syrian Orient*. Louvain: CorpusSCO, 1958.

Wadham, Ben. "Brotherhood: Homosociality, Totality and Military Subjectivity." *Australian Feminist Studies* 28, 76, (2013): 213–235.

Wainberg, Sara. "Teaching Pelvic Examinations under Anaesthesia: What Do Women Think?" *Journal of Obstetrics and Gynaecology Canada* 32, 1, (2010): 49–53.

Waldman, Ellen and Marybeth Herald. "Eyes Wide Shut: Erasing Women's Experiences from the Clinic to the Courtroom." *Harvard Journal of Law and Gender* 28, (2005): 285–324.

Wallis, Faith, ed. and trans. *Medieval Medicine: A Reader.* Toronto: University of Toronto Press, 2010.

Walsh, Nonee. "Bonuses Paid to Men in the Finance Sector are Worth Triple Those to Women." ABC News, Australian Broadcasting Commission, September 2, 2013.

Warren, Carol. "Genital Surgeries and Stimulation in Nineteenth-Century Psychiatry." In *Gendered Perspectives on Reproduction and Sexuality*, edited by Marcia Texler Segal, Vasilikie Demos and Jennie Jacobs Kronenfeld, 165–197. London: Elsevier, 2004.

Warsh, Cheryl. *Prescribed Norms: Women and Health in Canada and the United States Since 1800.* Toronto: University of Toronto Press, 2010.

Weigle, Marta. *Creation and Procreation: Feminist Reflections on Mythologies of Cosmogony and Parturition*. Philadelphia: University of Pennsylvania Press, 1989.

Weininger, Otto. *Sex and Character.* Vienna: Braumüller, 1903, Reprinted Munich, Matthes and Seitz, 1980.

———. *Geschlecht und Charakter, Sex and Character: An Investigation of Fundamental Principles.* Translated by Löb Ladislaus. Bloomington: Indiana University Press, 2005.

———. "The Main Parts of *Sex and Character* ." Selected by Kevin Solway from the 1906 English Edition." Accessed December 9, 2013. http://www.theabsolute.net/ottow/sexcharh.html.

———. *Collected Aphorisms, Notebook and Letters to a Friend, Ed 1.12.* Translated by Martin Dudaniec and Kevin Solway. Accessed November 6, 2013. http://www.huzheng.org/geniusreligion/aphlett.pdf.

Weir, Allison. *Identities and Freedom: Feminist Theory between Power and Connection.* New York: Oxford University Press, 2013.

Weldon, Brooklynn. "Restoring Lost 'Honour': Retrieving Face and Identity, Removing Shame, and Controlling the Familial Cultural Environment through 'Honour' Murder." *Journal of Alternative Perspectives in the Social Sciences* 2, 1, (2010): 380–98.

Whitmer, Barbara. *The Violence Mythos.* Albany: State University of New York Press, 1997.

Wiesner-Hanks, Merry. *Gender in History: Global Perspectives* 2nd ed. Chichester, West Sussex: Wiley-Blackwell, 2011.

Wiesner-Hanks, Merry. *Women and Gender in Early Modern Europe*. New York: Cambridge University Press, 2000.

———. *Working Women in Renaissance Germany*. New Brunswick, NJ: Rutgers University Press, 1986.

Wilkinson, Richard. "How economic inequality harms societies." TEDGlobal transcript October, 2011. Accessed July 9, 2013. http://www.ted.com/talks/richard_wilkinson.html.

Wilkinson, Richard and Kate Pickett. *Spirit Level: Why Equality is Better for Everyone*. Allen Lane/The Penguin Press, 2009.

Willer, Robb, Christabel Rogalin, Bridget Conlon, and Michael Wojnowicz. "Overdoing Gender: A Test of the Masculine Overcompensation Thesis." *American Journal of Sociology* 118, (2013): 980–1022.

Williams, Sophie. "Facts and Figures Relating to the Childhood Sexual Abuse and Sexual Assault of Males." Nungar, Queensland: Living Well, 2014.

Wills, Garry. "Scandal." *New York Review of Books* May 23, 2002.

———. *Why Priests? A Failed Tradition*. New York: Penguin, 2013.

Wilson, Robin. "Autonomy Suspended: Using Female Patients to Teach Intimate Examinations without their Knowledge or Consent." *Journal of Health Care Law and Policy* 8, 2, (2005): 240–63.

Winkler, John. *The Constraints of Desire: The Anthropology of Sex and Gender in Ancient Greece*. New York: Routledge, 1990.

Winter, Sam, Pornthip Chalungsooth, Yik Koon Teh, et al. "Transpeople, Transprejudice and Pathologisation: A Seven-Country Factor Analytic Study." *International Journal of Sexual Health* 21, (2009): 96–118.

Winter, Timothy. "Abraham from a Muslim Perspective." In *Abraham's Children: Jews, Christians and Muslims in Conversation,* edited by Norman Solomon, Richard Harries, and Timothy Winter, 28–35. New York: Continuum, 2005.

Wirth, Michelle and Allison Gaffey, "Hormones and Emotion: Stress and Beyond." In *Handbook of Cognition and Emotion,* edited by Michael Robinson, Edward Watkins, and Eddie Harmon-Jones, 69–94. New York: The Guilford Press, 2013.

Wong, Joel and Aaron Rochlen. "Demystifying Men's Emotional Behaviour: New Directions and Implications for Counselling and Research." *Psychology of Men and Masculinity* 6, 1, (2005): 62–72.

World Economic Forum. "Global Gender Gap Report." World Economic Forum, 2012. Accessed June 28, 2013. http://www.weforum.org/reports/global-gender-gap-report-2012.

World Health Organisation. "Global and Regional Estimates of Violence against Women." Geneva: Department of Reproductive Health and Research, World Health Organisation, 2013. Accessed June 22, 2013. http://www.who.int/reproductivehealth/publications/violence/9789241564625/en/index.html.

World Report, Saudi Arabia. Human Rights Watch, 2011. http://www.hrw.org/world-report-2011/saudi-arabia.

Wroe, David. "Army Shamed by Sex Videos." *Sydney Morning Herald,* June 14, 2013. Accessed June 22, 2013. http://www.smh.com.au/opinion/political-news/army-shamed-by-sex-videos-20130613-2o79z.html.

Young, Iris. "Gender as Seriality: Thinking about Women as a Social Collective." *Signs* 19, 3, (1994): 713–738.

Yousafzai, Malala. Speech to the United Nations. UN, July 13, 2013. Accessed July 13, 2013. http://www.guardian.co.uk/commentisfree/2013/jul/12/malala-yousafzai-united-nations-education-speech-text.

Yousafzai, Ziauddin. "My Daughter, Malala." TED.com transcript, March 2014. Accessed April 28, 2014. https://www.ted.com/talks/ziauddin_yousafzai_my_daughter_malala.

Zinn, M., P. Hondagneu-Sotelo and M. Messner. *Gender through the Prism of Difference* 3rd ed. New York: Oxford University Press, 2005.

Index

About the Author

A. L. Jones is a retired academic and psychologist whose published work includes three non-fiction books and numerous articles in scholarly and literary journals.

CPSIA information can be obtained at www.ICGtesting.com
Printed in the USA
BVOW05*0256220814

363558BV00001B/1/P

9 780739 190968